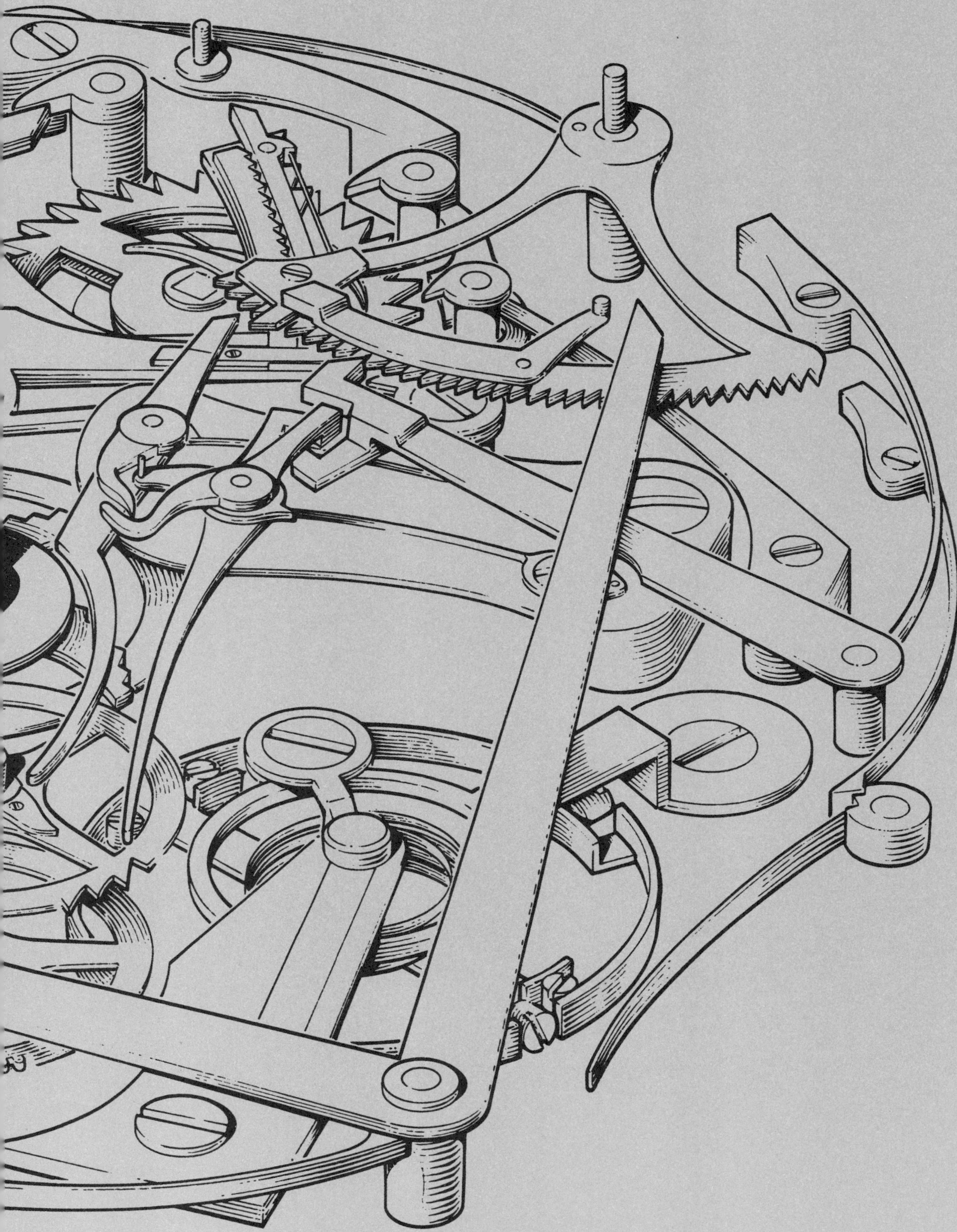

Watches

WATCHES

*A complete history of the
technical and decorative development of the watch*

REVISED AND ENLARGED EDITION

Cecil Clutton and George Daniels

PHILIP WILSON PUBLISHERS
Bloomsbury Publishing Plc
50 Bedford Square, London, WC1B 3DP, UK
29 Earlsfort Terrace, Dublin 2, Ireland

BLOOMSBURY, PHILIP WILSON PUBLISHERS and the PHIILIP WILSON
PUBLISHERS logo are trademarks of Bloomsbury Publishing Plc

First published in Great Britain in 1965 by B.T. Batsford Ltd
Second edition published in 1971
Third edition published in 1979 for Sotheby Parke Bernet Publications by Philip Wilson
Publishers Ltd
This reprinted edition published in 2022

A catalogue record for this book is available from the British Library
Library of Congress Cataloguing-in-Publication data has been applied for

ISBN: 978 1 78130 113 5

2 4 6 8 10 9 7 5 3 1

Printed and bound in China by RR Donnelley Asia Printing Solutions Ltd

To find out more about our authors and books visit
www.bloomsbury.com and sign up for our newsletters

Contents

Note: Roman numerals in the inner text margins refer to colour plates; arabic numerals refer to monochrome illustrations.

Note to Reprinted Edition

It was Cecil (Sam) Clutton, a great expert on antiquarian watches and clocks, who proposed to George Daniels that they should write a book together on watches. George, who knew Sam well, was astonished but very flattered as he remarked: 'I had only written birthday cards up to that point!'. This would be the first of several books written by the late Dr George Daniels CBE FBHI who was widely regarded as the greatest watchmaker in the twentieth century.

George had photographed all the watches that had been through his hands and had made drawings of mechanisms and in particular escapements which were always uppermost in both his and Sam's minds. The book was published to great acclaim in 1965 and George was a very proud co-author. Selling very well, the book was revised in 1975 and went to a total of three editions in addition to a German translation. Once out of print, George did not think the book would ever be reissued as the plates that were stored in a warehouse were ruined by damp. He did, however, raise the hope that the book might one day be scanned with new technology and this is precisely what Bloomsbury has achieved in this reprint.

Watches is a remarkable record that remains an invaluable reference source for watch, clockmakers and collectors of today. This reprinted edition with a foreword by Jonathan Betts addresses a long-awaited call to see this important book back in print. On behalf of the trustees of The George Daniels Educational Trust we would like to thank the Clutton family for assigning their share of the copyright of this publication to the Trust.

David Newman
Chairman
The George Daniels Educational Trust
2022

Foreword

It is a considerable privilege to be asked to write a foreword to this edition of *Watches*, as I have long been an admirer of the horological contributions of both Cecil (Sam) Clutton and George Daniels. The two authors were especially complementary, Clutton with his great knowledge of horological history and Daniels the horological technician, *par excellence*. When the first edition appeared in 1965 (Batsford in the UK, Viking in the US) there was very little already published on the subject save for G.H. Baillie's book of the same title (1929) and T.P. Cuss with *The Story of Watches* (1952). Neither of those titles dealt with the technical side of watch mechanism in the detail to be found in 'Clutton & Daniels', and the appearance of *Watches* was an important landmark in British horological bibliography; indeed, there is still no modern work dedicated to the history of watchmaking on the scale of this book. One area where subsequent research has revealed new insights is in the achievements of the great horological pioneer John Harrison, and the statements here that he 'made no final contribution to the science of horology', and that H4's design was '...a triumph of workmanship over design' are now recognised as conforming to the old-fashioned 'faint praise' view of Harrison's importance. Nevertheless, over 40 years after this edition saw the light of day in 1979, and in spite of much new research and discovery, the content usually stands up well to the scrutiny of we pedantic horological historians!

When the book was first published, there was an amusing contretemps in the journal *Antiquarian Horology* (*AH*), when, pre-publication date, the celebrated bookseller Malcolm Gardner (Charles Allix) pre-empted book reviews by publishing an advert in which he provided a critique (a review by any other name) of the contents. It was generally very positive, but included a few remarks which gave offence. Both Sam and George made their feelings very clear in the Letters to the Editor pages in the following issue of *AH*, but all remained friends! A second, extended edition of the book was published in 1975 and the third edition (Sotheby Parke Bernet), of which this volume is the reprint, appeared in 1979. As well as much new information, this edition brought the subject of watchmaking up to date and included photographs and descriptions of a number of Daniels' own, exquisitely beautiful creations.

George had begun making his own watches in the late 1960s and, with the publication of his seminal work *Watchmaking* in 1981, the name of Daniels surely crowns the pantheon of 20th century watchmakers. The last decades of the century witnessed a veritable renaissance in the art of mechanical watchmaking and, in the opinion of this horological historian, no single artist craftsman, from any country, inspired that rebirth more than George Daniels.

In the early 1970s Daniels began developing alternative watch escapements, with a view to seeing something better than the lever escapement introduced into large scale production by the Swiss. He would of course succeed with his celebrated co-axial escapement, which continues in production today at Omega and by his protégé Roger Smith. It is interesting to see that under the dust jacket of the 1979 edition of *Watches*, the cover of the book is embossed in gold with a drawing of the co-axial, presaging its future success.

Another beautiful creation of Daniels was his double-wheel chronometer escapement, inspired by Breguet's *échappement naturel*. Daniels specialised in the work of Abraham Louis Breguet (his book *The Art of Breguet* [Sotheby Parke Bernet, 1975] was reprinted in 2021) and his double-wheel chronometer escapement has also been very successfully put into production in recent years. Being George's only escapement truly needing no lubrication, the double-wheel is actually, in my view, his finest design. However, requiring two separate trains, and being very difficult to manufacture on a small scale, Daniels feared the Swiss would not contemplate putting such a design into production. It was only in the early years of this century that the firm of Frodsham's picked up the idea. Inspired and encouraged by George's great friend the late Derek Pratt, the firm developed their new wrist watch with the Daniels double-wheel escapement. Needless to say, George was very pleased; now both his creations were to go into production.

Practical Watch Escapement and *Watchmaking* were reprinted in recent years, and this new edition of Clutton & Daniels' *Watches* sees Daniels' best written works reprinted and available for the next generation of aspiring watchmakers and antiquarian horologists. With many young watchmakers now busy producing their own horological creations in this third decade of the 21st century, the renaissance in the making of mechanical watches is as strong as ever, and these volumes will undoubtedly continue to inspire.

Jonathan Betts MBE, FSA, FBHI
RMG Curator Emeritus
2022

Preface

When *Watches* was published in 1965 it was the first comprehensive history of the subject to have been written for many years, and with this historical essay was coupled a technical assessment of the principal escapements from the verge onwards. The book quickly gained for itself a considerable reputation, so that the authors were sad to see it fall out of print despite a continuing demand. They were therefore greatly encouraged when Sotheby Parke Bernet Publications expressed an interest in republishing it.

Since 1965 a vast amount of new information has come to light, especially concerning the development of the precision watch and also the work of Abraham-Louis Breguet. It has therefore seemed appropriate to re-write something like half the book. Although Ferdinand Berthoud made very few precision watches, he played an important part in the advances leading up to them which culminated in the superb productions of his nephew, Louis. Despite Ferdinand Berthoud's voluminous writings it is nevertheless difficult to assess how great his contribution really was and to ascribe precise dates to the important advances which may be attributed to him. An attempt has been made to put his work into perspective with the English pioneers, especially John Arnold.

Although an interest in watches of the Victorian era and the present century was beginning to develop in 1965 it has grown enormously in the intervening years, and the book has been brought up to date in this respect, particularly by the inclusion of a large number of additional monochrome illustrations. The number of pictures in colour has also been greatly increased. The technical section has been considerably revised and expanded and now includes a technical glossary.

This new edition is therefore no mere *rechauffé*, and the authors hope it may be seen as a presentation of the subject worthy of the date of its publication.

C.C.   G.D.

1979

<table>
<tr><td>A</td><td rowspan="2">E</td><td>G</td></tr>
<tr><td>B</td><td>H</td></tr>
<tr><td>C</td><td rowspan="2">F</td><td>I</td></tr>
<tr><td>D</td><td>J</td></tr>
</table>

A ANDREWS, London, England. *Circa* 1800. Cylinder escapement. White enamel dial. Gold arrow-head hands. Translucent enamel over engine-turning with applied filigree and pearl decoration.

B JOSIAH EMERY NO. 1057. London, England. *Circa* 1784. Lever escapement. See also fig. 22, for illustration of escapement, and fig. 180 for NO. 1289.

C SOLOMON PLAIVAS, Blois, France. *Circa* 1640. Verge escapement. Fusee and gut. Painted enamel dial with single gilt hand. Blois painted enamel case. The brilliant but delicate colouring is typical only of the earliest Blois enamelling.

D AUGUST BRETONNEAU, Paris, France. *Circa* 1640. Verge escapement without balance spring. Fusee and gut. Painted enamel dial with single hand. Blois painted enamel case. A watch of exceptional size and the very highest quality of Blois enamel painting of the school of Toutin. See also col. pl. XVIE.

E JAQUET DROZ. Swiss. See fig. 174.

F ABRAHAM-LOUIS BREGUET NO. 4038. Paris, France. 1824. Lever escapement. Two-arm compensation balance. Spiral balance spring with end-curve. Regulator operated through sector in dial. Going barrel. Half-quarter repeater on single gong. Engine-turned silver dial with eccentric chapter ring and subsidiary dial for seconds. Gold hands. Silver and gold engine-turned case. Breguet short chain and male key.

G ANONYMOUS. Swiss. Verge escapement. Quarter repeater on gongs. Three-colour gold dial with eccentric chapter ring and automata to simulate striking of hours and quarters. Gold hands. Engine-turned gold case.

H JOHANNES VAN CEULEN, The Hague. *Circa* 1670. Verge escapement originally without balance spring. Fusee and chain with long links, probably original. Painted enamel dial with blued steel hands. Painted enamel case with colours typical of the Huaut School. See also col. pl. XVIK.

I DANIEL QUARE, London. See fig. 121.

J JOHN BUSHMAN, London. See fig. 134.

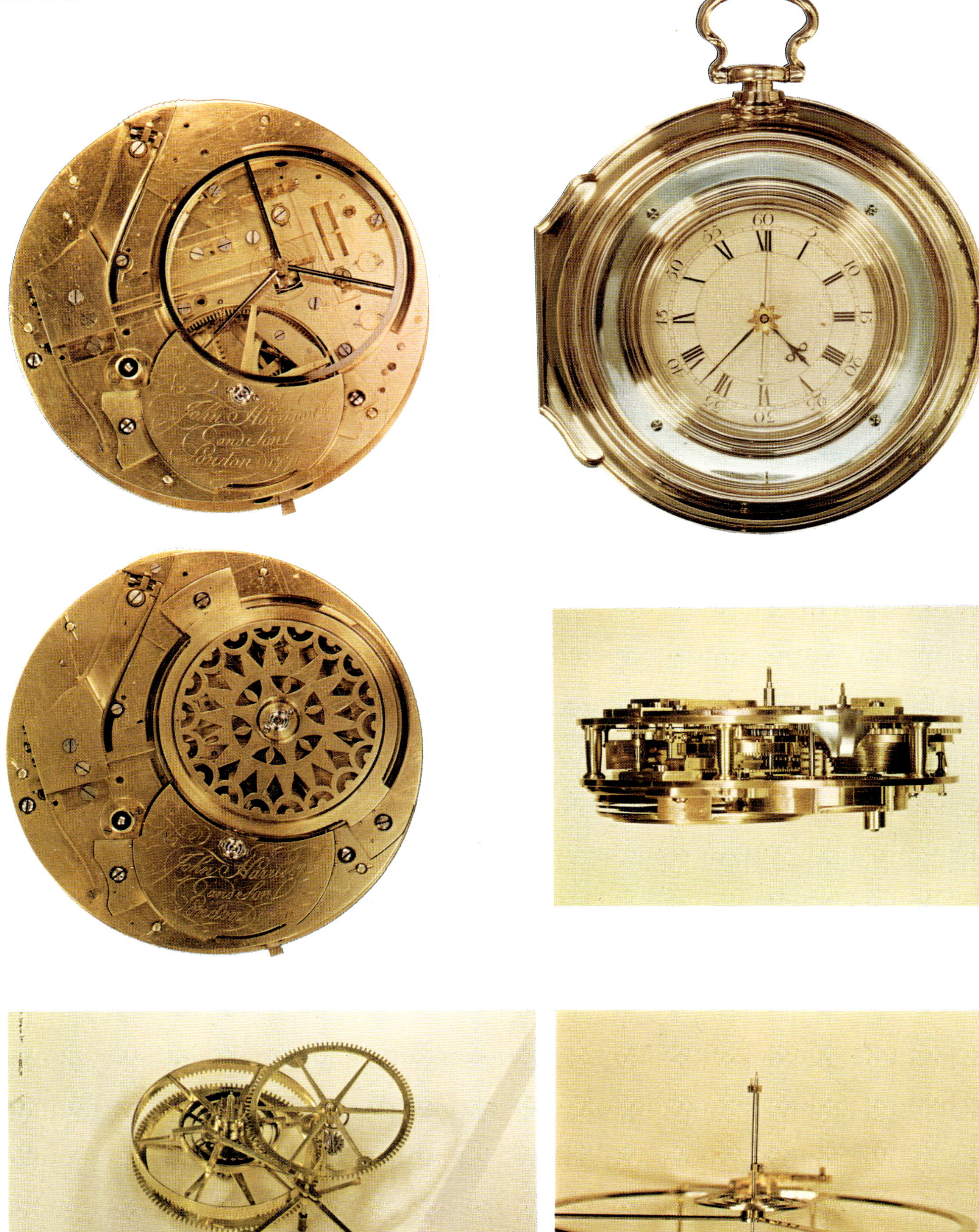

JOHN HARRISON, London, England. H.M. 1768. Verge escapement with specially shaped pallets of diamond (*bottom right*). Fusee and chain. 7½-second remontoir (*bottom left*). Plain balance with spiral spring and long bi-metallic compensation strip with adjustable mounting (*centre left*). Enamel dial. Steel hands with gilt hand-setting button. Silver pair case. This machine, known as Harrison's No. 5 timekeeper, is a near duplicate of his prize-winning longitude watch, which is the world's first practical and portable precision timepiece for navigational use during long voyages.

A

THOMAS MUDGE, London, England. H.M. 1769. Lever escapement with two-plane roller and jewelled fork and pallets (*right*). Plain balance with two balance springs, the lower spring with pivoted regulator activated by bi-metallic strips, the upper with adjustable cycloidal pins. Fusee and chain. Enamel dial. Steel hands. Gold pair cases. The first lever watch ever made, constructed for George III. *Illustrated by gracious permission of Her Majesty Queen Elizabeth II.*

B

THOMAS MUDGE NO. 574. London, England. H.M. 1764. Ruby cylinder escapement. Plain balance with spiral spring and regulator. Fusee and chain. Enamel dial with apertures for moon phases, day of the week and perpetual calendar with provision for leap year; outer circle for the date. Gold pair cases. Dial removed (*bottom right*). This is a rare and very early example of a watch with perpetual calendar.

A

CHRISTOPHER LOWNDES, Pall Mall, London, England. *Circa* 1682. Verge escapement. Steel balance wheel. Spiral steel spring with regulator. Fusee and chain. Silver dial, blued steel hands, tulip hour and poker minute. Tortoiseshell and silver outer case. The lack of a name cartouche on the dial, the irregular edge to the cock foot and the absence of a mask on the cock indicate an early date.

B

GEORGE GRAHAM NO. 6287. London, England. H.M. 1747. Cylinder escapement. Steel balance wheel. Spiral steel spring and regulator. Fusee and chain. Brass-gilt dust cover. Enamel dial. Blued steel beetle and poker hands. Single gold case.

A

ABRAHAM-LOUIS BREGUET NO. 44. Paris, France. 1791. Lever escapement. Bimetallic compensation balance. Spiral steel spring with regulator. Going barrel, wound-through dial. Ten-minute repeating (a double stroke for each ten minutes) on a block. Enamel dial with blued steel hands. Plain gold case. The flat hands, script signature and star-shaped minute divisions are typical of Breguet's earliest work. Part of a series of similar watches extending from 32 to 50, in the second numbering series and the first set of books.

B

ABRAHAM-LOUIS BREGUET NO. 3452. Paris, France. *Circa* 1819. Ruby cylinder escapement. Plain gold balance. Spiral steel spring with bimetallic compensation curb. Going barrel. Half-quarter repeater on a single gong. Silver engine-turned dial. Blued steel hands. Engine-turned gold case with gold cuvette. The movement is jewelled throughout and is typical of Breguet's 'ouvrage première classe'. In the illustration of the movement the repeater push piece is shown withdrawn and ready for use.

C

ABRAHAM-LOUIS BREGUET NO. 4775. Paris, France. 1828. Lever escapement. Bimetallic compensation balance wheel. Spiral steel spring with overcoil and regulator. Two going barrels. Engine-turned silver dial with eccentric chapter ring and subsidiary dial below showing position of regulator. Blued steel hands. Engine-turned savonette gold case, both the inner and outer covers to the dial being of glass. (It is likely that there was originally an alternative snap-in gold outer cover.) Gold cuvette signed (most unusually for Breguet's authentic work of this late date) 'Breguet à Paris'. The watch is almost the last of a series started in about 1812, representing Breguet's final work on the thin precision pocket watch.

A

JOHN ARNOLD & SON NO. $\frac{327}{628}$. London, England, H.M. 1786. Spring detent escapement. Steel balance wheel with two affixed bi-metallic compensation arms. Helical free-sprung steel spring with end curves. Fusee and chain. Enamel dial. Gold hands. Silver consular case. Arnold's 'second sort'. In this series there is a constant difference of 101 between the upper and lower figures of the number.

B

JOHN ARNOLD NO. $\frac{3}{44}$. London, England. 1778. Pivoted detent escapement. S-shaped compensation balance. Helical free-sprung steel spring with end curves. Fusee and chain. Enamel dial. Blued steel hands. Silver consular case. The 3 of the number probably refers to the type of balance and 44 is Arnold's chronometer series. As with all Arnold's detent watches prior to 1782 it is exceptionally large, 73 mm in diameter and 36 mm thick.

C

JEAN-ANTOINE LEPINE NO. 6100. Paris, France. *Circa* 1805. Double-wheel duplex escapement. Brass balance wheel. Spiral steel spring with bi-metallic compensation curb and regulator. Going barrel. Quarter repeating on one gong. Enamel dial. Blued steel hands. Gold engine-turned case. This watch is typical of Lepine's own form of the 'Lepine calibre'.

A

GLORIA À ROUEN, France. *Circa* 1700. Verge escapement. Brass balance wheel, spiral steel spring with regulator. Fusee and chain. Quarter repeating (without 'all-or-nothing' piece) on bell. Gold dial. Enamel plaques for hours. Enamel circles for hour divisions and minutes. Blued steel hands. Single gold case with shallow engraved arabesque decoration.

B

THOMAS TOMPION NO. 86 (later renumbered on the plate as 986, but the original number remains under the dial). England. *Circa* 1680. Verge escapement. Steel balance wheel. Spiral steel spring with regulator. Fusee and chain. Gold dial (the centre part is not original and probably dates from the addition of the present outer case). Blued steel beetle and poker hands. Gold pair case, not original. Hallmark for 1713, bearing a coronet and initials MV.

C

THOMAS EARNSHAW NO. 582. London, England. H.M. 1801. Spring detent escapement. Bi-metallic compensation balance. Helical steel spring with end curves, free sprung. Fusee and chain. Enamel dial. Gold hour and minute hands. Steel seconds hand. Gold pair case. The movement is gilt and engraved, typical of Earnshaw's more expensive watches.

A

JOHN WISE, London, England. *Circa* 1680. Verge escapement. Plain balance. Spiral steel balance spring of two turns. Fusee and chain. Silver champlevé dial. Blued steel tulip and needle hands. Silver inner case. Outer case leather with silver pinwork decoration. All the details of this watch, especially the unsigned dial centre, show it to be among the earliest balance spring watches. See also fig. 107.

B

TOMPION & BANGER NO. 233. London, England. *Circa* 1710. Verge escapement. Steel balance wheel. Spiral steel balance spring with regulator. Fusee and chain. Quarter repeater on a bell. Gold champlevé dial. Steel hands. Gold pair case, the outer case pierced, engraved and repoussé. See also fig. 130.

C

WILLIAM KIPLING, London, England. *Circa* 1705. Verge escapement, balance in the form of a pendulum visible in the dial. Spiral spring with regulator. Fusee and chain. Gold champlevé dial with sun and moon aperture for the hours below, central minute hand, aperture for the mock pendulum above and apertures for the regulator and for locking the movement into the case to the right and left respectively. Plain gilt-metal case.

DANIEL QUARE NO. 699. London, England. *Circa* 1680. Verge escapement, three-arm steel balance with spiral steel spring and regulator. Fusee and chain. Silver champlevé dial. Roman numerals for the hours 1-6 with Arabic numerals 7-12 superimposed, outer circle divided into sixty minutes for each hour. Single blued steel serpentine hand. Plain silver pair cases.

FRANCIS PERIGAL, London, England. *Circa* 1782. Ruby cylinder escapement. Three-arm brass balance. Spiral steel spring with spiral bi-metallic strip and pivoted regulator for temperature compensation. Mean time regulation by pivoted compensation bridge. Fusee and chain. Enamel regulator dial with hours above and seconds below. Gold hands. Later silver case (*circa* 1805) with spring catches for back and front covers.

JOHN HAMBLIN, London, England. *Circa* 1685. Verge escapement. Three-arm steel balance with spiral steel spring and regulator. Fusee and chain. Silver champlevé dial with inner circle of quarter-hour divisions, signed 'Hamblin London'. Single tulip hand. Plain silver inner case, silver outer case with tortoiseshell covering decorated with silver pinwork.

A

JOSIAH EMERY NO. 939. London, England. *Circa* 1782-3. Two-plane lever escapement. 'Double-S' bi-metallic compensation balance. Helical steel balance spring. Fusee and chain. Enamel dial, with subsidiary thermometer dial. Blued steel hands. Silver-gilt case. This is the lowest-numbered surviving Emery lever watch and is unique among them in having a thermometer with bi-metallic strip (see view of under-dial, fig. 158).

B

CHARLES VINER, London, England. H.M. 1814. Duplex escapement. Two-arm compensation balance, spiral steel spring with regulator, going barrel, quarter-repeating, alarum with Viner's patent setting crown in the band of the case. Gold champlevé dial with centre alarum hand. Subsidiary dial for seconds. Gold hands. Gold case with milled band and bezels.

C

ARNOLD & SON NO. 95. London, England. *Circa* 1787. Arnold spring detent escapement. Three-arm compensation balance. Helical gold spring with terminal curves free sprung. Fusee and chain. Enamel dial. Blued steel hands. Plain silver case.

A

DANIEL FLETCHER, London, England. Mid-seventeenth century. Verge escapement without balance spring. Worm-and-wheel set-up type regulator. Fusee and gut. Engraved silver dial, the centre part pierced over a gilt background. Single blued steel hour hand; fixed calendar chapter ring with moving annular ring between· hour and date chapter rings, with gilt pointer for the date, seen in the illustration pointing to the fourteenth day. Silver pair case, the outer case covered in black leather decorated with silver pinwork. Foliate-type decoration of the cock typical of mid-seventeenth century English design. This is a very early example of what was to become the typical English pair case.

B

LANGLEY BRADLEY CODE NO. HYY. London, England. *Circa* 1710. Verge escapement with spiral steel balance spring and regulator. Fusee and chain. Champlevé silver dial with beetle and poker blued steel hands. Silver pair case. This watch is unusual in that the dial is in mint condition, showing that the matt background of the silver champlevé dials were almost white in colour. See fig. 116a for unusual type of decorative pillars in which this maker seems to have specialised.

C

GUER À ANGULEM, France. *Circa* 1710. Verge escapement, three-arm steel balance with spiral spring and rack regulator. Fusee and chain. Movement without pins – the back-plate secured with a steel bayonet ring. Alarum with bell in the case. Silver champlevé dial with single hand and central alarum setting disk. Silver case pierced with roundels.

ABRAHAM-LOUIS BREGUET NO. 160. Paris, France. Started *circa* 1790. Completed *circa* 1850. Perpetuelle, called the 'Marie Antoinette'. Minute repeating, perpetual calendar, equation of time, 'secondes d'un coup'. Equal-lift lever escapement with two-arm compensation balance with recessed screws. Helical spring with terminal curves free sprung. Winding weight arbors supported in a parachute. Every ˙friction surface jewelled, the plates and wheels of gold. Rock crystal dial with gilt numerals and subsidiary dial for seconds with inset day of the week dial. At the top left sectors for the state of winding and the equation of time and at the top right the date of the month and thermometer. Blued steel hands. Plain gold case with rock crystal back to display the movement; repeating piston in the pendent.

A

GEORGE MARGETTS NO. 1098. London, England. *Circa* 1790. Cylinder escapement. Plain balance with spiral spring and regulator. Fusee and chain. Enamel dial with revolving subsidiary dials for indicating solar and sidereal time. Gilt-metal case with blue enamel back.

B

GEORGE MARGETTS NO. 311. London, England. H.M. 1783. Cylinder escapement. Plain balance with spiral spring and regulator. Fusee and chain. Astronomical dial with indications for sidereal year including sign and degree of the zodiac and visible stars in the northern hemisphere, mean time at the centre with moon phases and eclipses and time of high tide at named ports. Gold pair cases.

C

PATEK PHILIPPE NO. 198385. Geneva, Switzerland. 1932. Double-dialled astronomical watch with striking, repeating, chronograph, perpetual calendar, visible stars, sunrise, sunset, equation, sidereal and mean time. Gold case. This extraordinary watch contains 110 wheels, 430 screws, 90 springs, 120 pieces of mechanism and 18 hands. See also fig. 377 for views of the under-dial work.

A

ANONYMOUS, Swiss. *Circa* 1800. Cylinder escapement. Plain balance with spiral spring and regulator. Going barrel. Musical automata. Enamel dial set in an automated scene of Moses striking water from the rock; aperture below with automated putti striking a bell for the quarter repeating. Gold case.

B

ANONYMOUS, Swiss. *Circa* 1800. Cylinder escapement. Plain balance with spiral spring and regulator, pinned-barrel musical mechanism released at the hour, enamel dial, automated dancing scene set in the back. Gold case with pearl-set bezels.

D

ILBERY NO. 5891. London, England. *Circa* 1820. Virgule escapement. Plain balance with spiral spring and regulator. Going barrel. Enamel dial. Gold case with enamel decoration, automated blacksmith's workshop inside the back cover. For illustration of movement see also fig. 249b.

E

ANONYMOUS, Swiss. *Circa* 1790. Verge escapement. Plain balance with spiral spring and regulator. Fusee and chain. Enamel dial. Automated blacksmith's workshop set in the back of the gold case. Pearl-set bezels.

A

GEORGE DANIELS, London, England. 1972. One-minute tourbillon with Earnshaw spring detent escapement. Stainless steel balance with eccentric gold adjusting weights. Ni-span-C balance spring with terminal curve free sprung. Two going barrels. Silver engine-turned dial with eccentric minute circle, sector for retrograde hour hand, seconds below. Gold engine-turned case.

B

GEORGE DANIELS, London, England. 1974. One-minute tourbillon with fifteen-second remontoir. Daniels' spring detent escapement with detent in tension. Stainless steel balance with recessed adjusting screws. Ni-span-C balance spring with terminal curve free sprung. Two going barrels. Equation of time cam. Silver engine-turned dial with eccentric chapter circle, seconds below, sectors for the state of winding and equation of time. Gold hands.

C

GEORGE DANIELS, London, England. 1976. Daniels' pivoted detent escapement with two escape wheels driven by separate trains and barrels. Stainless steel balance with eccentric adjusting weights. Ni-span-C balance spring with terminal curve free sprung, seconds zeroing lever. Silver engine-turned dial with eccentric chapter circle, seconds above, sectors for the state of winding and thermometer. Gold engine-turned case.

A	E	I
B	F	J
C	G	K
D	H	L

A ABRAHAM-LOUIS BREGUET NO. 5. Paris, France. 1787. Lever escapement. Four-arm balance with two bi-metallic affixes. Helical spring with end curves. Screw-operated regulator at edge of dial. Pedometer winding by platinum weight. Two going barrels. Dumb repeater. Silver engine-turned dial with subsidiary dials for seconds, up-and-down, and moon phases. Blued steel hands. Engine-turned gold case with Breguet-type piston for repeating. This watch is one of the series of thirty watches described by Breguet as 'perpetuelles' and with which his surviving books commence in 1787.

B ALEXANDER HARE, London, England. *Circa* 1780. Verge escapement. Fusee. White enamel dial. Gold hands. Gold pair case, the outer decorated with a garland of translucent enamel over engine turning with central painted cartouche.

C SOLOMON PLAIVAS. *Circa* 1640. See col. pl. IC.

D P. GREGSON, Geneva and Paris. *Circa* 1795. Virgule escapement. Going barrel. White enamel dial. Gold hands. Gold case with translucent enamel over engine turning.

E AUGUSTE BRETONNEAU. *Circa* 1640. See col. pl. ID.

F J. MELLIE, Châtellerault, France. *Circa* 1620. Verge escapement without balance spring. Fusee with ratchet set-up regulator. Gold dial with champlevé enamel floral decoration. Single gilt hand. Gold case with champlevé translucent enamel floral decoration. Outer case of black leather with gold pinwork, possibly of later date.

G JOHN ELLICOTT NO. 2700. London, England. 1742. Cylinder escapement with spiral steel balance spring and regulator. Fusee and chain with bolt and shutter maintaining power. Complicated calendar and astronomical trains. Equation work. Seconds dial on the back of the watch. Gold case.

H ALLMAN AND MANGAAR NO. 354. London, England. *Circa* 1765. Verge escapement. White enamel dial. Hands not original. Pair case, the outer of Bilston enamel. The use of English enamel for watch cases was extremely rare.

I CHARLES FRODSHAM, London, England. 1914. Lever escapement. Compensation balance. Going barrel. Keyless winding. Silver engine-turned dial with gold cartouche for signature and gold seconds ring. Blued steel hands. Engine-turned gold case. A late example of high quality English work in the style of Breguet.

J SIMPTON, London, England. See fig. 136.

K JOHANNES VAN CEULEN. *Circa* 1670. See col. pl. IH.

L Probably by ILBURY, London, England. *Circa* 1800. Chinese duplex escapement. Going barrel. The movement of Lepine calibre, decorated overall with floral patterns in three-colour gold. Gold case with painted floral decoration, the pendent and borders inset with pearls. (For movement, see fig. 222.)

Historical

Mechanical 1500–1750

The invention of watches

No conclusive evidence exists as to the invention of a clock with a mechanical escapement. *Horologium* may mean a sundial, a mechanical clock or a water clock, and the latter go back to a very early date. Indeed, the Chinese had very complicated and monumental water clocks with a sort of very slow escapement, not unlike an automatic flushing tank of today. It may well be that this was developed into a weight-driven clock with a mechanical escapement in the Middle East and thence travelled into Europe.

It is certain that Villard de Honnecourt, a most observant and much-travelled architect in the middle of the thirteenth century, did not know of a mechanical clock, while by 1300 it is becoming reasonably clear that at any rate some of the literary references to an *horologium* mean a mechanical clock. It therefore seems reasonable to date the invention of the mechanical clock, or at any rate its appearance in Europe, in the last quarter of the thirteenth century.

The oldest clock surviving in any state of completeness is that at Salisbury Cathedral, made in 1386. It had been converted to pendulum, but in 1956 was reconverted to its original verge and foliot escapement. By its date quite small domestic clocks existed, although none survives; but these also were weight-driven.

No watch could be made until a reliable form of portable power was available, and it is once more difficult to fix exactly when a coiled spring was first applied as the motive power of a clock.

In the Bibliothèque Royale of Belgium is a manuscript copy of a fairly well-known work, *Horologium Sapientiae* (the Clock of Wisdom), which can be dated between 1455 and 1488. One of its illustrations shows a number of timekeeping and allied devices, including the movement of an octagonal spring-driven clock. This is accurately portrayed and a fusee is clearly visible. Some authorities prefer a 1455 rather than a 1488 date for the manuscript and thus infer that spring-driven clocks must have existed before 1450. There are, however, objections to the manuscript being so early, and the earliest dated reference to a spring-driven clock exists in a manuscript in the Augsburg Stadtbibliothek, by Paulus Alemannus, which can be dated accurately to the years 1477–8. It describes clock mechanism and several spring clocks, and a fusee is illustrated.

There is also a letter of August 21, 1482 from Comino da Pontevico to the Duke of Mantua, containing the following passage (as translated by Dr. Morpurgo). 'The clock is provided with a steel ribbon, hidden in a brass box around which a string is wound. . . . If this steel ribbon were missing the instrument would not be able to run, even though the string were there, the said string tied to the box around the steel ribbon, is inserted to cause the pulling of the screw or force to which it is attached; the purpose of all this is to make all the wheels of the clock turn by propelling the said screw by force of the steel ribbon; thus are my clocks made, which I showed Your Excellency, and thus all masters make them.'

In 1480, on April 4th, Jean de Paris, clockmaker to Louis XI, received 16 livres for a clock striking the hours 'to be carried with him everywhere'. This must have been a spring clock, and there are two portraits of the king standing by a table on which stands a little hexagonal clock, with a hexagonal domed top.

There seem therefore to be good grounds for dating the spring-driven clock at about 1470, but any earlier date calls for some degree of wishful thinking. No complete spring clock survives which can be attributed to the fifteenth century. However, the British Museum possesses the remains of a clock which seems to come from the fifteenth century and which appears originally to have been spring-driven. The oldest dated spring clock is that belonging to the Society of Antiquaries of London, made by Jacob the Czech, for Sigismund I, King of Poland, dated 1525. This is drum-shaped, 247 mm in diameter.

From the time of the first spring clocks it would not take long to compress the mechanism into a small enough case to be carried about the person, but there is no reference to a watch before 1500.

In 1511 Johannes Cocclaens writes of Peter Hele, or Henlein of Nuremburg (1480–1524): 'from day to day more ingenious discoveries are made; for Petrus Hele, a young man, makes things which astonish the most learned mathematician, for he makes out of a small quantity of iron, *horologia* devised with very many wheels, and these *horologia*, in any position and without any weights, both indicate and strike for 40 hours, even when they are carried on the breast or in the purse'. Henlein had a contemporary, Caspar Werner, who was said to have 'devoted himself with especial industry to the making of watches which he brought into great popularity as the result of continual study and the introduction of various new inventions, though he injured his memory and health thereby'.

No watch attributable to Henlein or Werner survives, and the earliest dated watch is French, of spherical shape, in the Louvre at Paris, by Jacques de la Garde, of Blois, dated 1551. A similar, larger watch by him, dated 1565, is in the Maritime Museum, Greenwich. It is not known exactly when the Blois industry started, but it cannot have been appreciably later than the Nuremburg school.

The Nuremburg watches appear also to have been spherical in shape and Henlein is mentioned by Dopplmayer as 'a locksmith artist who gained renown through the small watchworks which he was one of the first to make in the form of the musk-balls at that time in use', and in 1524 he received fifteen florins for a gilt musk-apple with a watch. Such a German watch, dating from about the middle of the sixteenth century, is in the Ashmolean Museum, Oxford. But although both Henlein's and de la Garde's watches were spherical, it is certain that drum-shaped watches existed before 1550, being miniatures of the Antiquaries' Jacob Czech clock previously mentioned.

The exact purpose of these small drum clocks, such as those of the Clockmakers' Company, is not very clear. Mostly 63 mm–75 mm deep and in diameter, they are rather large to be worn and in any case have no pendant from which to suspend them. On the other hand, if they were not intended to be carried about there seems to have been no point in making them so small. They are usually quite plain, with a little simple engraving. It may be that they were therefore carried in a bag, or regarded as travelling clocks. There is one such in the British Museum with its movement pivoted between plates; and two in the collection of the Worshipful Company of Clockmakers at the Guildhall in the City of London, which have skeleton movements and fusees, similar to those of the Jacob Czech clock. It may be therefore that the fusee and skeleton movement comes from the south, since the Henlein watches probably, and their immediate successors certainly, had plated movements with a stackfreed, a most inferior substitute for a fusee.

Fusee and stackfreed

56, 58 It is a matter of common experience that when a coiled spring is newly wound up it exerts more power than when it is nearly run down, and this is the great defect of springs as against weights, as the motive power for clockwork. As has already been shown, this problem had been overcome no later than about 1475 by the fusee; one of the most beautiful mechanical inventions of its own or any other time. Although described in detail in the technical section, the principle of its operation is self-evident; in effect it is a progressively-variable gear, in which the spring pulls a high gear ratio when it is fully wound, and a low ratio when it is nearly run down and at its most feeble. In this way an even torque, or nearly so, is applied to the train of wheels and the escapement, throughout the day or whatever the running period of the watch may be.

59, 60 By contrast, the stackfreed is a most brutal arrangement. It consists of a cam geared to the arbor fixed to the mainspring, so as to turn once in the going period of the watch. Pressing against the cam is a roller mounted on a strong spring. The contour of the cam is such that when fully wound the stackfreed spring works against the mainspring and so slows the watch; while towards the end of the run it assists the mainspring and so hurries things along.

The movement and its decoration up to 1675

The earliest German watch movements were made entirely of iron, although the pinions were made of harder iron (possibly steel) than the rest. After about 1580 the plates were increasingly of brass, but all-iron trains are found as late as 1625, and the stackfreed survived even longer. Where iron plates are brass-bushed this is a later repair.

The French realised early that brass is more easily worked than iron for wheels and plates, and wears at least as well.

57, 60 The balance wheel or foliot was always outside the plates and its top pivot therefore had to be pivoted into a cock fixed to the top plate. In the earliest German watches the bottom pivot ran in a hole in the bottom plate, or sometimes in a spiral-shaped cock cut out of the bottom plate. The earliest balance cocks were S-shaped, so that they could safely be bent to some extent, to adjust the depth of the engagement of the pallets with the escape wheel. The cock was pinned to a pillar riveted to the top plate.

It usually had a very short tail to steady it on the plate. Later, the tail was to grow into what is more properly called a foot, with steady pins to locate it more precisely. The crown or escape wheel was pivoted between two more pillars, or potences, also riveted to the top plate. In the earliest German watches there was no adjustment for depthing the escapement, except by bending the balance cock, as above.

In even the earliest French watches, and soon afterwards in German and all others, the bottom pivot of the verge had been supported in an extension of the crown wheel potence. The outer end of the crown wheel arbor was always pivoted in a small circular plug, fitted friction-tight in the potence. By this means, the end float and depthing of the escapement could be adjusted within fine limits, which is by far the best way of regulating a pre-balance-spring verge watch.

Especially in the skeleton movements, the makers were at considerable pains to place the crown wheel arbor so as to be exactly radial to the contrate wheel. In later watches, where the contrate wheel is pivoted between the plates, and the crown wheel is pivoted at its outer end in a small separate potence attached to the plate, this is no longer possible, and the teeth of the contrate wheel have to be cut on a skew so as to mesh smoothly with the eccentrically placed crown wheel pinion. In the early skeleton layout this was avoided by having one cock pierced with two holes, at right angles to each other; one carrying the contrate wheel pivot and one the crown wheel pivot. The advantage of this is obvious; its disadvantage is the impossibility, or at any rate considerable difficulty, of providing any means of adjusting the end float of the crown wheel staff.

It is very unlucky that we know so little about the fifteenth-century Italian watches, since the three quite different types of watch movement found in the early sixteenth century seem to have very little in common with each other and it is therefore difficult to say which, if any of them, stems from the Italian original. However, since the cultural connections between Italy and France were so strong it seems fair to assume that the French watches followed the Italian model. The French craftsmanship is so greatly superior to that of the German iron movements—whether the full-plate, stackfreed type, or the skeleton, fusee type—that it seems quite likely that the watch was separately invented in Germany, and that the German watches owe nothing to the fifteenth-century Italian school. Nor is it by any means improbable that the watch should have been evolved simultaneously in two or even three different places.

Only in the hands of quite exceptional artists, such as Jost Burgi and Hans Kiening, was wheel-cutting in iron brought to a standard which will bear comparison with that of the French watches.

Mean time regulation before the balance spring

A watch with a balance spring is readily adjusted for mean time, but the earliest watches had no such convenience and other, cruder means had to be found.

Fusee watches were provided with a ratchet and click mounted on the spring-barrel arbor, to which the inner end of the spring is attached. This arbor also ended in a square so that it could be turned by a key, and thus the initial tension of the mainspring could be set up by means of the ratchet. Since the amount of set-up operated as additional power, and since the rate of a verge escapement is directly related to the power applied to it, this set-up ratchet provided a rough means of regulating the watch.

In about 1640 the ratchet set-up began to be replaced by a worm and wheel. The wormwheel was mounted on a long arbor pivoted in two blued steel supports, screwed to the plate. The supports had elegantly pierced and shaped tails. After a short time the wheel, mounted on the mainspring arbor, was capped by a silver plate, usually numbered 1–6, as a guide when regulating the watch.

With a stackfreed it is not quite so simple to adjust the set-up, although it is not impossible, as is generally stated. The stackfreed incorporates stopwork which predetermines the full-wind and fully-run-down positions. As the pinion which drives the stackfreed wheel is squared on to the spring arbor there is no great difficulty in setting up the initial tension of the spring by winding it up to any desired extent, before meshing the pinion with the stackfreed wheel. In this way the mainspring can be set up in just the same way as a fusee. However, an additional means was usually provided, and this consisted of a pair of bristles placed in the path of the arms of the balance or foliot, and mounted on a pivoted arm. By moving the arm, the bristles could be moved closer to or further from the centre of the balance and thus limited the arc of its swing. A smaller arc would speed up the watch and a larger arc slow it down. It was a crude and unsatisfactory arrangement, although the fusee set-up is not much better as a means of regulation.

General characteristics and national styles

From the earliest times, watches had a striking train, and such watches are known as clock-watches. When Florimund Robertet, Treasurer successively to Charles VIII, Louis XII and François I, died in 1532, he left no less than twelve watches (presumably all French) which were listed as follows: 'Twelve watches, of which seven are striking and the other five silent, in cases of gold, silver, and brass of different sizes; but of these I (the widow) attach value only to the large one, merely of gilt copper, that my husband had made; it shows all the stars and the celestial signs and motions, which he understood perfectly'. The latter is also the first reference to a watch with complications.

French and German watches had a somewhat different system of striking, and when English watches began to appear at the end of the century, they followed the French system.

German watches had striking trains less frequently than the French, an alarum being preferred. Sometimes a watch had both striking and alarum. Sometimes, again, the alarum mechanism was separate, so that it could be attached to the watch proper.

The earliest plate pillars were plain, square or round in section, and riveted to the bottom plate. The top plate fitted over them and was secured by pins, as continued to be the practice until well into the nineteenth century. Even in this earliest period watches had at any rate one screw. In stackfreed watches it secured the stackfreed spring, and in fusee watches it secured the set-up click.

After about 1580 national styles began to develop more markedly and a native school began to appear in Britain.

German watches developed slowly. Brass plates began to appear about 1580, and sometimes brass wheels; but iron trains and the stackfreed continued in use throughout the first quarter of the seventeenth century. Finally, the Thirty Years' War (1618–48) so crippled the country that Germany ceased to be an important factor in horology, and France was left in undisputed supremacy, which she was to enjoy until successfully challenged by Britain in the last quarter of the seventeenth century.

By 1600, the watch movement had reached a fairly sophisticated state, yet it remained a villainously bad timekeeper. The daily rate might easily vary by a quarter of an hour. No means of improving this being found, it was natural that attention should turn to decoration, both of the movement and the case, and to mechanical complications, such as calendar and astronomical trains.

In French watches, the balance cock became larger in size and pierced and engraved in a roughly spiral pattern. Similar pierced and engraved decoration was used to cover the locking plate of the striking train, alarum stopwork and set-up ratchet. The maker's name also began to appear on the plate. — 63

The British affected a more floreate pattern of decoration for their balance and other cocks, and for a short time after 1600 put an engraved border round the edge of the top plate. — 65, 66, 71

In about 1620 the balance cock began to be screwed to the plate instead of pinned, although it continued for a time to be fitted over a pillar at the junction of the foot and the table.

As the century progressed, the cock tended to become less oval, and more circular, completely covering the balance. — 74, 85, 94

Pillars became increasingly decorative, with spiral and other fancy shapes; but the Egyptian tapering form was most common, sometimes pierced and engraved. After about 1660 the very decorative tulip shape began to appear, as also did other forms of decoration between the plates, such as turned arbors, engraved spring barrels, and a pierced mounting for the fusee stop arm. — 97 108

Especially in British watches, the maker's name was inscribed on the top plate in an elegant, tall, sloping script, among which Edward East's is outstanding. — 86

Three-wheel trains and a 14-hours going period continued to be almost universal up to 1650 and quite usual up to 1675. In rare cases, where 26-hours going was attempted, this was sometimes achieved by the questionable practice of using five-leafed pinions instead of six, rather than by introducing a fourth wheel.

During the third quarter of the seventeenth century it became fashionable to make thin watches, and some really surprisingly thin movements were produced latterly. To achieve this, the bottom plate was often cut away to allow the balance wheel potence to be recessed into it. — 86, 92

Fusee chains were introduced for table clocks by about 1630 and the earliest chains have very long links. But gut continued to be used in watches, and a fusee chain is rarely found before 1670.

For the escapement, there was no serious challenger to the verge. Richard of

Wallingford invented (*c.* 1330) and Leonardo da Vinci re-invented a variant of it with two parallel escape wheels with radial, pin-shaped teeth. The verge lay between them, with the pallets at 180 degrees to each other. One pallet engaged with one wheel and the other pallet with the other wheel. The effect is not different from that of a verge.

More significant was the 'cross-beat' escapement invented in about 1580 by Jost Burgi of Prague (1552–1632). This had two balances geared together with one pallet on each and a single escape wheel. The balances moved in a vertical plane and each consisted of a foliot with arms roughly vertical when at rest. Had Burgi made the lower arms longer than the upper he would, of course, have invented a very good compound pendulum escapement. But as it is, his escapement is not very much better than an ordinary verge. Several examples survive, made over a period of about half a century, including one in the British Museum. Although the escapement could have been applied to a watch, this is not known ever to have been done until a practically identical escapement was re-invented almost a century later by Robert Hooke.

Mechanically, the first three-quarters of the seventeenth century was thus a time of almost complete stagnation in watchmaking; the main interest lies in the variety of decorated cases, and for collectors more interested in the external appearance than in the mechanism, it is the most rewarding of any.

Clock and watchmakers' guilds had existed abroad from early in the sixteenth century, but despite various attempts, nothing was achieved in England until 1631, when the Worshipful Company of Clockmakers was founded, with David Ramsay (*c.* 1600–50) as its first master. From this time, under the leadership of such makers as Ramsay and Edward East (1602–96), the quality of British watchmaking improved steadily, until by 1675, with the development of the balance spring, the British makers were able to take the lead from the French.

The balance spring 1675–1700

The invention and development of the balance spring mark the turning point in the history of the watch. What had previously been little more than a rich man's ornamental toy became suddenly a scientific instrument, capable of running accurately within two minutes a day. In the story of it the names Robert Hooke, Christian Huygens, Thomas Tompion and Isaac Thuret figure prominently.

Robert Hooke (1635–1703), the first 'Curator of Experiments' to the newly founded Royal Society, had an extremely enquiring and inventive mind which ranged over most scientific problems of his day, including those of horology. In 1658 he began making experiments to improve the timekeeping of watches, by using a spring to control and regularise the oscillations of the balance. In 1668 Lorenzo Magalotti, of the Florentine Academy, visited London and saw Hooke demonstrating his invention before members of the Royal Society. He described it as 'a pocket watch with a new pendulum invention. You might call it with a bridle: the time being regulated by a little spring of tempered wire which at one end is attached to the balance wheel, and at the other to the body of the watch. This works in such a way that if the movements of the balance wheel are unequal, and if some irregularity of the toothed movement tends to increase the inequality, the wire keeps it in check, obliging it to make the same journey.'

Evidently this was a straight spring, anchored at one end to the plate, with its free end playing between two pins set vertically on the rim of the balance. Hooke claimed to know twenty ways of attaching the balance spring to a watch, but, withheld the best until he should have derived some benefit from it'. He also experimented with two balances geared together, with one pallet on each (presumably a revival of Burgi's cross-beat escapement); at another time he tried out a loadstone as a regulator.

Hooke was not alone in this field. As early as 1660 Christian Huygens van Zulichem (1629–95), the eminent Dutch physicist, mathematician and astronomer, saw a balance-spring watch being made by Martinet to the designs of Pascal and the Duc de Roannais. Huygens did not think their method practicable and said that he himself already knew of a better.

As often happened with Hooke, having had the seeds of an idea, he laid it aside before bringing it to fruition, and when someone else did so he accused them of stealing his invention. It seems that he abandoned his experiments in 1668, whereas Huygens continued his. Like Hooke, he had tried two balances geared together, but eventually chose a single balance with a spiral spring. However, it differed in one important way from what was to become standard practice. Instead of the balance wheel and spring being applied to the verge, the balance with its spring was on a separate staff, geared to the verge in such a way that instead of having the usual verge arc of about 100 degrees, the balance described several complete turns at each vibration. To make this possible, his balance spring had four turns, as against the one and a half or two turns found in the earliest surviving balance spring watches.

The year 1675 saw a splendid contest develop between Huygens and Hooke for priority in inventing the balance spring. It seems that both were active towards the end of 1674, as on October 4th Hooke notes in his diary: 'Tompion here all day. Discoursed . . . of the ways of springs etc.' Huygens brought his ideas to finality over Christmas in Paris, and on January 22, 1675 explained his invention to Isaac Thuret under the seal of secrecy. Thuret worked fast to put it into practice as only two days later Huygens notes 'Thuret showed me his trials at making a watch'. On January 30th he wrote to Henry Oldenburg, first secretary of the Royal Society, telling him of the watch, and enclosing an anagram to establish the priority of his invention. On February 20th he sent the solution of the anagram which is: 'Axis circuli mobilis affixus in centro volutae ferreae' (the arbor of the moving ring is fixed at the centre of an iron spiral).

Huygens next sought a patent in France, in which he was first opposed by Thuret who, fearing that he might get left out, laid a counter-claim. This Huygens defeated without difficulty and eventually forgave Thuret, who made an abject apology. But the Abbé Jean de Hautefeuille also laid claim to the invention, and although his experiments had been with a straight spring, or alternatively with a helical spring in tension (as later used by Harrison in his first three marine timekeepers), Huygens decided that it would be too troublesome and expensive to establish his claim, and so left his invention free to anyone who might care to profit by it. However, he did not despair of getting an English patent and offered Oldenburg an interest in the invention if he could make use of it.

In the meantime, on February 17th, Hooke had been 'At Mr. Boiles. He told me of Mr. Zulichem's watch with springs' (Hooke always referred to Huygens by the second part of his surname). He at once hurried round to Tompion to make a watch which should establish his own priority of invention. It is interesting to speculate how Robert Boyle came to have this early knowledge of Huygens' watch. It may have come from Denis Papin, best known in his own right as a pioneer in the evolution of the steam engine; but also employed as an assistant by Huygens. Papin, who was a Huguenot, left France in this significant year of 1675, to avoid religious persecution. He then settled in England and was employed by Boyle who may thus have learnt of Huygens' horological experiments.

On February 20th Hooke writes: 'Zulichem's spring not worth a farthing', but unless he had received detailed information from Boyle this can hardly be more than bravado, since no watch by Huygens seems to have reached England until April at the earliest, or quite possibly until the end of June, and Oldenburg could not yet have received the solution of his anagram. Indeed, this is borne out by Hooke's diary for March 18th, 'Saw Zulichem's Watch scheme and transcribed it' (presumably meaning the anagram). Already, on March 6th, he had been 'At Sir J. Mores. He told me of Oldenburg's treachery his defeating the (Royal) Society and getting a patent for Spring Watches for himself'. And on April 3rd 'Segnior told me of Oldenburg procuring a patent for Zulichem'.

In the meantime Tompion had been active and had produced a watch with a balance spring, though of what type is not disclosed. On April 7th Hooke was 'With the King and shewd him my new spring watch. Sir J. More & Tompion there. The King was most graciously pleased with it & commended it far beyond Zulichem's. He promised me a patent'. On April 8th the President of the Royal Society, 'Lord Brounckner and Oldenburg discovered their designe. R. Southwell told me of the King's refusing the warrant to Oldenburg after I had left the King. I vented some of mind against Lord Brounckner & Oldenburg. Told them of defrauding'. Nevertheless, on April 10th he was 'At Sir Jonas Mores who told me the King's message and that he said unless we made haste with the watch he would grant the patent'.

The watch Hooke had showed the King on April 7th was a hurriedly run-up affair in a brass case, and Tompion was thereupon set to make a finished article in a gold case for the King. It was this that the King was pressing for on April 10th, and it was finally delivered on May 17th when Hooke went 'with Sir J. More to the King who Received the watch very kindly, it was locked up in his closet'.

All this time Hooke and Tompion were at work on all sorts of springs. 'Tompion I shewd my way of fixing Double Springs to the inside of the Ballance wheel' (an impractical idea, consisting of no more than a balance with two curved, spring steel spokes). 'Tryd perpendicular spiral spring at Tompions . . . did well' (this was probably a helical spring in tension between two balances, which he later illustrated in one of the Cutlerian Lectures) and again 'the thrusting spring did best'. So Hooke had certainly not decided at any time during 1675 that a spiral spring was best. What is equally uncertain is when one of Huygens' watches first arrived in England. Charles II seems to have had one as early as April, but in the correspondence between Oldenburg and Huygens there is no mention of it, and Oldenburg does not seem to have received one until the end of June. On July 1st he thanked Huygens for it. This watch was destined for Lord Brounckner and had a regulator in the form of a sliding piece for shortening the spring. Oldenburg tells Huygens that 'no one except the King has yet seen Hooke's watch'.

No more is heard of Huygens' watch, but all the year the Hooke-Tompion instrument went backwards and forwards; thus on August 4th 'Gave Tompion King's watch' and on the 8th 'Tompion here. King's watch spring loose' and even a year later, on June 14th, 1676 'At Tompions he had mended King's watch'.

Hooke continued to vilify Oldenburg as 'a trafficker in intelligence'. Oldenburg wrote to Huygens on October 21st that Hooke, having learnt that Huygens had offered Oldenburg the benefit of any privilege granted to his watches in England, had publicly declared that this was a reward to Oldenburg for having disclosed Hooke's invention. Huygens dutifully wrote to Brounckner disclaiming any such disclosure by Oldenburg.

The last we hear of Huygens' watch is in Hooke's diary for December 14th, when Sir John More told him 'of Zulichem's new watch moving $\frac{1}{4}$ turn'. Historians seem previously to have overlooked these important words, which show that by the end of 1675 Huygens had finally abandoned his pirouette balance, and fixed the balance wheel and spiral spring on the verge itself, as has been the universal practice ever since.

Hooke continued to pester Tompion with experiments. Tompion, however, no doubt having decided that Huygens' lay-out was best, became increasingly tardy in their execution, so that over Christmas he appears in Hooke's diary successively as a 'slug', 'a clownish churlish Dog, I have limited him to 3 day & will never come neer him more'; despite which, after five days, 'Tompion a Rascall'. But they made it up and, so far as is known, continued friends throughout Hooke's life.

In the end, Charles II granted no privilege to anyone.

As to Huygens' priority in arriving at a practical watch with a spiral balance spring there can be no doubt. It is impossible to say if Hooke had ever experimented with such a thing before Huygens' invention became public. The evidence as set out above seems entirely against him, and the evidence of William Derham, in *The Artificial Clockmaker* was not only written twenty years later, but is manifestly partial.

Nevertheless, in 1675, when Hooke maintained that he had 'found out' the spiral spring long previously, Sir John More and Christopher Wren supported him and they were accurate observers of the utmost integrity, who were in a good position to know the facts. Finally, there is the opinion of Tompion himself. In a biographical sketch of Hooke appearing in 1740 in John Ward's *Lives of the Professors of Gresham College* Ward writes: 'I have lately seen a round brass plate, which was formerly a cover to the balance of one of Mr. Hooke's watches. It is cut through in the form of sprigs, and has on it this inscription "R. HOOK invenit an. 1658. T. TOMPION fecit 1675". This plate is now in the hands of the ingenious and accurate Mr. George Graham, fellow of the royal society, who informed me, that he heard Mr. Tompion say, he was imployed three months that year by Mr. Hooke, in making some parts of those watches, before he let him know, for what purpose they were designed; and that Mr. Tompion was likewise used to say, he thought the first invention of them was owing to Mr. Hooke.'

It seems unlikely that the matter will ever be settled conclusively.

The movement 1675–1700

An invention of Hooke's about whose authorship and immense utility there is no dispute, is his wheel-cutting or dividing engine, which he also evolved in about 1670. It is still used with little variation.

By the end of 1675, other watches had been made or commissioned on Huygens' and Hooke's principles and it is thus fairly certain that during 1676 the balance sprung verge watch as now known began to appear in increasing numbers.

Tompion made the first improvement in the form of his regulator. The curb-pins moved on a segmental rack corresponding with the outer turn of the balance spring. This was geared to a wheel with a squared arbor, carrying a numbered plate such as appeared on the later set-up regulators. This was preferable to Nathaniel Barrow's regulator, which looked more like the old set-up regulator, with its endless worm on a long arbor pivoted on the plate. A slide moved along this with two pins spanning the balance spring. The defect of this was that the spring had to end in a long straight section, for the operation of the regulator. Tompion's type of regulator became universal, and Barrow's type is now extremely rare.

Tompion was not of an inventive nature, but he was very good at putting an idea into execution. He was also a very good craftsman and later, when he had become successful, he had the ability to secure first-rate work from his employees. It was on this footing that he built up his reputation. Others, such as Quare and Knibb, might approach him at their best, and frequently their products are to be preferred to his in terms of elegance; but for consistently good work he was in a class of his own. Undoubtedly, too, he owed much to Hooke for introducing him into the Royal circle. Apart from his regulator, Tompion's only other innovation in the watch field was an escapement which is illustrated in Rees's *Cyclopaedia*. It has been described as a forerunner of the cylinder escapement, but in fact it is nothing so much as a prototype of the virgule (to be described later) since impulse is given in only one direction. Even the idea of this may have come from Hooke who on October 15th, 1676 'taught him the way of the single pallet for watches'. Tompion took out a patent in 1695 in conjunction with Edward Barlow and William Houghton; but none of them seems to have followed it up. It is described in the patent as 'a ballance wheele either flatt or hollow, to work within & crosse the centre of the verge or axis of the balance with a new sort of teeth made like tinterhooks to move the balance & the pallets of the axis or verge, one to be circular, concave & convex'.

As soon as the balance spring was established it became common to use four wheels with a going period of twenty-six hours. Some makers, both French and English, Tompion included, at first thought they could do without a fusee, but, in conjunction with the verge escapement, they quickly found that their optimism was unfounded.

The need for a set-up regulator did, however, become superfluous except as a means of once-for-all adjustment by the maker or anyone cleaning the watch. It was accordingly moved to between the plates in English watches, with a worm drive as previously; or became a ratchet between the plate and the dial, in most French watches.

An entirely new and most useful invention came ten years later, in the form of a repeating mechanism. It was invented simultaneously by Edward Barlow (1636–1716) and Daniel Quare (1647–1724) and both applied for a patent; James II favoured

Quare on the grounds that, on his watch, one push piece caused both the hours and quarters to sound, whereas in Barlow's version there were two push pieces, one for the hours and one for the quarters. The patent was granted in 1687.

The first repeaters struck on a bell filling the back of the case, like the earlier clock-watches. A single stroke denoted each hour and a double stroke each quarter. Somewhat later, some were fitted with a 'pulse piece'. When this was pressed and the repeater operated, the hámmer and bell did not sound; but the operator was informed of the time by a series of taps on the pulse piece. In the last quarter of the eighteenth century Le Roy and Breguet substituted wire gongs for bells, and Breguet often had only a block of steel which was struck by the hammer. These are known as 'dumb' or 'à toc' repeaters.

The work of Tompion and Quare was faked in their lifetime and throughout the first half of the eighteenth century, and latterly that of Graham; almost as freely as Breguet in the nineteenth century. Tompion and Quare fakes usually come from Holland and are easily detected. Dutch fakes almost always have an arcaded minute-circle on the dial. Quare had a very thriving export business. Some seemingly authentic Quares are signed 'Quarré' or 'Quaré'. It is thought that these are not necessarily fakes, if the quality of the watch is up to Quare's standards; and that he was simply making things easier for foreign customers who might otherwise find the pronunciation of his name insuperably difficult.

There are fortunately a good many aids to judging the validity and date of a Tompion watch (assuming it is not hallmarked, which silver cases seldom were before about 1740).

From 1701–1707 or 1708 Tompion was in partnership with Edward Banger and pieces are signed 'Tompion & Banger'. In about 1711 he took George Graham into partnership, but pieces with the joint signature are very rare (Tompion died only two years later, in 1713). For Tompion's system of numbering see p. 286.

The movement 1700–1750

In 1704 Facio de Duillier and P. & J. Debaufre, foreign artists domiciled in England, discovered a method of piercing jewels for the pivot holes in watches. At first, it was used only for the balance staffs of high-grade watches, and Graham and later Mudge used it for the balance arbors of their cylinder watches from about 1725 onwards. The jewelling of the train came increasingly into use with the development of the precision watch from about 1775. Even then it was entirely unknown abroad, and Breguet was the first foreign artist who succeeded in jewelling his watches, by using English workmen. The French instead used polished-steel end plates for their balance arbors, called 'cocquerets'.

Facio and his confederates applied for a patent, but were defeated by the production of a watch by Ignatius Huggeford with what was ostensibly a large jewelled bearing to the balance staff. This watch survives in the Guildhall Museum. Only in 1848 was it dismantled and the jewel found to be unpierced, and decorative only.

The next improvement came from France, where the English Henry Sully (1680–1728) had the idea in about 1715 of oil sinks for retaining oil. These were finally perfected by Julien Le Roy. About 1740 Le Roy also improved the method of mounting and locating the crown wheel in the verge escapement. This he did by screw-adjusted sliding plates containing the pivot holes, whereby the escapement could be adjusted with great accuracy, both laterally and as to depth. Le Roy was an artist of the highest quality who did much to raise the standards of work in the French industry to a level at which they could again rival British watches.

44   In 1704 Debaufre invented an escapement which goes by his name, or is known alternatively as a 'club-foot verge'. It consists of two co-axial escape wheels operating on a single steel or jewelled pallet on the balance staff. This pallet has 'dead' faces, on which the escape wheel teeth rest during the supplementary arc, and sloping impulse faces down which the points of the teeth slide alternately, giving impulse to the watch at each swing of the balance. This was the first of the 'frictional rest' escapements of which the cylinder is the best known. It is not known if any example by Debaufre survives, but in any case it was little used for nearly a century, until the Lancashire makers took it up, after which it became known more generally as the 'Ormskirk escapement'. It was long supposed that the recoil inherent in the verge escapement was a bad thing; but in fact in an escapement which is not at all isochronous, recoil can provide an automatic corrective to small variations in motive power, and thus serves to preserve a constant arc. Thus, although the frictional rest escapements are theoretically better than the verge, it is only in their later and most sophisticated forms that they produce better results; while the verge, as carried to its final perfection

II   by John Harrison, proved to be capable of an almost miraculous accuracy. In 1718

45   Sully devised a variant of Debaufre's escapement, which he used in his unsuccessful marine timekeeper, and in 1735 it was developed further by Julien Le Roy. It differs from Debaufre in having only one escape wheel. Its use in watches is, however,

205 exceedingly rare.

4   In 1725 came the cylinder escapement of George Graham (1673/4–1751). Graham was apprenticed to Tompion, married his niece and was for a few years in partnership with Tompion up to the time of the master's death in 1713. He was the outstanding British maker in the second quarter of the eighteenth century and fully carried on Tompion's standards of craftsmanship. He was, moreover, of a more original mind than Tompion, as is evidenced by his election as a Fellow of the Royal Society. In clocks, he made the first highly accurate instruments, when fitted with his dead-beat escapement and mercurial pendulum. It is difficult to see quite why the cylinder escapement acquired the pre-eminence it did, since it is exceedingly difficult to make (especially the escape wheel); much more fragile than the verge; wears rapidly; and does not (at any rate as made before Breguet) give a greatly superior performance to the verge. Probably it was Graham's established position, and the great elegance of his work, that did much to establish the cylinder escapement. Julien Le Roy saw it at an early date and preferred it to the verge, and Graham used no other escapement from about 1727 onwards. After his death, cylinder watches almost identical in every

151 way continued to be made by Mudge, then by Mudge and Dutton, until well into the 1770s. Graham and Mudge used a brass escape wheel and a steel cylinder, but John Arnold and John Ellicott sometimes used a steel wheel and a ruby cylinder. Ruby cylinders are, however, rare in English watches. James Ashley mitigated the

rapid wear of ordinary cylinders by staggering the teeth of his escape wheels at three different levels, thus reducing wear on the cylinder threefold.

Somewhat allied to the cylinder but really a development of Tompion's escapement is the virgule, which is a frictional rest escapement, like the cylinder, but gives impulse only on alternate swings of the balance. The impulse pallet takes the form of a long curved tail to the 'cylinder', giving it the appearance of a comma, whence the name 'virgule'. As far as can be ascertained, it was first devised by Jean-André Lepaute, but it was not much used until the last quarter of the eighteenth century, by Jean Antoine Lepine. A very rare variant, the 'double virgule', was invented by his brother-in-law Pierre-Augustin Caron, with two lots of teeth on opposite sides of the escape wheel, and two corresponding pallets, which thus secured impulses in both directions.

Although belonging more properly to the end of the century, and the precision period, the virgule escapement can hardly be regarded as one of precision, and it is therefore mentioned here because of its derivation from Tompion's prototype.

Another escapement of this period which did not come into use until much later was the rack lever, invented by the Abbé de Hautefeuille in 1722. This contained the seeds of the lever escapement, the escape wheel and anchor being identical. But instead of being detached, the lever was geared directly to the balance by a toothed sector on the lever, meshing with a pinion on the balance staff. ·It is not known whether Mudge was indebted to it for the idea of his detached lever escapement, and it did not come into general use until after 1791, when a considerably different form of it was patented by the Liverpool maker, Peter Litherland.

Despite this variety of escapements, the verge continued to be used in the great majority of watches throughout the eighteenth century. Although the arc is limited to about 120 degrees at most, so that there is no need for a long balance spring, this did gradually increase from its original one and a half turns .to as much as four. When it increased even more, during the nineteenth century, its timekeeping qualities suffered.

An immediate effect of the balance spring was to make watches much thicker. So long as they remained very inferior timekeepers there was a tendency to make them thin, and this applies particularly to enamelled watches. But it was always recognised that a large-diameter escape wheel was more conducive to good timekeeping; so that when timekeeping became the primary objective, the thickness of an average English watch increased from about 25 mm to 31 mm, or 38 mm for a typical French watch of the period which, from its almost spherical shape, earned the name of an 'oignon'. Later in the century the situation was reversed, and French verges became slimmer than the English, with detriment to their timekeeping. Otherwise, French balance spring verges only differ from English to any important extent in having very much larger balances; sometimes the full diameter of the plate. And these are supported in a bridge cock, instead of the single-footed English cock. These bridge cocks may be pierced, but sometimes they are solid and engraved, or covered by a painted enamel plaque. Oignons are generally wound through the dial, but this is very rare in English watches, except for clock-watches and repeaters.

Decoration of the movement 1675–1800

IV, VIIB, C, VIII   Soon after the introduction of the balance spring the decoration of English watch movements became fairly standardised. The floreate-patterned cocks of the earlier period gave way to an arabesque pattern, at first bold and open, but becoming increasingly fussy and perfunctory. The table, completely covering the balance, was circular, although not at first with a solid rim. The foot was more irregular in shape, but by 1690 both had acquired a well-defined rim, that of the foot following the edge of the plate. Most cocks have a mask engraved at the point where the table joins the foot, but prior to 1685 this is very small, and before about 1680 it may not exist. This mask survived in British verge watches until the middle of the nineteenth century. Up to about 1740 the pattern of the cock decoration is symmetrical about the centre line, but thereafter, with the rococo taste, it becomes generally asymmetrical.

At the invention of the balance spring, the application of the pendulum to clocks was within the memory of a great many people, and it was found that the newly accurate watches sold better if they appeared to have a pendulum. In France, watches were therefore made with solid cocks, with an annular slot through which a seeming pendulum bob was seen to swing, but which was in fact no more than a disk on one arm of the balance. This arrangement was applicable to French watches where the bridge cock could be made to conceal the balance wheel completely. With the English

114   form of cock, the illusion was not so impressive, and the matter was arranged differently. Here, the balance was generally placed between the back plate and the dial and the bob appeared through an annular slot in the latter. This silly delusion is not found much after 1690.

Pillars were made in considerable decorative variety in British watches, especially the Egyptian, baluster and tulip, and variants of them. The French seldom used anything but the Egyptian type in their oignons, and plain baluster pillars in their later, thin watches, where the movement did not hinge out of the case.

The cursive script of the maker's name becomes less bold soon after 1680 and, soon after 1690, increasingly appears in plain capitals, as it almost always did in France.

Dust caps fitting over the movement are very rarely found before 1715 and seldom before 1725, after which they become usual for watches of any quality; especially

120, 139   Graham's and Mudge's cylinder watches. They are sometimes of silver, but generally of brass-gilt.

After about 1725 the cock foot is increasingly solid and engraved, instead of being pierced. After the introduction of compensation balances in the last quarter of the century the table ceased to be circular and became wedge-shaped. Sometimes it was

139   solid and engraved, but pierced cocks, including Arnold's, continued up to the end of the century. Verge watches continued to have pierced circular tables until the middle of the nineteenth century.

The beginnings of the precision watch and the modern watch

The formative years of the precision watch

If the third quarter of the seventeenth century saw the greatest advance in the whole history of horology, with the practical application of the pendulum to clocks and the balance spring to watches, the third quarter of the eighteenth century produced advances no less spectacular in the realm of marine timekeepers and precision watches. There has certainly been no time when there was greater experimental activity among the leading artists in England and France.

In the formative period of the precision watch its development is so closely connected with that of the marine timekeeper that the two cannot be considered separately. It is from the marine timekeeper that the precision watch was developed, and in its first decade its only serious champion was John Arnold. However, in the quarter-century before he made his first precision watch in 1773 a complicated and fascinating web of history has to be unravelled and put into some sort of chronological order.

Five very different personalities are involved: John Harrison, Thomas Mudge, Pierre Le Roy, Ferdinand Berthoud and John Arnold. Each made an essential contribution to the final emergence in 1773 of a really accurate pocket watch. Each contribution may be clearly defined and it is only difficult to establish priority between Berthoud and Arnold who worked, quite independently, on closely similar lines, over the same critical period.

JOHN HARRISON

John Harrison made no final contribution to the science of horology but he did prove that a portable marine timekeeper could perform well within the limits of the famous Act of Queen Anne: roughly two minutes in six weeks. This he achieved by detailed refinement of the basically unsound verge escapement. However, he did make the first pocket watch with two of the prime essentials of a precision timekeeper: namely, maintenance of power during winding, and temperature compensation. This 146 was the watch used as a test piece for the successful H.4 and contained all the essential features of No. 4 except its remontoir. It was made for Harrison in 1753 by John Jefferys and is on display with the collection of the Worshipful Company of Clockmakers at the Guildhall in the City of London.

44

II H.4 went through its trial in 1761 and demonstrated convincingly that it was a practical and accurate marine timekeeper. This was upheld by H.5 and Kendall's copy, known as K.1.

Harrison's timekeeper represented the triumph of workmanship over design but it could never have formed the basis of a commercially-produced instrument. For that, five ingredients are essential:

1 Maintaining power during winding.
2 Temperature compensation incorporated in the balance wheel.
3 Immunity from position errors.
4 A detached escapement.
5 A simplicity of design and a standard of workmanship which will enable an acceptable standard of accuracy to be maintained over a period of years.

PIERRE LE ROY

Pierre Le Roy was the first man to comply with all of these except immunity from position errors which, indeed, never have been overcome entirely.

Galileo made the first detached escapement in 1641 and it seems remarkable that this device was not taken up for over a century. It was to be 120 years before another detached pendulum escapement was produced. Galileo's escapement used a pivoted detent but, as with all the early pivoted detent escapements, unlocking and impulse took place, with a recoil, at the end of the balance arc and not at or near its centre or dead-point, which is essential for good results.

Thiout designed a version of Galileo's escapement, applicable to a pocket watch with a balance spring, in about 1740, and Le Roy refined this in an escapement he presented to the Académie Française in 1748. This had the same defect (that unlocking and impulse took place, with a recoil, at the end of the balance arc) and before the detent escapement could become effective it needed some means of allowing the balance to disengage the detent either side of the centre line of each vibration. This refinement was the passing spring and Le Roy failed to think of it.

In a later escapement he realised the requirement by using an escapement with a pivoted lever which, being subsequently re-invented by Robin, is known universally 40, 41, 42 as the Robin escapement. In it the lever has two locking pallets and, in effect, a fork to engage the balance. Impulse is given at alternate vibrations. During the impulse vibration the lever is carried to one side to release the escape wheel and allows it to impulse the balance. The escape wheel is then locked on the second of the two locking pallets. During the return vibration the lever is carried back to its original position and the escape wheel advances by a small amount to transfer the locking, in readiness for the next impulse vibration.

Le Roy devised two methods of achieving temperature compensation. One consisted of a wheel with two bi-metallic rims of the kind which was to become standard for a century and a half. But he never developed it, because he found that it took on a permanent deformation in use. He therefore preferred his alternative of glass tubes attached to the balance, containing mercury and alcohol. This worked very well

45

although with its size and delicacy it could not have been applied to a pocket watch. Le Roy made two of these timekeepers which were tested at sea in 1765. One erred thirty-eight seconds in forty-six days and the other only seven and a quarter seconds. Finding small appreciation of his efforts in France, Le Roy then retired from the scene, and although he made some subsequent experiments with precision watches, which will be mentioned later, these did not lead to anything. (See p. 53.)

Of the five pioneers, Ferdinand Berthoud is by far the most difficult to assess, despite the high survival rate of his work, and his voluminous writings.

Berthoud was born in 1727 and his first attempts at a marine timekeeper were as massive as, and far more crude than Harrison's earliest machines of thirty years earlier. No. 1 was completed in 1761. It was weight-driven and had an undetached lever escapement, which the impulse pin was permanently imprisoned in a slot in the lever. Of all his archaic machines, the most successful was No. 8, which had a form of ruby cylinder escapement and gridiron compensation. It was first tested at sea in 1768 and again, for a whole year, from October 1771–2. During this time it performed amazingly well, its extreme variation in daily rate being only six seconds. For the most part, it kept within one second gaining. This, like H.4, merely showed how patient development and attention to detail can make an isolated, basically unsound machine keep time within astonishingly close limits. Berthoud understood very well the basic weakness that the escapement required oiling, although he states that he once ran it dry for three months, when the arc was greater than when oiled and did not vary during the test period. He concluded from this that a ruby cylinder does not need oiling; and from the regularity with which Breguet ruby cylinder watches will perform over a considerable period of years, it does appear that his remarkable conclusion was not entirely unjustified.

It was when Berthoud began gradually to move towards a detached escapement that his work began to become significant in the history of the precision watch. But even by 1773, when he had produced a quite sophisticated detached detent escapement, he showed how reactionary he was by writing that although these small, spring-driven marine timekeepers were all very well for battleships or the merchant navy, for serious navigation on a long voyage there was nothing to equal his weight-driven monsters.

In the meantime, he had made one or two attempts to develop an accurate pocket watch. The first, in 1764, had a verge escapement and gridiron temperature compensation; but in 1773 he admitted that it did not come up to his expectations, largely because he did not have workmen who could work to the fine limits necessary for a precision pocket watch. He therefore sold it to the King of England, and the watch is now in the British Museum. It was not until about 1784 that he was able to write of 'one of my workmen who has since become skilful enough to execute these small machines successfully'. This was almost certainly his nephew Louis Berthoud, born in 1750, of whom he wrote in 1787 that he was now solely in charge of developing and making the spring-driven 'montres marines' and pocket watches.

Berthoud continued to distinguish sharply between the big, weight-driven 'horloge marine' and the small spring-driven ' montre marine' and confused attempts to date his work by sometimes (but not consistently) using different sets of numbering for the two types. He returned at least twice to a new series, starting at No. 1.

However, it does seem clear that he produced a marine timekeeper with a somewhat crude detached detent escapement some time between 1771 and 1773. He approached it by what is evidently a rather clumsy version of Le Roy's 1765 Robin escapement.

The new timekeeper was a montre marine, apparently No. 3, which started life with a type of double virgule escapement, but Berthoud then modified it to receive his first detached escapement. The date is not clear but about 1768 seems likely. In it he returned to the lever but used it only to unlock the escape wheel, and to this extent he was only repeating what Le Roy had done three years earlier. The main defect of Le Roy's escapement was that unlocking and impulse took place at the rim of the very large balance, so that it was difficult for the escape wheel to accelerate sufficiently rapidly to impart an effective impulse to the fast-moving rim of the balance; in fact the balance arc was no more than a meagre 90 degrees. Berthoud evidently realised that he had got to move the action on to a small roller at the centre of the balance. But even so, he seems to have doubted if his escape wheel could move fast enough to catch up the impulse roller and give it an adequate impulse. He therefore conveyed impulse through a lever pivoted closer to the escape wheel than to the balance. When not required for duty this lever was held against a stop by a return spring. It seems he had first contemplated this escapement in 1754 but did not make it until about 1768.

It is interesting that at the time he was working on it, Berthoud had already seen Mudge's lever escapement, since he wrote in 1773 'M. Mudge, a very capable artist, an horologist in London, showed me in 1766 an escapement with free vibrations which he had made a long time previously'. This can only be Mudge's lever, and presumably what Berthoud saw was the original clock of 1754 now in the British Museum. It is therefore remarkable that he did not use it, instead of the complicated arrangement just described. Probably he was put off by the fact that the lever escapement needs to be oiled, to which Berthoud was greatly and rightly opposed (even if he did use oil in his own escapements when he felt like it).

The new escapement was not a success because Berthoud had not incorporated a safety action into it, and it was prone to trip. He therefore went back to his non-detached escapements, proposing a cranked staff holding a large cylinder made of a 'rubis de l'orient' (which he confusingly calls an anchor) with an escape wheel having hard steel vertical pins operating on the cylinder. But (writing in 1773) 'I confess that this escapement still has the defect of needing oil. Nos. 6 and 8 had this and went very well, but because of this defect I am working with all my energy to apply to No. 10 a new free escapement and I hope it will fulfil the desired effect'. It seems clear therefore that on his own showing he had not completed a satisfactory detached escapement by 1773. He does show a pivoted detent escapement although it can hardly have worked very well (*Traité des Horloges Marines*, 1773). This is because he had not yet thought of the passing spring, which had already been used by Arnold in his primitive detent escapement in the two chronometers of 1770–71, now owned by the Royal Society.

What made the pivoted detent into a satisfactory escapement was the passing spring and it is here that Arnold has the clear priority of invention.

Berthoud's detent escapement of some time between 1771 and 1773 does get the action into the centre line of the escapement, but by a rather clumsy device which falls short of the passing spring. The detent had a curved tail and when the unlocking pin on the balance roller came in contact with this curve it moved the detent aside. This unlocked the escape wheel which gave impulse. A spring returned the detent to its stop and locked the next tooth of the escape wheel. The detent, which was quite wide on plan, giving it the rigidity in that plane necessary to secure unlocking, was very thin in depth and chamfered. On the return swing the unlocking pin met the chamfered underside of the detent, deflected it upwards and so passed. This was applied to horloge marine No. 4.

Berthoud's next move was to introduce a passing spring onto the roller itself, which made the escapement work satisfactorily, and in many of his subsequent escapements he kept the passing spring on the roller instead of attaching it to the detent. This was applied to horloge marine No. 9 and as it worked well he applied it also to No. 3, which had already had, successively, the double virgule and then the Robin escapement with separate impulse lever. As it is mentioned in the 1773 book it must clearly be prior to that date, but subsequent to the deflecting detent on No. 4.

One might expect the whole chronological sequence to be set out in Berthoud's great *Histoire*, published in 1802, but in fact it is very short of specific dates and gives none for this important sequence of events.

Berthoud's next important move was to invent the spring detent which, on his showing, he did some time between 1780 and 1782. It was first applied to montre marine No. 24 which was started in 1780, completed in 1782 and went to sea in 1784. Nos 23 and 25 went through in the same batch. It worked in compression, in the Earnshaw style, but the passing spring was on the roller. Berthoud makes no claim for its merit except that of simplicity and he remarks that it would be suitable for pocket watches although it seems rarely to have been used in that capacity by him. Louis, in his *Entretiens sur l'Horlogerie*, makes it clear that he regarded it only as a cheap substitute for the pivoted detent, and, as he was in charge of the development and manufacture of the montres marines from about 1785 or perhaps earlier, this probably explains why it was so little used by the Berthouds: perhaps only in his first precision pocket watch made in 1786.

However, the evidence is that Berthoud arrived at the spring detent independently of Arnold and Earnshaw and at roughly the same time.

Berthoud's earliest attempts at temperature compensation used a gridiron, but as soon as he became aware of Harrison's bi-metallic strip he preferred this. He added the complication of making its leverage adjustable so that the degree of compensation imparted to the regulator was also adjustable. (Arnold did the same thing, probably in 1768). The 1773 book makes no mention of a compensation balance, but the 1787 book cited below says he was experimenting with one in 1773. It had two short, straight affixes with a heavy screw mounted on the free end of each; but he could not obtain enough compensation effect from it and had to supplement it by a bi-metallic curb of Harrison type. In the 1787 book a balance with four affixes is illustrated and was presumably in use by then. He says it was put on to No. 8; but No. 8 of what series and at what date is not vouchsafed.

Arnold made his first compensation balance in 1775, so if Berthoud was experimenting with one in 1773 he anticipated Arnold. Both were long preceded of course by Le Roy, but Arnold was the first to achieve complete compensation with a bi-metallic balance.

The foregoing is an attempt to place Berthoud's work in chronological sequence, but this involves correlating the *Traité des Horloges Marines* of 1773; the *De la Mesure du Temps ou supplément au Traité des Horloges Marines, et à l'essai sur l'Horlogerie* of 1787; the *Traité des Montres à Longitude* of 1792 and finally *L'Histoire de la Mesure du Temps par les Horloges* of 1802. It is not always easy to find consistency between them, and dating by instruments themselves is difficult because of the different numbering series he employed and the subsequent modifications introduced into various instruments, notably the original No. 3.

In quantity, Berthoud's production is small. By 1787 he claimed to have made only forty-five marine timekeepers, and Commander Gould estimated that his total production was only between seventy and eighty.

Turning to precision pocket watches, there was a long gap between the 1764 verge and the detent watches, all (it seems safe to suppose) made by Louis. Ferdinand did embark on a second version of the 1764 watch, with cylinder escapement, in 1766, but he was not satisfied with it and nothing further was completed until 1785 (when Louis was thirty-five). This 1785 watch is illustrated on plate V of the 1787 book. It had gridiron compensation and a spring detent escapement with passing spring on the roller. The diameter of the plate is 52 mm. The watch was sold to 'S.A.R. Monseigneur le Prince des Asturies'. It is clear therefore that so far as precision pocket watches are concerned, Berthoud was a very late starter.

John Arnold's chronology is fortunately less difficult to trace and all the necessary research has been done by Dr. Vaudrey Mercer in his book *John Arnold and Son*.

John Arnold was born in 1735 or 1736, that is, eight or nine years after Ferdinand Berthoud. He seems to have turned his mind to marine timekeepers in about 1769 and his first attempt had a detached cylinder escapement with a Harrison bi-metallic curb, adjustable for effect. This may be described as a type of Robin.

In 1770 he made the first of the series of instruments that was to lead him in only five years to the immediate ancestor of all subsequent marine chronometers, with a pivoted detent escapement and bi-metallic compensation balance.

The first 1770 model had a curious see-saw type of detent, its arbor being parallel to the plates. However, it did have a passing spring, thus establishing Arnold's priority in the invention of this vital ingredient of the detent escapement. Dr. Mercer considers that Arnold made five of this type, which he presented to the Board of Longitude in 1771. Two survive in the ownership of the Royal Society. Two went to sea with Admiral Sir Robert Harland at the end of 1770 and he spoke well of their performances, although when three of Arnold's machines accompanied Captain Cook on his second voyage in 1771 they did not perform well.

Arnold went on to arrive at an escapement which, subject to refinement, and when

49

mounted on a spring instead of a pivot, has been used in all marine chronometers   11
to the present day. Only the modern safety action is materially different.

Somewhat curiously, he seems to have used this pivoted detent escapement for the first time in a pocket watch and not in a marine timekeeper. Indeed, in the early years Arnold seems to have preferred his pocket instruments and certainly his early successes were obtained with pocket watches rather than boxed timekeepers. In 1773 one of these watches was worn by Captain Phipps on a voyage and out-performed the other timekeepers taken. Captain Phipps remarked 'timekeepers of this size are more convenient than larger, on several accounts; they are equally portable with a pocket watch, and being kept nearly in the same degree of heat, suffer very little or no change from the vicissitudes of the weather'. From the success of these earliest of all true precision watches it must appear that Arnold was outstandingly successful in minimising position errors.

In 1775 Arnold took out a patent for a helical balance spring and for a compensation balance, but at first these were used only in marine timekeepers. The compensation consisted of a bi-metallic spiral similar to compensation curbs which had been used by Mudge and others for several years, but now mounted on the   147 balance itself. The inner end of the spiral was fixed to the balance arbor and the outer end was attached to a movable cross-arm which moved weights at the rim of the balance inwards or outwards. This was in use for three years and, according to the valuable details of Arnold's balances in Rees' *Cyclopaedia*, about ten or twelve were made. None survives, although a replica was made by the late Mr. Gazeley.

In 1778 the spiral compensator was replaced by two straight bi-metallic strips loosely spigoted at each end into the balance rim. Attached to the centre of each strip was a rod terminating in a weight outside the balance rim. As the bi-metallic strip flexed under changes of temperature the weights moved in or out. These had the   156 merit of being easier to make and Arnold made about twenty of them between 1778 and 1780. This was also the first type of compensation balance to be applied to a pocket watch, the first probably being No. 36 which caused such a sensation when it was tested by the Astronomer Royal at Greenwich over a period of thirteen months. During this time its greatest error was 2 minutes 32·2 seconds and the greatest difference from its mean rate was never so much as four seconds. There were only four days in the whole trial when the difference exceeded three seconds.

This watch also had a helical balance spring, which Arnold had first used in 1775. Soon afterwards he found that by in-curving the two ends he could improve isochronism and this was covered in his patent of 1782. He began to use gold springs by about 1779 but not generally until 1784. Gold is less elastic than steel, but it is non-magnetic and does not rust.

In 1778 Arnold had therefore arrived at a precision pocket watch performing within limits which would be creditable even today; eight years before Berthoud had produced a pocket watch with a detached escapement and even then not with a compensation balance.

In 1780 came Arnold's famous 'double-S' balance of which about forty were made.   VIB, 172, 11 A further six or eight had the same type but with straight, instead of curved affixes. In 1781 or 1782 Arnold made his first spring detent watches, with virtually the modern type of balance, with two bi-metallic peripheral rims to the balance.

50   The pivoted detent watches measure 75 mm in diameter, which is about the

maximum for pocket wear, but the spring detent watches were much smaller, mostly about 50 mm.

17 18 It is not clear whether Arnold changed from pivoted to spring detents to reduce friction, or on grounds of cheapness. He could easily have taken his pivoted escapement and mounted the detent on a spring and this is precisely what Thomas Earnshaw did with his claimed 'invention'. However, this would have placed the necessarily delicate spring in compression which was an unattractive idea to Arnold. He therefore re-designed his escapement entirely so as to place the detent spring in tension. In the event, Earnshaw's simple adaptation is perfectly satisfactory and is the form that has survived universally, although both were in production for about forty years.

Earnshaw claimed to have invented his escapement in 1780, but to have been delayed by Wright and betrayed by Brockbank so that Arnold was able to learn of it and take out his patent in 1782 before Earnshaw's rights could be established. The point is not very important; the change from pivot- to spring-mounting is not so much an invention as a modification. Berthoud, who developed it at much the same time, certainly rated it no higher.

20 At about the same time Peto, one of Brockbank's workmen, devised an escapement which combined the merits of Arnold's and Earnshaw's escapements. He used Earnshaw's lay-out, but mounted the passing spring on the opposite side of the balance staff to the detent which he was thus able to place in tension. For general purposes the advantage is negligible, but where the detent has to be very small or under some additional pressure it has an advantage. When Breguet used a spring detent escapement in a tourbillon watch he employed the Peto variant.

VIIc The useful contribution Earnshaw did make was his method of fusing together the steel and brass laminae of his compensation balances, where Arnold had pre-shaped the two parts and then soldered them together.

Whatever may be said in Earnshaw's favour it must be remembered that he was not able to go into production on his own account until about 1790, by when Arnold alone had borne twenty years of continuous development of the precision watch and the marine chronometer.

In trying to assess the respective merits of Berthoud and Arnold, it is interesting to note that the dates on which they arrived at the pivoted detent with a passing spring and at the spring detent are not very far apart. But Berthoud was his own worst enemy. A compulsive experimenter, he could never leave well alone and was always trying to do better. This is a laudable quality but it militates against quantity production, and it is here that Arnold completely outstripped Berthoud, from the mid-seventies, with his simple, relatively cheap pocket and marine timekeepers. Apart from knowing when to stop experimenting, Arnold had two great advantages over Berthoud. First, he was a superb craftsman, and the state of the art in England enabled him to employ workmen capable of conforming to his standards. Berthoud admitted that until 1785 he could not command the necessary standards. Secondly, Arnold could use jewelled holes throughout his instruments, which were not available in France. Berthoud therefore had to rely on the complicated, bulky and expensive friction rollers or, latterly, on plain brass holes and steel end caps, about which he constantly complained.

It thus emerges that John Arnold was the sole pioneer of the precision pocket watch and for ten years no-one else could approach his standard of performance. 51

Chronometers continued to be the most accurate pocket instruments until the middle of the nineteenth century, although from the 1790s they began to be rivalled by Breguet's lever escapement watches, which were also thinner, and less prone to setting.

THOMAS MUDGE

This leads to a consideration of Thomas Mudge, the fifth of the pioneers of the precision pocket watch. Although he received almost no acknowledgment in his day, it is his work that has had the most enduring effect. This was largely his own fault, because instead of developing his first invention, the lever escapement, he preferred to concentrate on his remontoir-escapement marine timekeepers which were too complicated to have any lasting success and which none but he could persuade to operate with any degree of accuracy at all.

Soon after his invention of the lever escapement, in 1755, Mudge made for the King of Spain a pocket watch with a refined verge escapement, bi-metallic spiral compensation curb and a remontoir let off once a minute. The movement only survives and is owned by the Worshipful Company of Clockmakers. *147*

All the evidence points to Mudge having invented the detached lever escapement in 1754 and almost certainly his first finished product containing it was the bracket clock now in the British Museum. He subsequently made two other much smaller lever escapeement clocks, and at least one watch, or perhaps as many as three. The surviving lever watch belongs to Her Majesty the Queen and always has been in Royal ownership. Two other watches are mentioned in Mudge's voluminous correspondence with his patron Count von Bruhl; these, from the context, may have had levers. One was made for the Count, and one for General Caillard. The Royal watch was completed in 1769 and acquired by or for Queen Charlotte, wife of King George III. Mudge almost certainly made the 'ébauche' of an exact duplicate of it, but whether or not he completed it is uncertain. If he did not, his son had it finished after his death and it remained in the family. Unfortunately the escapement was changed completely in the early years of this century. That Mudge made more than one is established by his son's reference to 'the escapement for a pocket watch first applied by him to the one he made for the Queen'. *55 III 185*

Mudge made two types of lever escapement, the first in the 1754 clock and the second in the watch. The 1754 version has a fork in the form of a loop encircling the balance staff, and the action takes place on the opposite side of the staff to the escape wheel. The purpose of this arrangement evidently was to reduce engaging friction between the balance wheel pin and the fork of the lever. The idea evidently appealed to subsequent pioneers as it was used occasionally by Breguet (in the Marie-Antoinette watch among others) and by Pendleton until the end of the century. *55*

For the Queen's watch, Mudge used separate pallets on the balance staff, placed one above the other. The two halves of the 'fork' on the lever are at different levels, each coinciding with one of these pallets. There is no practical advantage in the arrangement, but in that it must have been tremendously difficult to make this escapement small enough to go into a pocket watch, for the first time ever, its use is understandable, as it did leave Mudge more room for manoeuvre. The temperature compensation is a form of Harrison's straight-line bi-metallic curb. This is the world's most historically important watch, since from it is descended every modern watch with a lever escapement.

Von Bruhl evidently had a keen mechanical brain and saw the great advantage of the escapement for pocket watches. He therefore ceaselessly badgered Mudge to make others, but by the 1770s Mudge was completely engrossed with his constant-force marine timekeeper and he resisted all the Count's blandishments. It seems likely that the lever escapement then almost lay fallow for over a decade after the completion of the Queen's watch, with one possible exception.

After he abandoned the development of his two successful marine timekeepers Pierre Le Roy continued to toy with the idea of precision pocket watches, with the notion that they should be worn by all ships' officers. By comparing them, they would be able to conclude the exact time. This type of watch was known as 'la Petite Ronde'. One complete watch and one or two movements survive. The escapement is Sully's frictional-rest arrangement and the compensation was by a gridiron. They were, there-fore, no advance over the Harrison-Jeffery's watch of 1753, Mudge's of 1755 or Berthoud's of 1764. However, there does survive one movement of size, type and compensation exactly similar to the 'Petite Ronde' but with a lever escapement. It is signed 'Julien Le Roy Ivenit et fecit No. 4757'. The researches of Signor Giuseppe Brusa and Mr. Charles Allix have established that Pierre Le Roy continued to sign watches 'Julien' long after his father's death in 1753 and the number 4757 suggests a date in the mid-1770s. It is possible that the watch started life with a conventional Sully escapement and was later converted to lever; but the visual evidence is against this—there are no signs of any conversion having taken place, and the escapement itself is so curiously constructed as to make it unlikely that anyone would have made it after about 1780. The impulse pin on the balance, and the lever inclines are all rollers, and while this achieves the object of reducing friction it produces an enormous 'drop' combined with a large escapement arc. Whatever the truth is, this remarkable survival does not alter the history of the lever watch, since Le Roy appears to have done nothing more to develop it; he had more brilliance but less stamina than his rival Berthoud.

Mudge finally provided Count von Bruhl with a rough model of the escapement, remarking only: 'I think, if well executed, it has great merit and will, in a pocket watch particularly, answer the purpose of timekeeping better than any other at present known; yet you will find very few artists equal to, and fewer still that will give themselves the trouble to arrive at; which takes much from its merit. And as to the honour of the invention, I must confess I am not at all solicitous about it: whoever would rob me of it does me honour.'

JOSIAH EMERY

Von Bruhl handed the model to Josiah Emery (1725–96), a Swiss workman who emigrated to England, who protested 'that Mr. Mudge was the properest person for such an undertaking, for to own the truth, I doubted whether it would be possible to ever make a common sized pocket watch with an escapement on so large a scale'.

However, in the end he consented to attempt the work and produced his first lever watch in 1782. He introduced various refinements into Mudge's proportions, with the result that his watches have a much livelier action than the Queen's watch. Also, he endowed them with a compensation balance, closely modelled upon Arnold's 'double-S'.

Emery's levers are superbly made and perform very well. But although they are more robust than Arnold's pocket chronometers they have never performed so accurately. Nevertheless, they won for themselves a good reputation and the Prince Regent owned one from about 1784. Emery seems to have turned out about three a year, selling them at £150 in a gold case. If he kept this up until approaching the time of his death in 1796 the inference is that he must have made about forty. Thirteen are known so far to have survived, but they are difficult to date, since most seem to have been re-cased so that the hall marks do not agree with the numbers. The earliest recorded is No. 939 and the latest No. 1379. In about 1792 Emery re-designed the escapement and replaced Mudge's complicated two-plane layout with a simple cranked roller.

According to Mudge's son, all Emery's levers were made for him by Richard Pendleton, and the three surviving Pendleton lever watches bear convincing evidence that such is the case. They have the same bridge cock, 'double-S' balance and helical spring. Two have the original Mudge loop-type fork and one has the Emery crank roller.

From surviving English precision watches with lever or ruby cylinder escapements it seems at least possible that most of them came out of Pendleton's workshop.

Other pioneer makers of the lever watch in England are Francis Perigal, John Leroux, William Dutton, George Margetts, John Grant and a mysterious Mr. Taylor whose initials are not even known.

None survives by Margetts and we have only the word of Rees' *Cyclopaedia* that he made them, with a crank roller escapement. They performed well, but were prone to stop in the pocket.

One watch by Perigal survives. It has a crank roller and Harrison compensation. It has been re-cased so that it cannot be dated by the hallmark, but a date in the late 1780s seems likely. Thomas Dutton produced one surviving watch (certainly made for him by Pendleton) with a crank roller.

Outstandingly the soundest, horologically, was John Leroux by whom two examples survive; one belongs to the Worshipful Company of Clockmakers and the other, a movement only, to the British Museum. The former is hallmarked 1785 at which date probably no-one but Emery-Pendleton was attempting the escapement. The most important feature of Leroux's watches is that they contain draw. This is the minute advance of the escape wheel after locking that ensures the lever moving to the end of its travel. This otherwise cannot be relied upon to happen. No other eighteenth-century maker of lever watches employed draw. The Leroux's also have a quite modern double roller escapement and compensation balances with four bi-metallic rims. The escape wheel is also unusual since all the lift is on the teeth and none on the lever pallets. They have no Pendleton characteristics and altogether are something of an enigma, since although any surviving work by him is of high quality, he seems to have made no other precision watches.

Taylor, at the end of the century, copied Leroux's escape wheel and generally copied different bits of the other pioneers. Two watches by him are known, the second one being made after he went into partnership with the Ellicott firm.

The most experimental of the English lever pioneers was John Grant, by whom eight examples are known to survive, with a variety of levers and compensation balances. One, a movement belonging to the Clockmakers' Company, has two geared-together balances. The escapement which Grant seems finally to have fixed on has been nicknamed the 'chaff-cutter'. In this, the escape wheel is vertical in the watch plates and parallel to the balance arbor. The lever pallets are at different levels so that they enter the path of the escape wheel from opposite sides alternately and receive impulse along their inclined faces. Grant liked to employ deep locking, which this escapement facilitates, and it also secures equal lift for the entry and exit pallets. This is now known not to be significant, but it loomed very large in the minds of all the pioneers of the escapement.

Grant's last-known lever watch is hallmarked 1805. By this date the escapement seemed pretty well to have died in England and it had almost been given up by Breguet, its outstanding continental protagonist. It was evidently very difficult to make and, in performance, was never equal to the best pocket chronometers. From the late 1790s the duplex escapement rapidly made ground against the lever, as it could be made up into a thinner watch and its performance at best was not far short of the lever. Probably not thirty English-made lever watches survive from before 1800. The escapement was not to be revived in England until the 1820s, in a much simpler, cheaper form, and it was only in the second half of the century that it finally achieved pre-eminence in the precision pocket watch.

On the continent, apart from Le Roy's solitary experiment, the lever escapement got off to a shaky start with some crude instruments by Jean Moïse Pouzait who was in charge of the first school of horology to be set up in Geneva, in 1788. There is a model by him, late 1786, in the Physics Museum of the University of Geneva, and several watches with balances the full diameter of the watch, beating seconds. The safety action, when it comes into play, introduces excessive friction.

Antoine Tavan (1749–1836), a distinguished Geneva maker, also made lever watches from about 1785 onwards, but his versions of the escapement are more curious than contributory. His best-known and most successful is the pin-wheel lever escapement.

Robert Robin (apparently after seeing one of Emery's watches which reached Paris) developed in 1791 a variant of it in which the balance is impulsed directly by the escape wheel, and the lever is only used for locking and unlocking. The escapement is known by Robin's name although, as has already been explained, it was actually invented by Pierre Le Roy as early as 1765.

ABRAHAM-LOUIS BREGUET

The outstanding pioneer work outside England was done by Abraham-Louis Breguet (1747–1823). Through his friendship with John Arnold he was in a good position to learn what was going on in England and probably saw one or other of the English lever escapements. Emery's are characterised by their massiveness, but all Breguet's work is of the greatest practicable lightness. He seems to have made his first lever watches in about 1786 and they owe little to English precedent. In particular, it is their elegant lightness which most clearly differentiates them from anything made previously. The lever has no fork, but two vertical pins engage the steel pallet (not a roller) on the balance. There is an ordinary crescent-and-dart safety action. There is no draw, but owing to the lever being so light and accurately balanced, it seldom fails to carry to the banking.

In 1787 Xavier Gide put a substantial amount of capital into Breguet's business, thus enabling him to lay down a stock of lever watches, fifty of which were to have a lever escapement (thirty with self-winding, or 'perpetuelle' repeaters, and twenty with ordinary key-wound repeaters). Unfortunately they had not got very far before the Revolution halted production and none was completed or sold before 1791, but from that date onwards, up to the end of the century, Breguet made more lever watches than all the other makers of all nationalities, put together. In conjunction with his beautifully light yet robust escapement he employed a lightened version of Arnold's final form of compensation balance and either a spiral or helical spring. He also abandoned the fusee in favour of a going barrel (two, in the case of the perpetuelles), yet despite this simplification his early lever watches perform with a degree of accuracy at least equal to Emery's heavy machinery and generally better. They are also only two-thirds the thickness of an Emery.

In addition to this normal type, Breguet occasionally used the loop-type lever.

Like some other pioneers of the lever, he was worried about the unequal action of the entry and exit pallets and, in the case of the famous Marie-Antoinette watch, he overcame the problem by using a divided lever, despite the additional friction involved. By about the time the English pioneers were abandoning the lever, Breguet also largely did so and for several years used mostly variants of the Robin escapement, apparently to avoid the entry pallet problem. When he again took up the lever escapement, in about 1812, it was in the form that has been used ever since, employing a double roller, and a modern type of compensation balance.

34

XII

42

36, 37

*   *   *

To sum up the formative years of the precision watch, the outstanding names are John Arnold, Josiah Emery and Abraham-Louis Breguet. From 1773 to 1782 Arnold alone was making pocket watches capable of a degree of accuracy which would be acceptable today. From 1791 Breguet provided a performance not far short of Arnold's in a much slimmer, yet robust form.

During the 1780s it was primarily Emery who pioneered the lever escapement and certainly his work prepared the way for Breguet.

By the 1790s, several makers, including Earnshaw and Brockbank, were making pocket chronometers equal to Arnold's in performance though none rivalled the elegance and finish of his work. The only person to do that was Louis Berthoud who from 1785 was responsible for all the best Berthoud work, including the unsurpassed deck and pocket watches with pivoted detent escapements.

Having traced the sequence of historical events in the development of the precision watch up to 1800 the collector will wish to know something more about the style of the makers concerned and the clues which enable their products to be dated.

John Arnold and his son John Roger come first in seniority, and for a detailed study of their work the reader is referred to Dr. Vaudrey Mercer's *John Arnold and Son, Chronometer-makers 1762–1843* published by the Antiquarian Horological Society. Nearly all the surviving early pocket and marine chronometers are enumerated in it with details of their present condition. A main criterion for dating Arnold's work is the year 1787, in which he took his son into partnership. Previously, work was signed 'John Arnold' and thereafter 'John Arnold & Son'. This continued until John's death in 1799 after which the signature on the dial is usually 'Arnold' and on the movement 'J. R. Arnold'.

In 1830 Edward John Dent joined Arnold and the work was signed 'Arnold & Dent, 84 Strand'. This partnership continued until 1840 after which Dent left to set up on his own. Arnold then once more signed his own name, 'J. R. Arnold' still at 84 Strand, until his death in 1843.

The business was then bought by Charles Frodsham and the work signed 'Arnold & Frodsham' until 1858 after which the 'Arnold' was dropped.

John Arnold's marine chronometers are particularly attractive to the collector since he was not very keen on gimbals and housed his instruments in elegant hexagonal boxes. The very earliest had enamel dials but nearly all have silvered metal dials.

Arnold said that he had made several No. 1s, but in fact his watch numbering is not so chaotic as might appear; the marines are not so easy. He used separate sequences for the marine and pocket chronometers, both starting at No. 1.

The marine numbering got off to a couple of false starts, but with the first compensation balance he went back for the third time to No. 1 and seems to have gone on until he took his son into partnership, when he started again. Some have plain numbers. Dr. Mercer noted that all have the balance spring below the balance, thus taking some of the weight off the bottom pivot. Those in which the spring is above the balance have a fractional number, always with a difference of ninety between the upper and lower figure. The upper alone appears on the dial.

After John's death, John Roger dropped the upper figure and continued with the lower. This carried on until 1858.

The pocket chronomoters also started with No. 1, but with the famous No. 36 he again introduced a fraction, $\frac{1}{36}$. These fractional numbers continued through the 'double-T' balances (as No. 36) and then through the 'double-S' balances, some of which were fitted to the first spring detent watches in 1782. The last recorded fraction of this kind is $\frac{34}{89}$. Thus, the two numbers do not progress equally and the probability is that the upper numbers refer to the balance. As Arnold took back several watches numbered prior to 36 and fitted compensation balances to them, it is natural that the upper numbers increased more rapidly than the lower.

With the two- and three-armed balances with compensation rims the lower number only appears but there are very few of these. There is also a very small number of watches 'of the second best kind' which have no seconds dial and a difference of 1000 in the fraction. By far the greatest number of surviving pocket chronometers 'of the second kind' are only 50 mm in diameter. The series apparently started in 1782 and continued until John's death. Throughout there is a difference of 301 between the upper and lower figures and the lower only appears on the dial. The balances are mostly Mercer's 'OZ': that is to say, a plain steel wheel with bi-metallic rims attached to the steel rim. Later, the steel wheel was left off and the bi-metallic rims were fitted direct to the arms of the balance. This, the 'Z' was John's final word and was carried on by John Roger. John seems constantly to have had early watches back to the works for modernisation, particularly regarding the balance. Thus, it may sometimes happen that the signature on the movement is plain 'John', but on the dial is 'Arnold & Son'. The watches before compensation balances apparently had spiral bi-metallic compensation curbs and a circular scoop out of the foot of the balance cock, now unoccupied, contained them. Chronometer repeaters are very rare. The workmanship is always of the highest quality and a joy to see.

159
202

THOMAS EARNSHAW

In about 1782 Thomas Earnshaw (1749–1829) developed the type of spring detent escapement which is now universally employed. When he could not afford to patent it he persuaded Thomas Wright to do so, on the basis that he would repay the patent fee by a royalty on the watches and chronometers as he made them. Wright kept putting off the patent proceedings and in the meantime Earnshaw claimed to have shown his design to John Brockbank who at once told Arnold about it. Arnold was thus able to get his patent in first. It is a rather unlikely story, since Arnold's spring detent escapement is totally unlike Earnshaw's or his own pivoted detent, while Earnshaw's consists of little more than Arnold's pivoted detent mounted on a spring. In any case, the question of priority is not very important and the likelihood is that they arrived at their two versions independently. Even when Wright finally did obtain a patent in 1783, Earnshaw still could not afford to set up on his own and supplied movements to established makers who put their name on his movement. He seems to have had his own establishment soon after 1790, but instruments by him, either marine or pocket, are exceedingly rare prior to 1800.

His marines are smaller in size than those of almost all other makers, and are fitted in gimbals. His watches are generally of handsome appearance externally but frequently the movements are completely plain and the plates not even gilt. This came about because someone said that his watches went well only on account of their high finish. Some however are gilt and the cock engraved, although in a somewhat coarse style. The train is far inferior in finish to Arnold's. Nevertheless, Earnshaw's pocket chronometers do perform very well. This is partly due no doubt to his method of fusing together the bi-metallic rims of his balance wheels, although he somewhat offset this advantage by using untempered balance springs made from soft-drawn wire hardened by rolling. These got tired and then lost on their rate. He used both helical springs with compensation balances and plain steel balances with a spiral spring and compensation curb. These curbs have two curved arms whose free ends embrace the outer end of the balance spring. By coming close together, or moving apart, they regulate the watch. From their appearance they have come to be known as the 'sugar-tong compensation'. It seems likely that these watches were supplied to hunting men or others with energetic habits, since Earnshaw's compensation balances were much heavier than Arnold's, which, if bumped about, might be expected to produce a crop of bent or broken balance pivots.

Towards the end of his life Earnshaw took up the lever escapement, usually in

VIIc
246
213
240

fairly cheap watches, for which he evidently regarded the escapement as likely to give adequate timekeeping, coupled with robustness.

In 1804 Earnshaw and John Roger Arnold were given £3,000 each by the Board of Longitude provided they would 'give in a full account in writing, with drawings of the principles and construction of their respective timekeepers, and methods of adjusting them'. They were also required to attend upon the Board and answer questions.

The results of these written depositions and interviews were published by the Board in 1806, and entitled *Explanations of Timekeepers constructed by Mr. Thomas Earnshaw and the late Mr. John Arnold*. This rare book makes both interesting and entertaining reading. In the interviews John Roger Arnold behaved with suitable dignity, but Earnshaw never lost an opportunity of turning his answers to Arnold's disadvantage.

Perhaps because of Earnshaw's persistent denigration of Arnold, John Roger has acquired a rather bad name. In fact nearly all his surviving work is to a very high standard of finish, and at least equal to Earnshaw's.

One of his workmen was Thomas Prest who in 1820 devised a very early system of keyless winding. Arnold applied it to a series of fine watches with ruby cylinder escapement and a compensation curb. These watches have a going barrel. The hands have to be set by a key, as previously; complete keyless operation did not come into general use for a further thirty years. The movements are signed 'Jn. R. Arnold. Patent'.

The birth of the modern watch: Lepine and Breguet

If the acquisition of a balance spring caused the English watch to grow fat, it resulted in the French watch becoming enormous. A common thickness of 38 mm won for it the sobriquet 'oignon' (see above, p. 42).

It was rescued from this inflated condition by Jean-Antoine Lepine (1720–1814). He trained with A. C. Caron and married his daughter. Caron rose in Court circles, took the name of Beaumarchais, and gave up watchmaking. Lepine then took over his business. Between them, Lepine and Caron invented the single- and double-virgule escapements which enjoyed a considerable vogue in France in the second half of the century. The virgule was a development of Tompion's escapement, although Lepine had probably never heard of it. It is a frictional rest escapement and has what amounts to a small cylinder with the exit lip extended into a tail, giving it the appearance of a comma (virgule), carried by the balance staff. The escape wheel carries vertical pins whose period of frictional rest takes place in the cylinder while impulse is given along the curve of the comma. It therefore receives impulse only in one direction. The double virgule consists of two commas facing opposite ways. The escape wheel has two sets of pins, on opposite sides of its rim, each engaging with one of the commas. Impulse thus takes place on both swings of the balance. The escapement is attractive in theory but in practice it proved impossible to retain the oil that it needed for its successful operation. Examples are quite rare and much sought after.

A curiosity peculiar to Lepine was his habit of mixing Roman and Arabic numerals on his dials in a style which he presumably found aesthetically satisfying but which to most people appears more perverse than decorative. He also showed his perfectionist approach by making the wheels and pinions in some of his best watches with 'wolf's teeth'. The difficulty of cutting such rachet-shaped teeth is not justified by their slight theoretical superiority. Lepine's real claim to fame rests on the movement layout which he devised, apparently in about 1770. In all previous watches, of whatever nationality, the train wheels were all pivoted between two plates with the balance   VIc separately mounted outside the top plate. Lepine abandoned the thick fusee and relied on a going barrel, and separately pivoted most of the wheels in the train. This not only made his watches easier to work on, but greatly reduced their thickness. 12·5 mm is a usual thickness for his watches.

The 'Lepine Calibre' was used subsequently by Abraham-Louis Breguet for all but his precision watches and enabled him to produce the thin and very thin watches with which he so largely captured the British luxury market.

Lepine retired from active participation in the business before the French Revolution, and handed it over to his son-in-law Claude-Pierre Ragust. Lepine died in 1814.

Ragust continued in his father-in-law's tradition and produced watches comparable in elegance and execution to Breguet's. For some of his best watches he used the double-wheel duplex escapement.   9, VIc

While Lepine paved the way for the advent of the modern watch, it was Abraham-Louis Breguet (1747–1823) who brought it to full maturity. He and John Arnold between them have contributed more to the development of the watch than any other maker. They were good friends and after John Arnold's death Breguet continued a steady correspondence with John Roger. But whereas Arnold confined his activities to the precision pocket watch with a detent escapement, Breguet explored every aspect of horology and loved to invent ways of solving the most intransigent mechanical problems. As he coupled this mechanical ability with highly-refined taste manifested in the appearance of his watches, and with a well-developed commercial instinct, he was assured of the success he so richly enjoyed. The work of no other maker is so eagerly sought or so expensively acquired, and because of the variety and size of his output, the study of his work can become a full-time horological occupation. It is a study whose rigour has been greatly eased by George Daniels' *The Art of Breguet*, published in 1975.

For the completeness of any book such as this, seeking to cover the entire history of watches, Breguet's work must be discussed in considerable detail.

Breguet was born in 1747 in Neuchâtel, Switzerland, and came to Paris to learn watchmaking at the Collège Mazarin, where his ability attracted the attention of the Abbé Mairie, who brought him to the notice of the king.

Breguet may subsequently have worked for Ferdinand Berthoud although there is no positive evidence. He was in business in a small way on his own account by   171 1780. Surviving examples of his work prior to 1787 are few and do not give much indication of what was to follow. Probably most of Breguet's sales at this time were of watches bought from the trade and even watches upon which he seems to have worked himself had some trade parts, such as repeating trains. These early watches have their serial number followed by a fraction, of which the most likely interpretation is that the upper figure represents the month and the lower the year of completion. If Breguet kept any books at this time they have not survived, but the earliest-known number is $2\frac{10}{82}$, which was a self-winding watch (perpetuelle) belonging to Queen Marie-Antoinette. Probably it was much like $8\frac{10}{83}$, which belongs to the Worshipful Company of Clockmakers. Also self-winding, this has an ordinary steel cylinder escapement. (The self-winding watch had been invented, or at any rate brought to a practical state, by Abram-Louis Perrelet by about 1780.)

By 1786 Breguet had probably seen an Emery lever watch and formed his own ideas of how the escapement could be made on very much lighter lines.

In the same year he went into partnership with Xavier Gide who put a considerable amount of money and stock into the business. This gave Breguet the opportunity to collect a team of workmen and lay down a stock which enabled him to reveal his capabilities to the full. The first thirty-one watches are the most sought-after of all Breguet's output and, considering their date, must be acknowledged as the most advanced and sophisticated watches ever made. Their construction was interrupted by the Revolution and none were completed before 1791. All were self-winding repeaters and all but one (a survival of the pre-Gide era) had lever escapements.

Despite having to contain the self-winding mechanism and a repeating train these
V A   perpetuelles were only 18 mm thick and 50 mm in diameter. All had plain cases of the reddish-coloured gold Breguet used all his life. The dials, with Arabic numerals, were enamel, but because they were subject to cracking, Breguet subsequently rebuilt most of the early perpetuelles, using engine-turned gold cases
XVI A and engine-turned silver dials with Roman numerals. He continued to employ the latter kind of dial, either in gold or silver, throughout his career, with little alteration.

Also in 1787 the famous books were started which recorded every watch made by the firm, the workmen employed in its construction, and the purchaser.

Despite this wealth of information, the numbering of Breguet's watches is not always easy to interpret. At times a manifestly early watch has a late number, in which case it had probably remained in stock for many years in an incomplete state. Sometimes Breguet bought an early watch and added a digit to its number before re-selling it. However, as a rough guide the number will usually provide a reasonably close indication of the date of the watch. The fractional numbers were virtually abandoned in 1787, although there are a few later ones whose purpose is not known. The second series, which started with Gide's partnership in 1786, ends at No. 295 in 1793 when the partnership terminated.

In 1793 Breguet was in danger of arrest under the revolutionary 'reign of terror' and fled the country. It was only safe for him to return in 1795. The intervening period was spent largely in Switzerland, and it is from 1795 that Breguet's triumphant career got into its stride.

A third numbering series had been started in 1793 and continued in use into the 1830s, finishing at No. 5121.

As a rough guide to dating, but with frequent exceptions, the following may be used:

up to No.	*up to Year*
500	1800
1,000	1803
1,500	1805
2,000	1809
2,500	1812
3,000	1818
3,500	1820
4,000	1823

Until about 1795 all work was signed on the movement 'Breguet à Paris', and the enamel dials had the signature 'Breguet' in script. Later, the 'à Paris' was dropped and the dial signature was in Roman capitals. The signature 'Breguet à Paris' was later widely used by fakers and the paradoxical situation therefore exists whereby a genuine watch before about 1795 should be signed 'Breguet à Paris', but any evidently nineteenth-century watch so signed is almost invariably a fake. Fakers also sometimes accented the first 'e' thus—'Bréguet'. In 1807 Breguet took his son Louis-Antoine into partnership and work was then signed 'Breguet et fils'. On the death of Louis Berthoud in 1812, Breguet was appointed 'Horloger de la Marine Royale' and these words often appear on cuvettes in first-quality watches. For a short time, in about 1833, after Breguet's death, some work was signed 'Breguet Neveu Compagnie'.

By 1800 Breguet had developed all the features which were to characterise his subsequent work and the principal among these, from the watch-collector's point of view, are:

The 'montre perpetuelle'
The rotating escapement ('regulateur à tourbillon')
The 'garde-temps'
The 'montre à souscription'
The 'montre à tact'
The ruby cylinder ('échappement isolée')
The shock-proof mounting for the balance ('parachute')
The ratchet-winding key
The secret signature
The certificates

The perpetuelles have already been mentioned. These are immediately recognisable by the up-and-down dial ('développement du resort') which is an essential adjunct to a self-winding watch.

Early perpetuelles have no provisions for manual winding, and must be shaken up and down a few times to be set going. The ordinary amount of unavoidable walking undertaken during a day will then keep them working. When the springs are fully wound by the movement of the platinum weight, a bolt is raised to lock the weight and prevent further winding. In the earliest examples this locking arrangement sometimes did not work, and there would occur what was known in the Breguet workshop as 'l'accident habituel'.

Breguet sealed his cases and as a result the watches will go longer than almost any other sort between cleanings.

Nearly all the perpetuelles have a lever escapement, but there are three known examples made between 1812 and 1817, with a Robin escapement.

The tourbillon was a radical invention typical of Breguet's approach to horology. Watchmakers before him and since have minimised position errors by a number of tricky adjustments which are only effective as long as the balance arc remains roughly constant. Breguet eliminated position errors by putting the escapement on a revolving platform, with a rotating period usually of one minute, but, in rare examples, of four and even six minutes. Thus, all the position errors are repeated in the rotating period of the platform. Breguet regarded this invention as his final word in timekeeping and the tourbillons are his only watches to rival Arnold's pocket chronometers in accuracy. Also, they keep their rate longer than any other kind of watch.

Most have a lever escapement, but his first tourbillon (No. 282) has an Arnold-type spring detent. Although Breguet later used an Earnshaw-type escapement he never did so in a tourbillon watch. When these have a detent escapement they always (with the exception of No. 282) have the Peto cross detent arrangement.

Tourbillons have a three-quarter plate ébauche and a fusee (except No. 282 which has two going barrels). They come in the category that Breguet described as 'garde-temps', which may also have straightforward detent escapements, of Arnold type in his earliest examples and Earnshaw later. Breguet seems never to have been perfectly at home with the detent escapement and except for the tourbillons his garde-temps do not perform with a degree of accuracy comparable to that of the London makers.

The montre à souscription was the nearest Breguet ever came to series production although no two even of these have ever been found to be identical. They were Breguet's way of making the cheapest possible simple watch that conformed to his high standards, and omitted every possible complication. By laying down a batch when enough orders had been obtained, the price was reduced further. The customer subscribed half the price with the order and the remainder upon taking delivery.

The souscription watches are nearly 62 mm in diameter and 15 mm thick. The reason for the large diameter is that one of the simplifications was the omission of a minute hand. The large diameter enables the division markings to be placed at five-minute intervals so that the time can be read easily to within one or two minutes. The dial is almost always enamel, metal dials being exceedingly rare. The mainspring is contained in an immense going barrel wound through the centre of the hand. The escapement is always Breguet's ruby cylinder, in conjunction with his parachute and bi-metallic compensation curb. The hand is set by pushing it round and care must be taken to push it from fairly close to the centre, with an orange-stick or similar wooden object that will not scratch the enamel dial.

The flat-sided band of the case, and the back, are usually made of engine-turned silver. The pendant loop and the rims holding the glass and the back are gold. However, all-gold and all-silver cases are sometimes found. Breguet also made souscriptions in smaller sizes.

As with all Breguet's ruby cylinder watches, a highly-acceptable degree of accuracy is provided and, owing to the parachute, a souscription will sustain a heavy blow without damage (but the authors offer no guarantee for this assertion).

The montre à tact was based on the souscription. Often described as a 'blind man's watch', it has a rotatable arm pivoted in the centre of the watch case; by moving it round clockwise until an obstruction is felt, the time can be found by touch alone.

While these watches could therefore be useful to a blind man, there is no doubt that Breguet regarded them as an alternative to the repeater. Studs are placed at the hours round the edge of the case by which the time can be judged; but it is difficult to 'read' to nearer than half an hour.

Montres à tact of souscription size are severely simple but smaller examples are often put into cases which might be decorated with transparent enamel over engine turning and have the 'tact' lever set in diamonds.

Breguet probably developed his ruby cylinder escapement while he was an exile in Switzerland during the Revolution. He called it an 'échappement isolé', and employed it in by far the greatest majority of his watches.

All other cylinder watches have the cylinder, (whether steel or ruby) between the pivots of the balance staff. Breguet used a perfectly plain balance staff between jewelled holes but he attached a crank to it which came round below the bottom bearing and contained the ruby cylinder at its extremity. This enables the cylinder to   2, 3 be of the simplest possible shape and also does not call for the complicated escape wheel of the typical English cylinder watch. Breguet evidently considered—with justice—that this very robust escapement, in conjunction with his compensation curb, provided a sufficient degree of accuracy for all ordinary requirements; and it will continue to do this for very long periods between cleanings.

The highest grade of cylinder escapement watches, described as 'ouvrage première classe', are repeaters, and usually distinguish the half-quarters. They sound sometimes on a metal block 'à toc', and sometimes on a single gong. These watches have an   VB easily identified Lepine calibre and are jewelled to the centre hole. They have gold cuvettes (as opposed to brass-gilt on cheaper watches). The cases are gold and the dials engine-turned silver. The independent seconds dial is nearly always placed near the IV.

The first souscription watches appeared in 1796 but the traditional Lepine-type lay-out was not finalised until about 1799. The première classe repeaters were finalised in about 1804 and both types were then in use until the end of Breguet's life with only small modifications being made.

Some of the repeaters have Breguet's 'jumping hour hand' which seems to be entirely peculiar to him. In these, the hour hand remains stationary for most of   VA the hour, and this must be borne in mind when reading the time, otherwise it is easy to mistake ten minutes to eleven for ten minutes to ten. The hour hand begins to move at five minutes to the hour, and during the following five minutes moves about half-way to the next hour. As the hour comes up it then jumps to the hour and remains there for the next fifty-five minutes.

Breguet's shock-proof mounting for the balance staff, which he called a 'parachute' consists of a spring suspension for the balance pivots and is fitted to almost all his watches. It needs a shock of extraordinary violence to break or bend a balance pivot.

The ratchet-winding key is simply an arrangement to ensure that if it is inadvertently turned the wrong way, no damage can be done. For obvious reasons, they are also known as tipsy keys. Breguet's very elegant keys are usually attached to the watch by   228, 238 a chain about 112 mm long, with circular double links, dividing into a double chain for the 50 mm nearest the key.

Breguet said he devised his 'secret signature' to defeat fakers of his work. It is certainly true that these signatures cannot be reproduced exactly without the

210, 227 original master plate. They are written in cursive lettering and formed by means of a pantograph. On enamel dials they are to be found below the figure 12. The name Breguet is subjoined by the number. Souscription watches also have the word 'souscription' between 'Breguet' and the number. On metal dials the secret signature is on the chapter ring, repeated on either side of the XII. On gold dials they have usually survived, but on silver dials, which eventually tarnish and have often been stoned clean, they are seldom found.

Breguet issued certificates with his watches, giving full details of their construction, the date of sale, and the name of the purchaser. Where an original certificate survives it is of great interest, but duplicates were issued on payment of a fairly modest fee, as long as the firm survived. The practice continues, on payment, with the present owners of the name. Breguet also supplied his watches in elegant red leather cases, usually with a compartment for a spare glass or crystal.

2, 3, 6, 8, 9, 34, 35, 36, 37, 42, 43 Breguet used almost every escapement then known, except the Debaufre and rack lever, and added some of his own invention. The escapements he used include verge; cylinder; virgule; double-wheel duplex; detent (of various sorts); lever (of various sorts); Robin; and 'échappement naturel'.

Although the early lever escapements, already described, performed outstandingly well, Breguet, like all the other lever pioneers, seems to have been worried by the almost unavoidably unequal operation of the entry and exit pallets. Therefore, soon after 1800, he almost gave up using the escapement and took up instead various forms of the Robin, in which the lever is used only for unlocking, and impulse is delivered direct by the escape wheel (but only on alternate beats). In order to achieve

214 impulses on both swings he developed his 'échappement naturel' in which two escape wheels, geared together on either side of the lever, impulse the balance alternately on separate swings. This complicated escapement was used almost exclusively on tourbillons of the highest quality and is the rarest of all Breguet's escapements. These tourbillons perform very well, but no better than the standard lever type. Breguet described as 'naturel' an impulse delivered direct by the escape wheel, rather than through an intermediate lever, and this explains his name for this escapement.

35 For his lever tourbillons Breguet devised the single roller lever which was to become the standard English lever escapement. By about 1812 he appears to have overcome his inhibitions about the lever escapement as a whole and devised his final, double roller, version with draw, which came into standard use on the Continent and is

36, 37 now employed universally. By about 1805 he had also arrived at virtually the modern form of compensation balance. His elegantly slim late lever escapement watches (less

Vc than 12 mm or ½ in thick), with two going barrels, represent his final word on pocket timekeepers and their distinguished appearance is fully matched by their precision performance.

In addition to the watches described, which cover Breguet's standard products, insofar as any of his watches were standardised, he made a number of more or less complicated watches to special order. This category was known as 'montres de fantaisie et grande luxe'. They were, and are, enormously expensive. The most complicated of all is the watch he started to make on order from an officer of the

XII Royal Guard, as a present for Queen Marie-Antoinette; it was not finished (and probably not even started) in her lifetime. Nevertheless, Breguet went on with it after the Revolution as a monument to eighteenth-century watchmaking. It was not completed until 1820. The complications include self-winding, perpetual calendar, equation of time, centre seconds hand marking whole seconds, up-and-down indicator, thermometer, minute repeater, equal locking and lift lever escapement.

Breguet probably always bought and sold watches made in the trade, but he sold one category, partly made outside, which was fitted with an escapement and finished in his own workshop. These he described as 'montres mixtes' and sometimes 'mixte' appears as part of the secret signature. These watches were not entered in the firm's books.

Breguet was no stranger to the idea of keyless-winding, which he employed in a very few watches; he evidently preferred his short chain and key which, indeed, are almost a Breguet trade mark. Apart from this, his late lever and late, thin, lever perpetuelle watches anticipate the modern watch in every significant particular. In elegance, robustness and precision they have never been surpassed and very rarely equalled.

Many subsequently famous makers trained with and then worked for Breguet. When they left and set up on their own they seem always to have done so with Breguet's good will, and were proud to sign their work with their own name and, frequently, the words 'élève de Breguet'. Among these were Audemars, Benoit, Fatton, Firche, Ingold, Jacob, Urban Jürgensen, Kessels, Laissieur, Lopin, Moinet, Mugnier, Oudin, Rabi, Renevier, Robert, Tavernier and Winnerl.

224

208, 232, 243 244, 271, 291

The development of the watch since 1800

The first half of the nineteenth century saw the development of the British watch to a fairly standard lay-out which continued unaltered for the next eighty years; nevertheless the verge escapement was made without alteration until the end of the century. The rise of the Swiss industry commenced, culminating in the bar lay-out of the movement, with a cylinder escapement for cheap watches and lever escapement for those of higher quality. Switzerland and America vied with each other in quantity production, but employed entirely different methods. In terms of factory production America was first in the field by many years.

The work of Abraham-Louis Breguet was a vital factor in setting the pace for the nineteenth century, but because his formative years lay in the eighteenth century, his work as a whole has been dealt with in the preceding section.

208, 273, 284, 289, 296, 298

Alongside the general trend towards standardisation, the work of a few individualists such as Urban Brunn Jürgensen and subsequently his sons Louis and Jules, James Ferguson Cole and Sylvain Mairet was able to prosper. Even after 1850, and the Great Exhibition of 1851, a handful of such men continued to make their mark up to the end of the century, but by that time, originality had come almost to imply eccentricity and from the second half of the century only the metallurgical researches of Charles-Edouard Guillaume were of significance.

Although Breguet had determined the shape of the modern watch by 1812, other escapements beside the lever continued to enjoy popularity for the first half of the nineteenth century, the duplex escapement, in particular, remaining a strong challenge. The British lever watch took until the 1850s to arrive at its final form, with a single-roller lever escapement and three-quarter plate movement.

The principle of the duplex escapement had been devised by J. B. Dutertre in the second quarter of the eighteenth century, but nothing came of it until the end of the century, and it was only in the early years of the nineteenth century that it enjoyed popularity.

In 1782 Thomas Tyrer of London patented an 'horizontal escapement for a watch to act with two wheels'. One wheel is for frictional-rest locking. Its teeth bear against a jewelled roller forming part of the balance staff. A narrow notch is cut in the roller, through which the teeth of the locking wheel are able to pass. As a tooth passes through the notch, a tooth of the impulse wheel imparts impulse to the balance by striking a pallet attached to the balance staff. On the return swing of the balance, the notch brushes past the locking tooth with a barely-audible click and a minute movement of the wheel.

The Continental makers who used the escapement were, notably, Lepine, Breguet and Jürgensen, all of whom used the two wheels postulated by Tyrer. Probably none of them employed it before 1800 and all used it only in high-grade watches. The locking wheel is steel, the impulse wheel is of gold, and the roller of sapphire.

The English developed a single escape wheel with normal teeth for locking, and vertical teeth, standing up from the rim of the wheel, for impulse. In the early 1790s, Josiah Emery was among the first to use the escapement with two wheels, but by 1800 the single wheel had been developed and several leading makers were using it. Brockbank was an enthusiastic exponent, but some of the finest and most handsome duplex watches were made by the Vulliamy firm. They did not use a compensation balance but adhered to a compensation curb. D. & W. Morice (the excellence of whose work is insufficiently recognised) developed a light, and quite modern, compensation balance, which they were using with the duplex escapement by 1815 at latest. Robert Pennington, primarily a chronometer maker, was another who arrived at a remarkably modern, light, bi-metallic balance, in the first decade.

Probably the most prolific of the early nineteenth-century English duplex makers was James McCabe. He came from Belfast, was received into the Clockmakers' Company in 1781, and died in 1811. He was succeeded by his sons Robert and Jeremy who continued in business until 1883. His best work was signed 'James McCabe', and the second grade 'McCabe'.

With the rise in prosperity of Liverpool as a port, and Lancashire as a centre of the Industrial Revolution, it was natural that an horological trade should grow up there.

Such an industry had indeed existed from the seventeenth century, and the well-known Thomas Aspinall was working in Toxteth, a suburb of Liverpool, in the early years of the century. It had been stimulated from time to time by injections of Swiss talent, fugitives from the revocation of the Edict of Nantes in 1685. More arrived a century later, attracted, no doubt, by the greater opportunities for skilled workmen in industrially-prosperous England. Lancashire had long supplied parts to London makers, but from about 1800 a recognisable style of Lancashire watch is met increasingly. Two escapements are particularly characteristic; the rack lever and the Debaufre.

The idea of the rack lever occurred to the Abbé Hautefeuille early in the century but it was not taken up. Peter Litherland of Liverpool probably knew nothing of this when he took out a patent in 1792 for a rack lever escapement. This consists of an ordinary Graham dead-beat escapement, but at the end of the lever a toothed rack engages with a pinion on the balance staff. The balance therefore swings through a very large arc, but while the escapement is dead-beat it is totally undetached. Litherland probably regarded the fact that it beat seconds as its principal merit, since he describes his patent as 'Improvements in escapements, by which watches are made to beat once only in a second and also to point dead seconds'.

From 1800 to 1820 rack levers remained very popular, and only gradually faded out as the detached lever began to re-assert itself in the '20s. Its manufacture was widespread but its main sources were Prescot, Ormskirk and Liverpool. A strange variant, 'Smith's five-tooth lever escapement', was patented in 1812 by Samuel Smith of Coventry. The escape wheel is of contrate form, with five very long, pointed teeth. These impulse the inclines of the lever which is mounted on a wheel of ninety-nine teeth and engages with a pinion of six on the balance staff. Another dead-beat, non-detached escapement much favoured by the Lancashire makers was a variety of the Debaufre escapement. Like the original Debaufre escapement of 1704, it has two co-axial escape wheels with ratchet-shaped teeth bearing upon a single, chamfered roller on the balance staff. It became so popular that it was known as the 'Ormskirk escapement'. Despite its popularity, it has now become a considerable rarity. The escapement is capable of many variants, the best-known being Sully's escapement which uses one escape wheel and two rollers. A particularly large variety of watches and movements with this type of escapement may be studied in the group of 'watches with miscellaneous escapements' in the collection of the Worshipful Company of Clockmakers at the City of London, Guildhall. Earnshaw devised an interesting variant, which was a compromise between the cylinder and Sully's escapement; he seems to have made few examples.

The Lancashire makers also developed an individual style of jewelling. These jewels were colourless and of enormous diameter, causing them to be known as 'Liverpool windows'. They are particularly characteristic of the second quarter of the century.

When the detached lever escapement began to come back into use in the 1820s it was taken up enthusiastically by the Lancashire men, who were conditioned to it by their use of the rack lever.

As has been said, the lever escapement in its pioneer form was driven from the field in England by the duplex for thin watches and the detent escapement for precision watches. John Grant was the last of the pioneer group and he seems to have given up after about 1805. For another ten years it is doubtful if any further lever watches were made in England, and there were certainly very few for ten years after that.

Edward Massey, a London maker who moved to Coventry in 1814, introduced in 1815 a type of detached lever escapement evidently derived from the earlier cranked roller escapement. He did make some cranked rollers, but in most of his watches the roller is not cranked. Although it used to be called 'Massey's crank roller escapement', the more correct and gradually accepted term is 'Massey's escapement'.

In it the balance staff roller may be made of solid steel, or the impulse nib may be a jewel let into the roller. On each side of the nib, a slot or notch is cut into the roller. The lever ends in a simple fork whose ears fit exactly into these slots. Thus, as the balance swings the nib enters into the fork and moves it, unlocking the escape wheel which then impulses the balance through the lever. If the lever is moved during the free swing of the balance wheel, the fork will butt against the roller and so be unable to move further until the slots register with the ears of the fork. The escapement thus needs no separate safety action. Watches signed by Massey are rare but his escapement was made by many London and Lancashire makers from the early 1820s for a period of twenty years.

George Savage devised an escapement in about 1820 which came to be known as the 'Savage two-pin escapement'. It enjoyed a following until around 1880 and was capable of giving good results if well made. The balance staff had a large roller with two pins on it, instead of the usual one, and these engaged with a wide fork on the lever. On the end of the lever another pin engaged with a very narrow safety slot in the roller. The two pins were only concerned with unlocking the escape wheel, their wide spacing being calculated to ensure that unlocking was completed with minimum friction by the centre line. Impulse was imparted by the safety pin and hardly, if at all, by the fork.

Concurrent with these two escapements, the single-roller lever appeared in England during the early 1820s. No-one seems to have claimed its invention, but the most likely inventor was Breguet who had introduced it into some of his tourbillon watches by 1810.

Among the London makers, including Earnshaw, the single-roller lever seems to have been regarded as a useful, robust escapement for medium-class watches. It was superior to the verge or cylinder but inferior to the detent. The single-roller gradually drove all the other types from the field in England and by 1850 was firmly established as the norm to be followed by British makers.

Probably the first English maker of the second generation of lever makers who regarded it as a precision instrument was the little-known T. Cummins of London. 266, 270 The earliest known lever watch by him is hallmarked 1822. All his watches have Massey's escapement with a jewelled pallet, and have a large roller, a modern type of compensation balance and a helical balance spring. The movement has the old-fashioned full plate lay-out, and at least one example has resilient banking pins. The purchaser's name is usually engraved on the balance cock. A tiresome feature is that there is no motion work on the regulator dials between the hour and minute hands, thus necessitating separate setting. Cummins's output was evidently small. About five of his watches are known to survive and there is also one hallmarked 1837 by his successor (his son?) Charles Cummins. Nevertheless his influence in pointing the way 288 to the re-establishment of the lever escapement as orthodox for high-grade precision watches should be seen as important.

Another maker probably influential in re-establishing the lever escapement for precision timekeeping in Britain was Sylvan Mairet, a pupil of Breguet who worked 284 for a time for the London makers Hunt & Roskell. He favoured Leroux's escapement 298 in which all the lift is on the escape wheel. Watches attributable to him are of the utmost elegance and superb finish and are remarkably slim. Hunt & Roskell were responsible for some of the finest British watches made during the second quarter of the nineteenth century. Continental influence is strongly evident in all their work.

After 1820 the use of draw became increasingly common. Although Breguet's final form of the lever escapement suggests that he intended to employ draw, it is in fact so minimal in most of his watches as to be almost non-existent; yet he had clearly contemplated using it some years earlier. Urban Jürgensen, who worked with him around 1800 and was much in his confidence, wrote in 1804 (speaking of Breguet's early lever escapement): 'the impulse faces of the anchor should be concentric to the pivot point of the lever, and one can make it slightly angled [*en talon*] to render the escapement as safe as possible; but this will then cause a slight recoil in the wheel'. Considering this comment and the even earlier use of draw by Leroux, it seems remarkable that it was not used more enthusiastically by Breguet and his immediate successors.

All the early British lever watches had full-plate movements, which made for a thick watch. The Continental makers, using the Lepine calibre, were able to make their watches substantially thinner, which led in turn to a demand in England for slim watches.

The British answer to this was the three-quarter plate movement, in which only the 256, 316 balance wheel and escape wheel were separately cocked. This arrangement continued the standard lay-out for all subsequent British watches.

253, 260 From about 1840 the leading London makers, notably Frodsham and Dent, made lever watches of high quality, with compensation balances. The earliest keyless-wound watches required key-setting for the hands; a patent for performing both operations by the winding knob was taken out by Adolphe Nicole in 1844. Dent was making 301 keyless-wound and keyless hand-set watches, with going barrels, by 1850 at the latest.

263, 295 Another winding variant is known as the pump-wind, patented by Robert Leslie in 1794. At the top of the pendent is a button, like some of Breguet's repeater buttons, which is pulled in and out to effect winding. It was employed by Viner, among others, in some early, high-grade levers.

In contrast with this history of steady development towards a standardised product must be set the work of such individual artists as Louis Berthoud, Urban Brunn Jürgensen and James Ferguson Cole.

Louis Berthoud has already been mentioned, but his early work was almost completely overshadowed by his uncle Ferdinand and he enjoyed only a brief period of independence between Ferdinand's death in 1806 and his own in 1812, when he was 'Horloger de la Marine Royale'.

His output was small, since he seems to have liked to do most of the work himself even down to jewelling, where he was severely critical of jewels supplied by the trade. 225 His workmanship is of an order that has never been surpassed and, without making any attempt at contrived elegance, his pocket chronometers, deck watches and marine chronometers are, by their sheer proportion and brilliance of craftsmanship, in a class of their own.

At the request of the French government he trained four pupils of whom Henri 262 Motel in particular and Pierre Saulnier carried on his standards in their own work. On the completion of their apprenticeship he wrote them a charming, largely non-technical little book, entitled *Entretiens sur l'Horlogerie*.

Urban Brunn Jürgensen (1776–1830) was a Dane, born in Copenhagen, and the son of a watchmaker. Having served five years in his father's workshop he was sent to Le Locle to study under Frédéric Houriet. He moved to Paris in 1800 where he was made free of the Berthoud and Breguet workshops. In 1801 he went on to London where he was received kindly by John Roger Arnold, who surprised him by saying that as he had worked in the house of Breguet it would help him little to work in London, 'for it is well-known that he is the most distinguished artist in Europe'.

After further wandering, and marrying Houriet's daughter, he returned to Copenhagen at the end of 1801. Apart from running the family business after his father's death in 1811, his great interest was in making pocket and marine chronometers. He made seventy-two chronometers out of a total of eight hundred 232 watches. The chronometers had a double number, the upper being the chronometer series and the lower the watch number. Nearly all have silver cases; probably only twenty-three are gold.

For his chronometers he favoured the Arnold pattern of spring detent and gold helical balance spring and compensation balance. To reduce air resistance he made his balance arms with sharp edges and oval weights. In addition to chronometers he made double-wheel duplex watches.

So insistent was he upon being independent of outside help that he persuaded two pupils of Mallet, who had brought the still secret art of jewelling from London to Geneva, to come to him in Copenhagen.

He trained his brother George Frederic and his own sons Louis Urban and Jules Frederic. Other famous makers who worked for him were J. T. Winnerl and Victor Kullberg.

Jules Jürgensen set up a separate establishment at Le Locle, in Switzerland, in 1834 and exerted a powerful influence upon the design and execution of the highest grade of Swiss watches. The Le Locle factory made both lever and pivoted detent watches and their train wheels were frequently of gold. The dials are characterised by very attenuated Roman numerals and Breguet hands.

The third generation of the family to be active in the business was Jacques-Alfred Jürgensen (1842–1912).

Paul Chamberlain said of Urban that 'he was a sound scholar, a master workman, a man who was satisfied only with the best and one who set such standards of excellence that their influence is seen in every first-class shop in the world where Scandinavians are employed—and that means the great majority'.

James Ferguson Cole (1798–1880) has been described as the English Breguet. This is perhaps an over-statement, but he did resemble Breguet in his keenness to try out new ideas and novel escapements. His workmanship was of a high order, and he owned to being much inspired by Breguet.

For elegance of appearance and brilliance of workmanship he perhaps never surpassed his 'masterpiece' completed in 1821, when he was twenty-three; it is now in the British Museum. The escapement is Cole's pivoted detent, but he went on to such complicated escapements as his 'double-rotary detached chronometer', 'constant-force double-escape wheel detent escapement', 'rotating detent escapement', 'resilient lever' and 'repellent lever escapement'.

None of these contributed anything to the development of watchmaking yet it is enquiring minds such as Cole's that make the second and third quarters of the nineteenth century such a constant source of interest and surprise.

Although the British makers arrived by 1850 at a fairly standardised version of the lever escapement watch with a three-quarter plate lay-out, they did little to rationalise its manufacture. Whether in London or Lancashire, the man who was eventually to put his name on a watch and sell it had only a small workshop. He relied on out-workers to supply most of the components of his watches. Specialisation was extreme, individuals making case-hinges, hands, dials, wheels, fusee chains and so forth. These were collected up and assembled at the head workshop where probably only the escapement was made. In London, the centre of all this activity was Clerkenwell, where it continued to be with diminishing prosperity until 1939. It was an uneconomic arrangement and the reluctance for change among all the leading houses, despite the insistent advice of Lord Grimthorpe, eventually ensured the utter collapse of this once thriving British industry.

Much the same happened in France, although fine watches continued to be turned out by the house of Breguet throughout the century, but with increasing reliance on Swiss ébauches. The same applied to the firm of Le Roy. It was in the manufacture of carriage clocks that the French showed the greatest initiative throughout the nineteenth century, starting in about 1830 with the elegant little clocks of Paul Garnier, with Debaufre escapement.

The Swiss are commonly credited with the introduction of mass-production methods into watchmaking but this is not true, for the real pioneers worked in America.

The Swiss concentrated more upon the organisation of skilled labour on roughly English lines, and on the making of specialised tools for the machine production of watch parts. It was not until the present century that full-scale factory production was achieved. The first step was taken in about 1780 by Frédéric Japy, who developed a range of cutting and drilling tools and lathes for making specific watch parts.

Various members of the Humbert family made attempts at quantity production, all of which failed. Joseph, in partnership with Borel & Poterat, started a watch-making factory, and Veuve an ébauche factory. More in line with Swiss requirements, a machine factory for making watch parts was set up in 1820 by Humbert et Darier, but the ultimate goal of interchangeability eluded them, and it in turn failed.

Full flexibility was not achieved until about 1840 by G. A. Leschot, working for Vacheron & Constantin, who by 1845 had designed and made a complete range of machine tools.

Despite these eventually useful advances, the Swiss industry was held up for a long time by sheer bad workmanship, and even as late as 1830 it seems that there was no-one in Switzerland capable of making a balance spring.

The ability of the merchant watchmakers to organise the outworking industry and assemble watches in significant numbers did most to build up the export trade which became the foundation of Swiss prosperity.

Throughout the first half of the nineteenth century, Swiss makers relied almost exclusively upon the cylinder escapement which they made up into fashionably thin watches. Their exhibits at the Great Exhibition of 1851 won them a vast export market in Britain.

In the early years of the century the Swiss made considerable numbers of musical and automaton watches. Some presented scenes of an innocent nature but others, described in the workbooks as 'salacious', revealed at the release of a secret shutter, a lively enactment of connubial intimacy. These are much sought after by collectors whose mental equipment renders them incapable of understanding anything of greater mechanical interest or merit.

Swiss watchmakers throughout the first half of the century revived the enamelling of watch cases, and while some of these are meretricious, many are of considerable beauty and technical merit.

Although none of them was destined for a long run, it was in America that real factory methods were first applied to watchmaking. Luther Goddard opened the first watch factory at Shrewsbury in 1809. His watches had silver cases, enamel dials, gold hands, verge escapement and full-plate movements with a decoratively-pierced balance cock. The whole effect was of a typical English verge of the early nineteenth century, and it may be that English ébauches were imported in the rough. In any case, the venture was a success, even if the total production was not spectacular. Goddard retired from business in 1817, and in the run of eight years only five hundred watches had been produced. Tompion had produced at a markedly greater rate.

It was, nevertheless, a successful start, but no further attempt was reported until 1838 when a factory was opened at Hartford, Connecticut, by James and Henry Pitkin. Some parts were made with power-driven machinery and in that the factory produced about eight hundred watches in its three years of life it may be accounted a success. It ultimately collapsed through failure to sell its products, but it is not known if this arose through lack of initiative by the Pitkins or from lack of performance by their watches. These are signed 'H. & J. F. Pitkin'.

Mass production on a far more grandiose scale was envisaged independently by Aaron Dennison and Edward Howard, in the 1840s, but it was not until 1853 that they got together and obtained the necessary capital to set up their factory. In the meantime, the pioneer efforts of Luther Goddard and the Pitkins had shown, long before anyone in Europe, that factory production was at any rate feasible. 354

The Great Exhibition of 1851 in Hyde Park was the first of many international exhibitions which encouraged machine-operated production. Horological exhibits were well represented, including an electric clock. At this time, America and Switzerland were both entering into the quantity production of watches, with the American makers concentrating on a full factory system. As this committed any factory to one model or very few, it was prey to changes in the market and too inflexible to follow them. The Swiss system of specialised production of individual watch parts for assembly by the named maker, possessed the necessary flexibility, prospering increasingly where American companies failed.

The production by either method might be of the cheapest possible watch, or of the highest quality consistent with machine production.

In Switzerland G-F. Roskopf (1812–89) was the first to persuade the various out-workers to make parts sufficiently simple and crude for him to be able to market the assembled product for twenty francs. The watches had a pin-pallet lever escapement and keyless winding. The hands were set by pushing them round digitally. Roskopf had launched into production by 1865–7, establishing his success in 1868 at the Universal Exhibition in Paris. 310

The cheap American watches came only a little later. Jason Hopkins developed the Aurburndale watch and obtained finance to put it into production in 1875; but the enterprise lasted only until 1883. More successful was the Waterbury venture which survived in one form or another for over forty years. The Waterbury had a crude form of duplex escapement and went through a rapid series of modifications, indicated by the letters A to W, from 1880–90. The firm continued to do business as the New England Watch Company, but could not sustain the Waterbury success and was taken over by the Ingersoll Company. The brothers R. H. and C. H. Ingersoll ran a mail-order business in which they incorporated in 1892, as a side-line, a one-dollar watch. This and successive cheap watches gained a considerable and deserved success until the termination of the company in 1922. 321

America also produced watches of quality and high performance; notable were the Waltham, Elgin, Illinois, Howard and Hamilton companies. Although unable to compete with the more flexible Swiss system, their products are rightly prized. No pains or expense were spared to achieve the highest quality and they may take their place in the most select horological company. Outstanding is the highest grade of Waltham, the 'Riverside Maximus'. 350, 360, 36 367

The Hamilton Company achieved its greatest success in the Second World War, when there was a sudden and heavy demand for very accurate timekeepers. The company responded with outstanding ability and its marine chronometers and deck watches, in particular, achieved a very high standard of performance. 379

America also had its individualists. Pierre Frédéric Ingold was born in Switzerland in 1787, trained under Breguet in Paris, and tried to set up a business for the machine production of interchangeable watch parts. He was first in Paris, then London, and finally, by 1845, in New York. He also made complete fine watches and evolved an escapement which has much in common with the better-known example by Fasoldt. It is a form of Robin, having a lever whose anchor pallets are concerned only with locking the escape wheel. Instead of impulse being delivered to the balance by the teeth of the escape wheel, a third pallet on the tail of the lever receives impulse from the teeth of the escape wheel on the side furthest from the balance. This imparts impulse to the balance via the lever. To a considerable extent, therefore, it is a reversion to Berthoud's 1768 escapement (see page 47). The movement has a half-plate, with separate cocks, for the fourth and escape wheels.

While little is known about Ingold, Charles Fasoldt (1818—98) is a more tangible character. He was born in Dresden, came to New York in 1849, and set up as a watchmaker, producing about fifty watches. In 1861 he moved to Albany and started a small factory making precision clocks and watches and other instruments. He is known mainly for his so-called 'chronometer' escapement. This is similar to Ingold's escapement in that impulse is delivered on alternate beats only, via the lever, but differs in that there are two co-axial wheels of which one engages with the anchor pallets, and is concerned only with locking, and the other engages with an impulse pallet on the tail of the lever. The movement is of bar type, and of imposing appearance.

The Fasoldt and Ingold escapements have the merit that they do not need oiling, but the disadvantage that impulse is delivered only on alternate swings of the balance. Fasoldt's version, like Berthoud's, has the lever pivot closer to its impulse pallet than to the balance pivot, so that a brisk impulse is imparted to the balance-wheel impulse pallet. The Ingold has no such mechanical advantage, and it is difficult to see what merit the escapement possessed as against an ordinary Robin. Nevertheless, unusual escapements of this kind, featured throughout the second half of the nineteenth century, can provide a most interesting and attractive field for the connoisseur-collector.

Albert Potter (1836–1908), of Chicago, was a most prolific contributor to the field and invented no less than fourteen different escapements. He excelled in the lever escapement and one version had an escape wheel of only one tooth. The movements had a somewhat unusual form of three-quarter plate lay-out and his watches were all of high quality. He provided the New Haven Watch Company with designs for a good timekeeping medium-priced watch, called the 'Charmilles'. The company failed in 1896.

Edward Howard (1813–1904) was perhaps the most important American watchmaker of the second half of the nineteenth century. His first attempt, at Waltham, during 1853–57, was not successful financially, but he was soon in business again and prospered increasingly until his retirement in 1882. His ambition was to build the best watch in America, and considering the very large output of his working life he came close to achieving it. Animals may often be found stamped on Howard movements; a deer signifies that it has been adjusted for temperature and position; a horse for temperature only; a hound, not at all. Keyless winding was employed from 1878. Long after Edward Howard's retirement the firm continued to make watches of high quality.

The Swiss do not seem ever to have been much interested in escapements, perhaps because they were practical enough to realise that Breguet had said virtually the last word with his double-roller, divided-lift lever escapement, or perhaps because they were commercial enough to see that there was no money in experimental escapements. The interest in Swiss watches, after 1850, therefore lies in their immense variety of presentation.

The cheap Roskopf watch has been mentioned already, but the Swiss system of specialisation enabled every other sort of watch to be made in the most efficient and economical way.

Among all this proliferation, it is difficult to choose individual types and indeed, since most of the big firms produced a wide range of models, it is the type of watch, rather than the name on its dial, which is more important to the collector.

Although the Swiss makers on the whole concentrated on the double-roller, divided-lift lever escapement, the best makers used the pivoted detent escapement for some of their finest work. Also, the cylinder and duplex escapements hung on into the present century.

Decorative watches have always been a feature of Switzerland and enamel painting continued to be practised until about 1930. A particularly durable technique used in Switzerland was 'Geneva enamel', invented in 1760. This consists of enamel under a flux, or transparent layer, mechanically polished after firing. Decorative engraving was practised throughout the nineteenth century but died out by 1914. Bracelet watches date back as early as 1806, but they were rare before 1850 and not common until after 1900. After the Zurich Exhibition of 1883 there was a return to the seventeenth-century fashion for form watches.

The Lepine calibre was introduced into Geneva in about 1790 and has been in almost universal use ever since, usually in the form of parallel bars across the movement. Extra-thin watches became fashionable from about 1840, when Tissot-Gounouilhon produced a movement only 2·5 mm thick. The thin watches might employ either a Lepine calibre or the reversed calibre, in which all the wheels are between the plate and the dial and none is visible on opening the back of the watch.

Complicated watches began to become fashionable from about 1850 and mechanical complication appeals to a sufficiently wide range of collectors to ensure that such watches command very high prices when they come on the market. A repeating train is a feature of every complicated watch, and the Tavannes Watch Company, founded by Henry-Frédéric Sandoz, of Le Locle, was the first to make the complicated equipment necessary for the mechanised production of repeating trains. L-E. Piguet followed with minute repeaters in 1853. From about 1870 Patek Philippe & Co. have been pre-eminent for their highly complicated watches.

Pioneers in the use of keyless winding and hand-setting were Louis Audemars & Sons (possibly as early as 1838), Adrien Philippe (in 1843, pioneering hand setting by pulling out the pendent) and Antoine Lecoultre (in 1847).

The fly-back chronograph was produced by Nicole et Capt in 1862.

Switzerland, until quite recently, has always produced some very high-precision tourbillon watches. Outstanding among makers of these was Albert Pellaton-Favre (1832–1914) who made eighty-two tourbillons during his long life.

It is in the field of metallurgical research that Switzerland has contributed most to the real advancement of horology. Magnetism and rust were the first evils to be attacked, although John Arnold had used gold balance springs as early as 1784. J. F. Houriet made compensation balances of white gold and brass in the 1820s and at the age of eighty-five showed a non-ferrous chronometer at the Swiss National Exhibition of 1828.

Palladium, derived from gold ore, is non-rust and non-magnetic, and was produced by Charles-August Paillard in 1877, but the most valuable work was done by Charles-Edouard Guillaume (1861–1938). He evolved 'Invar' in 1899, composed principally of iron, nickel and manganese. As Invar has almost no co-efficient of expansion it rendered the cut, bi-metallic balance unnecessary and thus eliminated middle-temperature errors. In 1913 Guillaume developed a refinement for balance springs called 'Elinvar', consisting of steel, nickel, chromium, manganese and tungsten. It has, however, the disadvantage that it is soft and easily deformed, and is also incurably variable in quality.

An improvement on Elinvar was evolved by M. R. Straumann. It is a complex amalgam incorporating the metal beryllium, which he called 'Nivarox'. This is more elastic than Elinvar, non-rust, and almost non-magnetic. A completely non-magnetic watch can now be made employing beryllium-copper alloy for the escapement, with a Nivarox balance spring.

Although terminal curves for balance springs were first used by Arnold in 1782, and Breguet had applied the principle to spiral springs, their conformation was empirical until Edouard Phillips (1821–89) worked out their mathematics and thus facilitated a great advance in the minimalising of position errors.

Although Paul Ditisheim (1865–1945) spent much of his youth in England, and worked for Rotherham's until 1891, he returned to Le Locle and it was in Switzerland that he did his most contributory work in the advancement of horology. As a watchmaker his interests were catholic, he delighted equally in precision 369 watches on the one hand and jewellery and form watches on the other (but only of the highest quality). Using Guillaume's inventions in metallurgy, he applied very small bi-metallic affixes to Invar balances, thus eliminating their minute residual temperature errors. Ditisheim's tourbillons took first place at the Kew and Teddington trials on eight occasions. However, his greatest contribution to horology was in the unglamorous but highly important sphere of oils. As recently as 1900, it was desirable for a high-precision watch to be cleaned every two years, but now an interval of eight years is acceptable, and this advance is due largely to the experiments carried out by Ditisheim.

In these inconspicuous ways, and in the perfection of machine tools, the Swiss made a marked contribution to the advancement of horology, but perversely it is the decorative, the ultra-thin, and the ultra-complicated aspects of Swiss watches that most appeal to collectors.

Sadly, it must be admitted that France has contributed almost nothing to horology since 1850, its most important production being that of carriage clocks. The firms of Le Roy and Breguet nevertheless continued to produce fine and elegant watches, although relying increasingly on Swiss parts.

Germany produced perhaps only one outstanding artist, Adolphe Ferdinand Lange 325, 326, 328, (1815–75) of Glasshütte. He trained in Paris under Winnerl, returning to Saxony 330, 331, 332, in 1845. The robust but superbly-detailed quality of his work set the style for the 365, 373 considerable number of makers who followed him in Glasshütte.

Although the British industry has not done much to advance horology since 1850, it has at least handed us down a large number of very fine watches. The firms of Dent and Frodsham maintained their pre-eminence in producing watches of the highest quality until 1939, although latterly they both obtained their ébauches from Nicole Nielsen until the firm's closure in 1935. Nicole Nielsen in turn had much of their material from Switzerland, so that a Smith or Frodsham ultra-complicated watch of the 1920s or 30s was not very different from a Swiss one. However, to British tastes, at any rate, the cases, dials and hands were greatly superior. These firms also produced superb lever-tourbillon watches. Swiss-based movements had double-roller, divided-lift escapements, but the English single-roller lay-out survived until well into the present century. 301, 313, 323, 329, 342, 343, 344, 351, 352, XVII, 356, 35 370

The London trade remained faithful to the fusee for as long as possible, delaying the universal adoption of keyless winding until the 1880s. However some efficient keyless-wound fusees were developed. As against this, from 1850 Dent's produced high-grade keyless wound and hand set watches, with going barrels. 316

The firm of Benson also made good watches, but like all the London firms, relied upon the highly-concentrated complex of out-workers in Clerkenwell for most of its supplies, perhaps making only the escapements in its own workshops. Regrettably, this logical system of producing watches, in which the emphasis was on quality and accurate performance, could not contend with the far more commercial Swiss industry, and by about 1935 the British industry had effectively died. 333

Against this rather sad overall picture, some individualists continued to make a living. Ferguson Cole lived until 1890 and the firm of Jump continued to produce watches and carriage clocks in the Breguet tradition, almost up to Breguet standards of finish. 311

The Lancashire industry went on throughout the nineteenth century but its decline was discernible by the 1880s. An outstanding firm, Richard Dover Statter and Thomas Statter, made watches of the highest quality and often, individuality, until at any rate the 1880s. The Lancashire Watch Company of Prescot was founded in 1889 in an attempt to revive the industry and survived until 1910. It was then continued on a small scale until 1960 by A. J. Huckle. The firm of Joseph Preston and Sons, marine chronometer and watch movement manufacturers, also survived until 1952. 306
338

An escapement often mentioned is the 'Union Chronometer' but very few examples have survived, or perhaps were ever made. C. G. Kelvey and W. Holland of Birkenhead took out a patent in 1859 for their 'Patent Union Chronometer'. It consisted of something between a Robin escapement and Ferguson Cole's repellent lever. Locking was by a lever, but with repellent pallets, so that the safety dart pressed permanently against the safety roller of the balance. Presumably Cole thought that the resultant friction was preferable to the effect of draw, whose use he avoided by this device. Impulse was imparted to the balance by the escape wheel teeth and therefore took place only on alternate swings of the balance.

Between 1825 and 1860 the characteristic heavy engraving of the Liverpool balance cocks became less pronounced, the jewelling became smaller, and the full-plate lay-out was replaced by three-quarter-plate movements. Massey's escapement was much used and even the rack lever survived for quite a long time. The earliest rack lever watches had fifteen-tooth escape wheels with radial leading faces to the teeth. Later, a three-wheel train with a thirty-tooth escape wheel became generally used. Rack lever 263, 269, 30 watches had 'Patent' engraved on the balance cocks, and on watches with Massey's escapement the cock was engraved 'patent improved' or 'detached'. (This is interesting, as it lends some support to the view sometimes expressed that Massey's escapement is a development of the rack lever.) At the end of the century, the fine Liverpool firm of Thomas Russell was making karrusel and tourbillon movements; the family is still in business in Liverpool.

Returning to London, Sidney Better was an artist who suffered, like so many others, from the world depression of the 1930s. A somewhat lonely figure, working on the 376 fringe of the Clerkenwell area, he contracted to make tourbillon watches of high quality for a retailing company called the Northern Goldsmiths. He obtained the considerable outfit of precision tools necessary for his purpose and had turned out some very fine watches when the agreement was terminated. He was thus left without employment and, broken emotionally and financially, died soon afterwards.

Among this confused pattern of events, two significant, productive developments 329, 342, 371 occurred in the British industry. The first was the development of the karrusel, patented in 1894 by Bahne Bonniksen (1859–1935). It is a less exacting, more robust, version of the tourbillon, its object being to minimise position errors. Whereas, in a tourbillon, the rotation of the carriage is an incident in the escapement and, because of its high rate of rotation, has considerable inertia, the karrusel has a much slower and less violent speed of rotation, the rotation being nothing to do with the escapement. The carriage is driven by the third wheel and the fourth and escape wheels mounted on it. It is, therefore, an arrangement hardly more exacting to make than an ordinary watch. The first watches had a rotation speed of fifty-two and a half minutes, but in 1903 Bonniksen produced a model with a more rigidly-mounted carriage, a centre seconds hand driven off the third wheel, and a rotation period of thirty-five minutes. This he called the 'Bonniksen Tourbillon'.

These watches proved their merit by consistently taking the first place at the Kew trials and, like all revolving escapements, are able to hold their rate for longer than almost any fixed-escapement watch.

Bonniksen's factory was in Coventry, where he employed about twenty men. It is very rare to find his name on a watch, as for the most part he supplied the trade, and several makers' names may be found on karrusel watches. The watchmaking industry in Coventry was a thriving concern throughout the second half of the nineteenth century, but decline set in by the 1890s, and by the 1930s it ceased to exist.

The now almost-universal self-winding wrist watch is also a British development. John Harwood made his prototype in Douglas, Isle of Man, in 1922 and went into 375 production in 1928 as the Harwood Self-Winding Watch Company Ltd. The ébauches were supplied by the Swiss firm of A. Schild & Co. but in 1932 the company was forced into liquidation. Had a British firm of substance had the vision to take up the invention, the future of the industry might have presented a very different picture.

Most recently of all, a British development of the precision watch has been 385, XV manifested in the watches now made in London by George Daniels. Cecil Clutton was the recipient of the first of a new series of tourbillon watches started in 1968. At his request the watch was fitted with a pivoted detent escapement but all subsequent watches have the spring detent escapement. The escapements have large-diameter stainless-steel balances to give high inertia for low-pivot pressures. The balance springs, with terminal curves, are nickel-steel alloy to compensate for temperature. Changes to the proportions of the escapement have resulted in increased balance amplitude and self-starting during winding from the run-down condition.

A later development has a one-minute tourbillon carriage isolated from the watch train by a remontoir mechanism winding at fifteen-second intervals. This was constructed to study the effects on isochronism of pivot friction and power variation without the confusion of positional errors. The watch also shows equation of time. 386

The results of tests carried out on this watch have encouraged the manufacture of an entirely new series of watches. These have two separate trains and a double escapement impulsing directly to the balance axis at each vibration, without need of oil. The wheels are unlocked alternately by a pivoted lever and the whole of the impulse is delivered after the line of centres. The resultant escapement loss is automatically compensated by the short arc-gaining characteristics of the Ni-Span-C alloy balance spring so that isochronism is continuous throughout the running period of the watch. In the hands of their new owners the watches have demonstrated their ability to keep a remarkably close daily rate although the principal object of the design is to improve long-term reliability of rate. With modern oils at the pivots and no oil at the escapement they are expected to run for a minimum of ten years without regulation. 387

The manufacture of individual watches by hand methods now demands that the constructor makes every component himself. The numerous small workshops that once skilfully interpreted the instructions of the watchmaker no longer exist. The makers of cases, dials, hands, springs and other parts, are no longer readily available to help the watchmaker complete his task, which is therefore made more complex and difficult. The reward lies in the continuous process of development of the watch towards a better and more reliable timekeeper and the authors see no obvious reason why the two hundred years of development of the precision watch since the pioneer work of John Arnold should not be continued.

Decorative

Types of decoration

There are three principal ways of decorating a watch case.

The great majority of cases are made of metal, which may be decorated in a variety of ways. Otherwise, cases may be enamelled, or they may be made of, or decorated with, precious or semi-precious stones. Cases may also be covered with leather, and decorated with pinwork; or they may be covered in tortoiseshell and inlaid with patterns in gold or silver.

Metal cases may be decorated in five ways:

1 casting
2 chiselling
3 engraving
4 repoussé
5 engine-turning.

62   1 As a preliminary, a decorative shape or pierced pattern may be cast, but this will need finishing by chiselling or engraving.

91   2 Chiselling is carried out on the surface of the metal by a hammer and punches.

78   3 Engraving is a finer form of decoration, executed by a 'graver' which produces fine V-grooves on the surface of the metal.

92   Chiselling and engraving are often combined as, for example, in 'champlevé engraving', where the background is chiselled back to a common level and then engraved with fine grooves or punch-marks all over, to give it a dark appearance. The foreground, or pattern, will also be finished with engraving.

144, XVIj   4 Repoussé is a very ancient art, but it does not appear much in watches before the end of the seventeenth century. It is sometimes difficult to distinguish on the surface from casting, since both produce a moulded surface. Repoussé is executed in fairly thin, pliable sheet metal (almost invariably silver or gold), worked on by hammering it out from the back into the desired shape.

170   5 Engine-turning, or guilloche, dates from about 1670 but is rarely found on watch cases before 1780. It is done on an engine-turning lathe which can produce a variety of repeated patterns. 'Barley-corn' is the most popular and usual form for watch cases.

Enamel is a glass composed of silica, red lead and potash. It is coloured with various metal oxides, but very occasionally it is used without colouring, when it is a transparent flux called 'fondant'.

Enamel may be hard or soft, according to its composition. Hard enamel retains its surface and colour indefinitely but can only be applied to metal with a high melting temperature. It cracks much more easily than soft enamel which, however, is more readily scratched.

Enamel decoration may take six forms:

1 plain enamel
2 painting in enamel
3 painting on enamel
4 basse-taille
5 champlevé
6 cloisonné.

1 This is self-explanatory and is hardly ever found alone.

2 Painting in enamels was developed during the sixteenth century after which it is seldom found. It was hardly applicable to curved surfaces, but it is found in its simplest form on watch cases, and these are some of the most beautiful enamelled cases in existence. Small blobs of enamel were deposited on the enamel base, in the form of a pattern, usually floral. After firing the blobs remained in relief. The method was used on English watches as frequently as any; perhaps the finest example is the watch by Edward East in the Victoria and Albert Museum, where a light blue ground is decorated with applied enamel flowers in white and pink. Another specimen is in the British Museum.

3 The technique of painting on enamel was evolved in about 1630 by Jean Toutin, a French goldsmith (1578–1644). The painting is applied on a plain enamel ground. The colours are metal oxides with a little fondant to make them vitrifiable. After firing, the picture was covered with a layer of fondant and again fired.

Blois was the centre of enamel-painting throughout the seventeenth and eighteenth centuries after which it moved to Geneva. Early Blois masters who painted watch cases were Henri and Jean Toutin (sons of the original Jean) and Isaac Gribelin. Pierre Chartier and Christophe Marlière specialised in flower painting. The early painted-enamel cases are of the greatest splendour as to design, technique, and colour. Many of them are miniatures and the colours are very brilliant. After about 1650 the quality fell off noticeably and the colours used were less brilliant. The favourite subject then became fat women in négligé attire, closely observed by eager-looking old men or satyrs. The most renowned artists of this style are the Huaut family. The members of the family who painted watch cases were Pierre, Jean and Amy. Pierre, the eldest, signed himself 'Huaut l'aisné', 'P. Huaud primogenitus', 'Pierre Huaud' and 'Petrus Huaud major natus'. Jean signed himself 'Huaud le puisné' but mostly worked with Amy when they signed their work 'Les Deux frères Huaut les jeunes' up to 1686. They then moved to Berlin after which they signed themselves simply 'les frères Huaut'. In 1700 they settled in Geneva after which the signature might be 'Les frères Huaut' or 'Peter et Amicus Huaut' or 'Frères Huaut'. It will be noticed that the terminal 'd' appears only in the German period. The Huauts' signatures have been retailed in some detail because they are so renowned. Many other competent artists specialised in watch cases; but since enamelling is a quite separate subject, the watch collector who wishes to specialise in it is recommended to study the standard works on the subject, rather than rely on the superficial account given here. The Geneva school continued throughout the eighteenth century while the Blois school declined. Latterly the Swiss artists specialised in scenes of an intimate nature, delicately referred to in their accounts as 'salacious', whose verisimilitude was sometimes heightened by moving automata. That very high prices are paid for these need not be a matter of concern to the horological collector, as they are invariably found in pieces of no horological interest or quality.

4 Basse-taille is perhaps the most elegant form of enamel decoration as applied to watch cases. It consists of a coloured, transparent enamel on a chiselled or engine-turned gold ground. Examples over chiselling survive in very small numbers from before 1650, but basse-taille did not become fashionable until after 1780 and surviving examples are mostly Swiss.

5 Champlevé is the commonest and earliest form of enamelling found in watches. It consists of cutting cells out of the metal and filling them with enamel.

Allied to champlevé enamel is the technique known as niello, which is found, rarely, in early seventeenth-century cases. Here the cells, instead of being filled with enamel, are filled with an alloy of silver, lead, copper and various sulphides.

Champlevé chapter rings are found even in sixteenth-century watches. It was a popular form of case decoration during the seventeenth century. The cells may be so close together that the walls between them are so thin as to be hardly visible, in which case champlevé is quite difficult to distinguish from cloisonné.

6 In cloisonné work, the pattern to be followed is outlined in thin strips of metal, or 'cloisons', almost always gold, fixed to the metal ground, and filled in with enamel. The enamel is then ground down to the level of the cloisons and polished.

Watches decorated with precious and semi-precious stones cover such a wide field of the lapidary's art that to cover it comprehensively here would not be practicable. It is sufficient for the horologist to recognise the principal types.

The earliest and commonest form is the case of rock crystal. This consisted of two parts, one of which was hollowed out to take the movement; the other formed a lid. Each was mounted in a narrow metal frame, or bezel, of which the two parts were hinged together. The crystal was decorated by faceting in a variety of ways. Very rarely, there was no metal frame, as in the case of a watch by Michael Nouwen (1582–1613) in the British Museum. In this type a simple hinge is riveted to the crystal body and lid. The crystal of this superlative watch is smoky in colour and the dial is decorated in red and green champlevé enamel. During the seventeenth century, the cases of small watches might be made of semi-precious stone; larger watches might have panels of cornelian and similar stones. Lapidaries' work is seldom found during the eighteenth century, but during the early nineteenth century Swiss watches in particular were often decorated in brilliants or split pearls, usually in conjunction with basse-taille enamelling.

Styles of watch cases and dials up to 1750

Having thus covered the main forms of decoration as applied to watch cases, and briefly described the techniques involved, they may now be considered in chronological sequence. This is important horologically up to 1650 in particular, as during this early period it is generally easier and safer to date a watch by the style and decoration of its case, than by its mechanism.

Cases other than plain metal are practically never found before 1600. It is known from records that cases were made in gold and silver, but none of these survive, and only brass-gilt cases are found. These cases may be cast, or made of soldered sheet-metal. The latter is obviously applicable to simple shapes, such as the plain, flat-sided drum, and engraving (and, to a limited extent, piercing) is the only form of decoration which can be applied to it.

The commonest use of sheet-metal was in the small drum-shaped clocks or large watches, already discussed, which usually have a pushed-on back and no cover to the 56, 57 dial. However, it is only by a slight stretching of the imagination that these can be regarded as watches at all, and nearly all other sixteenth-century watches had a hinged cover to protect the hand. It might be solid; or pierced, so that the hand and numerals 60 could be seen through it.

Since most watches were either clock-watches or alarums the case, as well as the lid, had to be pierced, to let the sound of the bell out; and these were usually cast. Apart from the very early French and German spherical watches (of which any modern collector is most unlikely to find an unknown specimen), and the very early, tall drum-shaped clocks or watches, the mid- and late-sixteenth-century watches had 65 circular, flat-sided cases. The bell was fixed inside the case, and the movement, fitted inside the bell, and was hinged to the case. Since the bell was a fixture, the movement had to be swung out of the case for winding (winding through the dial, or drilling a hole through the bell, did not come until the second half of the seventeenth century). The dial cover was also hinged to the case.

As has been said, the case and cover were cast, pierced and chiselled in quite high relief. As the century drew to an end the chiselling became less deep and engraving 62, 65 began to take its place. During the last quarter of the century, the cover began to take on a slightly domed contour, and the sides became curved, instead of flat, in section. It is very rare for a sixteenth-century watch to be anything but circular, but in the last quarter, octagonal and elongated octagonal watches were occasionally 64 made.

The dial was almost invariably engraved gilt-metal. An applied silver chapter ring is seldom if ever found before 1600. Bohemia, Italy and south-west Germany used a twenty-four hour day, and calibrated their dials accordingly. The first twelve hours were marked in Roman numerals, and prior to 1650 these are always so stubby that the IIII is wider than it is tall. From the very earliest times, IIII was used instead of IV. This was done for purely aesthetic reasons. The IIII is heavier in appearance than IV and is thus a better balance for its opposite number, the VIII. (The truth of this is seen in the rare late-seventeenth-century clocks with Roman striking. These chapter rings, necessarily using the IV, have a distinctly lop-sided appearance.) The second twelve hours were marked in Arabic numerals, and of these the 2 invariably 60 appears as a Z. When English watches began to appear after 1600 they had a normal 2, but the Germans continued to use Z for another quarter-century.

The centre of the dial is usually decorated with a star-shaped pattern, of which the 56 centre may be embellished with a little champlevé enamel.

So long as the watch had to be set to time by pushing round the hand itself, this was nearly always of iron or steel. It might be quite elegantly shaped and rounded, but it was always sufficiently robust to withstand direct handling. Despite their robustness, the early hands are seldom clumsy in appearance; and when they are (usually of unrelieved square section) it is almost certain that they are replacements. Practically the only exception to the iron hand is found in enamelled watches, when the hand is always gilt-metal.

Prior to the first appearance of pockets, about 1625, pendents were usually fixed, and pierced from front to back, and fitted with a loose-fitting ring-shaped pendent. This 70 was therefore at right angles to the watch, indicating that the watch was worn suspended on a cord, round the neck.

Surviving watches of the sixteenth century are very rare, and the majority of them are already in museums. After 1600 watches become increasingly common, and are found in the greatest variety of shape and decoration. For the first three-quarters of the seventeenth century there was no significant technical advance in watchmaking, and the timekeeping of watches continued to be of the most casual nature. Interest was therefore concentrated upon decoration, and it is during this period that the greatest variety of high-quality decoration is found, both in the cases of watches, and in the movements. So far as the decoration of movements is concerned, this has already been dealt with in the mechanical section.

In case work, the most curious development during the first half of the seventeenth century was that of the 'pair case'. As soon as watches began to have expensive and delicate cases it became a sensible precaution to provide them with a protective outer case for use when travelling or when not in use. At first these were made 70 of stiffened leather, with the outer side sometimes decoratively tooled.

As time went on, it apparently became usual to wear the outer case. This practice probably started with the introduction of pockets, but by the middle of the century, outer cases are found increasingly which are definitely part of the watch and meant to be worn with it. For this purpose, the dial side is open and the back of the case becomes increasingly decorated, generally with a pattern of silver or gold pinwork. 85, 86 Thus, the emphasis moved from the decorated inner case to its protective outer, so that by 1670, especially in English watches, the inner case had generally become completely plain and it was the outer case that was completely decorated. Sometimes this decoration was of such a delicate nature as to need a third case to protect it. 84 In the ordinary watch such as most people owned, from 1680 onwards, both inner and outer case were perfectly plain, the outer serving merely to prevent dirt entering the winding hole in the inner case and to make the watch a good deal thicker than it need have been. The French rarely used this illogical arrangement, preferring a single-case watch which was wound through the dial.

This trend was therefore one which was continuous during the whole of the seventeenth century, but other fashions came and went more quickly. A newcomer 61, 67, 78 after 1600 (or possibly just before) was the form watch, made in a number 83 of irrelevant shapes such as those of books, skulls, dogs, birds, crosses, flower-buds and the like. Particularly popular were forms in the *memento mori* class, such as the skull, and the cross. The skull was hinged at the jaw to open and reveal the dial. With such trivialities did early seventeenth-century watchmakers amuse themselves, and keep their customers' minds off the lamentable performance of their watches. It must be said, however, that British makers did not go in much for these frivolities, but concentrated on trying to make their watches keep time for which they were rewarded in the last quarter of the century by finding themselves in a position of unassailed supremacy.

The traditional sixteenth-century round, octagonal, and increasingly, oval, watches were made throughout the first quarter of the seventeenth century. Flat-section sides continued, but the rounded section became increasingly popular. In British watches, particularly, this trend eventually developed into an almost unrelieved, oval egg-shaped watch, devoid of decoration.   84

Brass-gilt remains the most usual metal, at any rate in surviving watches, but silver survives increasingly, and gold very occasionally, although at first only in conjunction with enamel decoration.

French watchmakers employed increasingly the elongated oval or octagonal shape, and Swiss watches began to appear, closely following the French style.

After the third quarter of the century, German watches become rare and relatively unimportant owing to the disastrous impact of the Thirty Years' War, although a surprisingly large number of fine German hexagonal table clocks were made from about 1660 onwards; but these highly decorative and functional instruments never won much favour in other countries.

For plain metal cases after 1600, engraving fairly rapidly ousted chiselling, no doubt as watches became more widespread as items of normal attire, and so were required to be smaller and lighter. Engraved landscape scenes containing figures and animals are more usual than hitherto, and generally fill the space inside the chapter ring (unless the watch is an alarm, when the setting dial may be decorative, in the form of pierced gilt metal over blued steel or pierced silver over gilt). The applied silver chapter ring becomes common as the twenty-four-hour dial tends to disappear.   68, 77, 78, 79, 92

Mechanical complications, in the form of astronomical movements and calendar-work, become fashionable for more luxurious watches after 1600, and this brought in a fashion for multiple subsidiary dials. One of these might be what amounts to a minute hand, especially in quarter-striking watches, where it was easy to fix a hand to the quarter-striking wheel, which revolves once an hour. However, the dial is not marked in minutes, but only in quarters. True minute hands are seldom found before 1680. But the fashion for multiple dials continued until after 1680 and, like so many seventeenth-century watch fashions, it was revived in the early nineteenth century.   66, 71, 85, 98, 99

During the first quarter of the seventeenth century enamelling is pretty well confined to champlevé decoration (occasionally also champlevé niello), but the middle of the century saw the very finest period of enamel-painting. This was short-lived, and after twenty or thirty years there was a marked deterioration of taste in the form already described. The enamel cases are single, painted inside and out.   Ic, D, H, XVIc, E, K

Glass covers over the dials appeared during the second quarter of the century. The first move in this direction was a round window of rock crystal, held to the case cover by tabs, but plain glass appeared soon after, held either by tabs or split bezels. It is, however, difficult to say with certainty that an ostensibly early glass has not been added at a later date. When the cover is engraved it is often quite clear that this has happened.   70 76 79

At about the middle of the century the completely plain, almost egg-shaped British watch, already mentioned, and now generally known as a 'Puritan watch', was developed. This is almost always made of silver. The dial cover always has a round glass window and the dial is also plain, usually with an applied silver chapter ring and fairly plain, but elegantly-shaped steel hand, with a tail. The only decoration   84 is on the plate of the movement, although the dial may sometimes be engraved.

These egg-shaped Puritan watches are quite different from the so-called 'Nuremburg Egg'. This meaningless term used to be loosely applied to any big old watch, pretty well regardless of its shape. The term is never met with before the eighteenth century and its use has now been quite discontinued, except by the most ignorant.

During the third quarter of the century tastes sobered down and the watch assumed increasingly the form it was going to retain for the next hundred years. A few form watches were still made, and enamel-painting continued in profusion although in deteriorating taste. The pair case became usual for British watches, the outer case being either plain, or covered in shagreen or fish-skin, with patterns of silver or gold pins.

The matt-surface silver or gold dial began to appear, and the figures of the chapter ring rapidly grew in length, by 1675 to such an extent that very little room was left for the hand in the centre.

Hands in general became simpler, sometimes being perfectly plain pointers. They continued to have tails, although not such long ones as previously, and it is quite common to find no tail at all.

The pendent hole gradually changes from its old back-to-front position to one where the pendent ring is parallel with the watch. This no doubt followed the introduction by Charles II of waistcoats and waistcoat pockets into England.

The case-piercing of alarums and clock-watches became increasingly floral in pattern and attained a very high standard of design and engraving. Watch glasses became universal and the snap-in bezel began to appear by about 1660, although the split variety may be found almost until the end of the century.

In French watches especially, a perfectly plain white enamel dial, with plain black Roman numerals, is very occasionally found.

The commercial application of the balance spring to watches from 1675 brought about a rapid change in decorative styles. As watches suddenly became capable of sensible timekeeping, the emphasis moved rapidly from exterior decoration to mechanical superiority. From now on for the next hundred years, by far the commonest form of British watch has a pair case completely devoid of decoration. Before 1675, the finest decorative cases contained the finest movements, but by soon after 1680, the finest movements usually have the plainest cases, and after about 1750 it is almost certain that an extravagant case will have an inferior movement, or at best one of no horological interest.

During the third quarter of the century, watches had become surprisingly slim (especially the enamelled cases), but as soon as the emphasis shifted to good timekeeping they plumped up wonderfully, the French watches particularly so, to the extent that they rapidly earned for themselves the name of 'oignon'—and indeed they have the almost spherical shape of a large onion. By far the most common form of French case after 1680 is a plain, single case of brass-gilt, decorated all over in an arabesque pattern. This might be formed by chiselling, or possibly in some cases by casting, simply cleaned up with a chisel. There is rarely any engraving, and indeed at this time engraving almost disappeared as a form of watch decoration.

The usual French dial was of brass-gilt, decorated similarly to the case, with the hour numerals on separate enamel plaques stuck on to the dial. Sometimes there is a narrow enamel ring inside the hour plaques, with hours and half hours marked on it.

87

Minute hands were unusual and did not become at all universal until well after 1700. When there is a minute hand, there is sometimes another, outer enamel chapter ring for minutes. The hand or hands long continued to be moved manually, and even after 1700 they may be found without motion work, so that the two hands are not positively geared together, but have to be set separately. Winding is through a hole in the dial plate, occasionally through the centre of the hand.

However, some French watches had silver dials with a champlevé chapter ring, not unlike the British, and, as already stated, a very few had plain white enamel dials.   113

Hands might be of steel, but equally often of brass-gilt, usually pierced and of rather clumsy design.

In fact, although French watches just after the introduction of the balance spring are by no means devoid of a certain robust charm, they are manifestly inferior, certainly in accuracy of performance, to their British rivals. There ought to be some very fine Dutch watches of this period, but for some reason very few seem to have survived, and such that have either follow French or British styles. In the latter case, British provenance is even sometimes claimed, Tompion and Quare being great favourites for this purpose. Without question, the outstanding watches between 1675 and 1750 were British.

As has been said, the finest watches have the plainest cases. Gold watches are hallmarked and were of 22 carat gold. Silver cases were seldom hallmarked before 1740. Casemakers stamped their initials, so these should be identical on inner and outer cases. Watchmakers also took to numbering their products and the watch number was usually repeated on the cases.

Repoussé began to appear as a decoration, but before about 1715 it is usually confined to such simple forms as radial fluting. Repoussé figures and scenes hardly   130, XVIj appear before 1715 and rarely before 1725. The great period of repoussé was 1725 to 1750, after which it again declined in favour, until by 1770 it is rarely found. At its most exuberant the repoussé became so deep that it could not be worked out of one sheet, but a second sheet had to be worked and then soldered to the main case.   136 Some of the finest repoussé cases are signed by the embosser.

A new metal used for watch cases was 'pinchbeck', so-called after its inventor, Christopher Pinchbeck. It is an amalgam of three parts of zinc to four of copper, but the secret was jealously guarded from the time of its invention, about 1720, throughout most of the eighteenth century. It was described as 'so naturally resembling gold, as not to be distinguished by the most experienced eye, in colour, smell and ductibility'.

Pinwork leather and tortoiseshell cases continue, and the tortoiseshell may also be   100, 105, 112 decorated with an inlaid pattern of very thin silver or gold strips. The shape of these   IVa is cut out of the shell, which is then heated and the strips are pressed in while the shell is still hot. This decoration may take the form of an arabesque, or sometimes a scene in the new Chinese taste.

Outer cases were also covered in shagreen. Real shagreen is the skin of a shark, but in commercial form it consisted of the skin of a horse or donkey, treated by pressing   109 hard, round seeds into it until the skin was perfectly dry. It was then rubbed down to a smooth surface, polished, and stained green.

Otherwise, apart from a few enamel cases, any other form of decoration was quite unusual from 1675 to 1775.

Pendents continued to be loose-fitting rings until about 1690, and sometimes after 1700, but soon after 1690 a hinged, stirrup-shaped pattern became almost universal. The hinges of the outer cases underwent a change at about the same time. The earlier ones are square-ended, but by about 1690–1700 the ends became curved, to merge less sharply into the rim of the case.

108, 110

Some unusual forms of dial

Dial design became standardised by 1700, after a most interesting, formative quarter-century, almost wholly confined to British watches. This arose from an uncertainty as to how best to indicate minutes as well as hours, now that the new standards of accurate performance rendered this information of practical value.

Although the concentric hour and minute hands had appeared by 1680 their acceptance was by no means universal, and four other principal variants were tried out, admittedly not in very large numbers, which makes them a considerable prize for any collector, especially as all of them are highly decorative. The four variants are:

1 the six-hour dial
2 the wandering-hour dial
3 the differential dial
4 the sun-and-moon dial

The first three seek to tell hours and minutes with a single indicator, each in a most ingenious way.

1 In the six-hour dial, a single, centre-pivoted hand rotates once in six hours. The dial is divided from I to VI in Roman numerals. Therefore, starting at twelve o'clock, the hand has performed a complete circle by six. Super-imposed over the Roman numerals are a set of smaller Arabic numerals from 7 to 12. So from 7 to 12 the owner must read the Arabic numerals. Owing to there being only six hours, instead of twelve, to the full circle, the space between each pair is twice as large as on an ordinary dial. This gives enough room for a minute ring round the edge of the dial with two-minute divisions. Also, between the hour and minute rings is another circle divided into quarters and half-quarters. The time can therefore be read with considerable accuracy, although not without a good deal of practice for a modern owner, since none of the divisions appear in their accustomed positions. He must also decide whether he is in the 'Roman' or 'Arabic' sector.

104, 108

2 The 'wandering hour' achieves its objective rather more effectively. An annular slit extending through half a circle appears in the upper half of the dial. Round its outer edge, on the dial plate, are numbers 0–60, usually at five-minute intervals, and minute graduations. Under the dial plate is another, rotating plate, the full diameter of the dial, usually engraved and gilt on the visible part. Two holes, the size of the annular slit, are cut out of the rotating plate, exactly opposite to each other. Pivoted to the rotating plate are two smaller rotating plates, one carrying the odd numerals from I to XI and the other the even numerals II to XII. These numerals appear each through one of two circles. By means of a detent, the numbered plates are advanced

125, 134, IJ

once an hour. Thus, at one o'clock, the figure I will appear through the hole in the large rotating plate opposite the O of the minute markings. XII will be visible at 60 through the other hole. The I and its hole then advance round the slot, the XII almost immediately disappearing from view, under the fixed dial plate. As it advances, the I will successively pass all the minutes of an hour, so that the hours and minutes are simultaneously visible at any moment. When I comes to the end of its journey the other hour plate will have advanced and will appear again as II, opposite O.

A peculiarity of the wandering-hour watches is that nearly, if not quite all, surviving English specimens have a royal attribution, such as a royal portrait (James II, Mary, or Anne; no Charles II is known to the authors) on the dial; or the royal arms engraved on the cock; or both. Also a considerable majority have a fluted repoussé outer case, which is rare at this time. 134

These distinctions have never been explained, and are additionally mysterious since several makers produced the type, not all of whom held a royal appointment. As one surviving example has the portrait of Queen Anne it is evident that what is, after all, a very sensible arrangement continued in production well after the turn of the century.

3 The differential-hour arrangement is very rare indeed. An ordinary minute hand goes round once an hour. The centre of the dial is occupied by a revolving disk with the hours from I to XII numbered on it in the ordinary way. This is geared to rotate $\frac{13}{12}$ of a full circle in an hour. Thus, the current hour is always immediately under the minute hand. 123, 132, 198

4 The sun and moon indicator is not quite such a serious affair as the rest. A minute hand revolves once an hour and is concerned with the minutes only. A semi-circular hole is cut out of the upper half of the dial plate, numbered VI to XII, and then I to VI, round its outer edge. Visible under the hole is a disk which rotates once in twenty-four hours. One half is decorated with the moon and the other with the sun. The sun first appears at VI and moves round the twelve hours and eventually sets again at the opposite VI, at which time the moon appears at the left-hand VI and moves in turn across the visible area. Apart from the questionable utility of distinguishing (within broad limits) between night and day, the arrangement is more decorative than useful; but decorative it certainly is, the moon and stars usually being silver against a blued ground, while the sun and its rays are gilt on a silver ground. It is a troublesome dial by which to tell the time. 112, 115

In addition to the above, there were some 'false pendulum' watches. These have an eccentric chapter ring, and what appears to be a pendulum (but is really a disk attached to the balance wheel) swings through an aperture in the lower half of the dial. As the significance of this arrangement is mechanical, it will be found discussed in the mechanical part of the book. 114

Apart from these elegant flights of fancy, the modern arrangement of two concentric hands quickly came into general use, and fairly soon with motion work under the dial, by which they are positively geared together. To set to time, it is then only necessary to apply a key to the square on the end of the minute hand arbor, and on turning it both hands will move progressively. Even seconds dials, in the ordinary modern position, are occasionally found from an early date. The hands are either of the 'beetle and poker' variety (from a fancied resemblance of the hour hand to a beetle and the minute hand to a poker), or else, the hour hand may be of the more elegant 110 'tulip' pattern. The 'tulip' did not survive long after 1715, but the beetle and poker went on until after 1800.

I1, VIIIA, B   Dials were silver or gold, with a matt ground, and champlevé hour and minute numerals. The maker's name appears on a polished cartouche in the centre part of the dial, except prior to about 1680, when the dial centre has a plain, radiating pattern.

Enamel dials appear on English watches only very occasionally, early in the eighteenth century or possibly even earlier; but they are not at all common before about 1725, at which time Graham adopted them as standard. They exactly follow 139, 151   the lay-out of the metal dials, but in Graham's and, later, Mudge's extremely elegant cylinder watches, they are further embellished by a polished steel, sweep-centre second hand. Some, however, have a single case and are wound through the dial; in such examples there is no seconds hand.

Styles of watch cases and dials 1750–1830

After Graham introduced the cylinder escapement about 1725, the century became one of a stagnation almost as complete as the century before, until the last quarter, which again was a time of tremendous activity in the development of the true precision watch. This stagnation is reflected in watch cases and dials, which continued the old fashions in an increasingly perfunctory sort of way. This applies both to Continental and British watches.

Once the balance spring had been fully mastered and its improved timekeeping taken as a matter of course, watches once again became smaller and thinner, although the British makers would never go so far as Continental in this direction, which contributed to their eventual downfall.

Enamelling continued, occasionally of high quality, especially when heraldic. But it 153, XVIB   was usually confined to a framed panel in the back of the watch. Such enamelled, or highly repoussé watches, sometimes had an additional outer case, with the back glazed, through which the repoussé or enamel was visible.

Dials remained with little alteration, but the old concentric circles gradually disappeared; and as people increasingly read the time at a glance, without having to look at the numerals, the minute numerals almost completely disappeared by about the turn of the century.

Engine turning began to appear quite generally from about 1790, probably owing 170, 211, 228   to Breguet's influence, and this led to a wide use of basse-taille enamelling, in superb translucent red and blue enamels over an engine-turned gold base. It sometimes appears in conjunction with a small central painting on enamel, either heraldic or a miniature; and it may be further embellished with diamonds or split pearls.

As the new school of precision makers (predominantly Arnold and Emery in 156, VIB   England and Breguet in France) put fresh life into watchmaking, so did they give a new look to the visual aspects. As at the birth of the balance spring, so again an elegant severity became the keynote. In England, Arnold's early dials and hands have an almost brutally functional quality. But they have so much character that they make a strong aesthetic appeal despite their lack of superficial elegance. For his new lever escapement watches, from about 1784, Emery used a form of dial copied from the 'regulator' clock, in which only the minute hand sweeps the whole dial, with subsidiary dials at twelve and six o'clock for hours and seconds. These, with Emery's special form of spade-ended hands, have an elegance seldom if ever surpassed, and were much copied in precision watches for the next forty years. Both Arnold and Emery, and some of the other early precision makers, used the 'consular' case, in which the front bezel and back cover meet on the centre line of the watch, so that no body is visible at all.

In common watches, the pair case remained current in England until the extreme end of the nineteenth century. Enamel dials had painted centres, often of some topical scene relevant to the owner, such as ploughing, or a railway train. These honest, unassuming, ungaudy watches, have an endearing quality quite their own.

In the semi-precision class of watch, generally with a cylinder or duplex escapement, a heavily handsome form of dial became fashionable in England from soon after 1800. This consisted of a perfectly plain matt-gold dial with highly-polished gold numerals, either Arabic or Roman, applied on to it. At their most elegant, these dials had serpentine or wavy hands, either of gold or blued steel. The cases were plain or engine-turned.

With the increasingly florid tastes of the time in England, the last type of watch later developed a heavy cast and engraved case, with a heavily encrusted body. The matt dial had applied decorations and figures in four-coloured gold. These watches, which may be found roughly through the twenties and thirties, have a ponderously handsome quality, allied to superb workmanship, which is by no means without appeal.

On the Continent, tastes generally were as uncertain as in England, with the exception of the basse-taille enamelling already mentioned. Painted enamel dials of poor quality became usual, but they can have a certain charm when allied to simple automata, such as a revolving windmill. Automata movements were often coupled with the repeater mechanism, whereby a figure on the dial struck a bell in time with the strokes of the repeater.

Cheap watches, especially Swiss, sometimes have a case of transparent horn, pleasingly painted on the inside with a delicate pattern of flowers or ferns.

Better-quality Swiss watches are sometimes of an alarming thinness, not surpassed at any subsequent time.

Form watches also came back into fashion in Switzerland soon after 1800, and have remained so, in varying degrees of popularity, ever since. Most of them are no sillier than their seventeenth-century predecessors.

163, Ib

297

219, 235, 236

245, 249, 250,
251, 252, XIV

252

Performance of early watches

There is not a great deal of contemporary evidence as to the performance of early watches when they were new, although all the figures survive of the trials of the earliest marine timekeepers. Thus, we know that in the trial of Harrison's No. 5 by George III its total error or mean time during a trial of ten weeks was only 4·5 seconds. Mudge's first timekeeper was tested at Greenwich for twenty-nine weeks from April to October 1777, during which time the difference between the greatest and least daily rate was only 1·62 seconds, and the vast majority of days come within one second's variation. As against these, the performance of two of John Arnold's early pocket chronometers shows up favourably. No. 36 was tested at Greenwich over thirteen months, in 1779–80, during the whole of which time it was worn in the pocket. Its total error amounted only to 2 minutes 33 seconds, and its daily rate never varied by more than 3 seconds on two consecutive days. Its position errors, too, were remarkably small. In the normal positions in which a watch operates, between IX, XII and III-up, and horizontal, dial up, its daily rate varied only 1·43 seconds. If the never-used positions of VI-up and horizontal dial down were added the error went up to 3·5 seconds. This watch had Arnold's 'S-shaped' balance and pivoted-detent escapement.

Another watch of Arnold's, No. $\frac{21}{68}$, made in 1780, was tested by Mr. Everard of Lynn pretty regularly between 1785 and 1790, and during the whole four and a half years its fastest daily rate was $+2\cdot7$ seconds and its slowest $-2\cdot5$ seconds. This watch forms part of the Ilbert Collection and also has an 'S-shaped' balance. It originally had a pivoted detent but now has a spring detent. This conversion may well have been carried out by Arnold before the Everard trials.

Some of Earnshaw's early watches also put up fine performances, but over long periods they tended to suffer from the inferiority of his balance springs.

Against this background it is interesting to see how a number of watches have performed in daily wear when between 150 and 250 years old. At this age a good deal of wear has inevitably taken place and this shows up particularly in position errors. Therefore it is not fair to test old watches in positions, and in all the tests they were restricted to the positions likely to be gone through in the pocket. Compensated watches are likely to show up relatively worse than others, since their delicate compensation balances are sure to have suffered some violence during their life, and in no case were they subjected to fine readjustment equivalent to the adjustments certainly carried out by their makers before they were first sold.

The oldest watch tested by the authors was a verge by Quare, No. 4465, hallmarked for 1713. In the course of one month the difference between the greatest and least daily rate was two minutes, and if the five worst days were excluded, it ran within one minute. This performance is not exceptional and disproves the widely-held belief that early verges are hopelessly bad timekeepers. Curiously enough, no nineteenth-century verge tested has approached the above performance. For some reason, too, verges are surprisingly little affected by temperature changes, especially the early ones with short, untempered springs. A watch by Andrew Dunlop of about 1710 was tested fairly continuously over a year, and the average midsummer and midwinter daily rates varied by only half a minute.

Next in date was a cylinder watch by Mudge & Dutton, No. 580, hallmarked for 1762. The duration of the test was again one month during which the difference between the greatest and least daily rate was 1·5 minutes. If the five worst days were excluded it ran, like the Quare, within one minute. This performance is not exceptional, one way or the other, and both the Quare and the Mudge & Dutton were in as good condition as could be hoped for considering their age, not much inferior to when they were new. It bears out how little superior the typical English mid-eighteenth-century cylinder escapement was to a verge of 1700.

Arnold's pivoted detent No. $\frac{17}{67}$, with 'S-shaped' balance, probably dates from 1781 (it has been re-cased so that it cannot be dated by its hallmark). It was tested for three weeks during January and subsequently for a fortnight in a very hot June. During the June test the greatest variation in daily rate was six seconds, but excluding one exceptionally hot day it was only four seconds. During the January test the greatest variation was as much as 16 seconds. Taking January and June together the greatest difference between daily rates was still only 16 seconds. It is noteworthy that the June test came after the January one, when it had only recently been restored after being brought from America, where it had been grossly abused, and it is the experience of the authors that where compensation balances have been mishandled, or even not used for a long period, they take quite a long time to settle down again. The June run may therefore be regarded as more typical and matches up reasonably with the original performances of Nos 36 and 68, allowing for the passage of 180 years.

Twenty more years bring us to John Roger Arnold's spring detent chronometer No. 1869, made in 1802, but this showed disappointingly little improvement, the greatest difference in daily rate over a month's trial being 13 seconds. After further adjustment, over a subsequent fortnight, this came down to 5 seconds. On yet a further test, when the watch was not worn, the difference was only 2·5 seconds a day.

A 1799 Earnshaw pocket chronometer, which once belonged to Neville Maskelyne, and is in good condition, was able to be tested for only a fortnight, during which time its maximum variation in daily rate was 8 seconds.

Urban Jürgensen's pocket chronometer No. $\frac{24}{58}$, dating probably from 1813, had a greatest variation in daily wear of 8 seconds over a month, and in a further three weeks, when it was not worn, the greatest variation was only 4 seconds.

Breguet was represented by two specimens. A souscription in fine condition, tested over a long period, showed maximum daily variations up to one minute, and the difference between the average winter and summer rates was 45 seconds per day faster in the summer, showing that the compensation curb was acting too strongly. Despite its single hand a Breguet souscription watch can be read accurately to within 15 seconds, and despite these not very remarkable figures, its performance over a long period of wear is always found to be a good deal better on average, and well up to the requirements of modern life.

At the other end of the Breguet scale is the four-minute tourbillon No. 1890, with 'échappement naturel', made in 1807, which may be regarded as Breguet's last word in pocket timekeeping. This has been repeatedly tested, and in daily wear regularly shows a maximum variation in daily rate of 5 seconds, but generally very much less, seldom exceeding 2 seconds. Over one period of three weeks, during which it was not worn, its daily rate never varied by so much as one second. However, even this performance was exceeded by one of Breguet's ordinary lever escapement tourbillons, No. 2572, which was tested in daily wear for a month, during which time the difference between the slowest and fastest daily rates was only 3 seconds.

Eighteenth-century English levers have been disappointing on the whole, doubtless owing to their lack of draw, but Emery No. 1057 tested in May kept within 7 seconds a day. When again tested in August it ran within 5 seconds a day. The difference between the May and August average was 12 seconds a day faster in August.

The best modern results to be obtained from any watch known to the authors were achieved by a Barraud pocket chronometer, No. 183, hallmarked for 1797, and owned by Cedric Jagger. This has John Arnold's type of escapement and balance, and over a period of 302 days, with one or two unavoidable gaps, it was timed on 211 days. Of these, fifty-five days showed no variation; ninety-two days showed plus or minus one second; sixty-four days showed plus or minus 2 seconds. The above excludes two periods of two or three days when the instrument appeared to become deranged and then settled down again: a malady not unknown even in the best regulator clocks. It is thus about the equal of the Breguet tourbillon No. 1890, but the tests are more conclusive for having been carried out over such a long period.

Technical

Introduction

The counting of the hours has been of interest since men found the need for regulating their affairs. The advent of the first mechanical means of achieving this started an almost continuous exercise in the development of an improved mechanical timekeeper and to this day the process continues and will always continue. The difficulty of indicating exact seconds inspired many talented men to devote their lives to the development of a mechanism to control exactly the energy of the prime mover.

The prime mover of the watch is its mainspring. It is well known that the torque of the spring increases directly as the angle of winding and, conversely, proportionately diminishes as it unwinds.

The verge escapement is particularly affected by variation in power and the first developments in timekeeping were directed towards smoothing out the power of the mainspring. One method, employed by early German makers, was the stackfreed. This device is extremely crude and cannot have had very much, if any, improving effect on the rate. In brief it consists of a cam geared to the arbor of the mainspring, so as to turn once in the going period of the watch. Pressing against the cam is a roller mounted on a strong spring fixed to the plate. The contour of the cam is such that when fully wound the stackfreed spring works against the mainspring and so slows the watch; while towards the end of the run it assists the mainspring and so hurries things along.

The earliest watches had unprotected springs hooked to a pillar between the plates at the outer end and to the great wheel arbor at the inner end. The great wheel was mounted freely on the arbor and the power transmitted to the wheel by a pawl and ratchet. Turning the arbor wound the spring and stopped the train in the process. To protect the spring, or more probably to protect the movement in the event of breakage of the spring, it was enclosed in a barrel or pierced cage screwed to the plate. This is known as a 'resting barrel'.

It is not known who invented the fusee, but its use spread rapidly for it gave the watchmaker a sound and practical method of controlling the power of the mainspring. It is used in conjunction with a barrel mounted freely on an arbor locked to the plate. The outer end of the spring is attached to the barrel and the inner end to the arbor. Wound round the barrel is a cord, the free end of which is attached to a spirally-grooved pulley of constantly-decreasing diameter fixed to an arbor running in the plates. Freely mounted on the arbor is the great wheel and the power is transmitted to the wheel by means of a pawl and ratchet. Turning the fusee arbor winds the cord from the barrel on to the spiral pulley and in so doing winds the mainspring so that the increasing power is matched by the decreasing diameter of the pulley. By careful shaping of the fusee curve it was possible to make a reasonable approximation to uniformity of torque at the pinion engaged by the great wheel, but beyond this the power would fluctuate to an amount depending on the uniformity of the wheel-cutting.

In order to make the first turn of winding equal to the subsequent turns it is necessary to put a little power on the barrel before winding commences. The amount of this 'set-up' as it is called depends entirely on the torque of the spring when wound into the barrel. A spring requiring to be wound through six turns to get it into the barrel will require less set-up than a spring requiring only four turns to achieve the same end.

With the rapid deterioration of the torque curve of early springs the watch soon began to lose on its rate so that it was necessary to re-set up the first turn of winding of the spring. Embarrassed by the frequent appearance of his customer to have this adjustment made, the watchmaker devised a worm and wheel set-up and fitted a dial to the wheel with a pointer fixed to the plate so that the owner could make the necessary adjustment himself. Thus when the watch began to lose it was only necessary to screw up the adjuster and the original charmingly erratic rate was restored. 75, 76

The gut line used to couple the fusee to the barrel was in about 1675 replaced with a chain. The only other improvement in the system was the introduction of maintaining power to prevent reversal of the train during winding. This was done by placing between the great wheel and the fusee cone a ratchet-toothed wheel of the same diameter as the fusee cone. The power was then transferred to the ratchet wheel instead of the great wheel by means of the pawl and ratchet. Between the ratchet wheel and the great wheel is a spring coupling the two together and kept in tension by the pull of the cord on the fusee. During winding the ratchet wheel is held by a pawl pivoted in the plates, so maintaining the coupling spring in tension and ensuring the continuous rotation of the great wheel.

To prevent overwinding a projection of the last turn of the fusee is caught by a pivoted steel arm raised into its path by the cord or chain.

Occasionally, in high-grade watches, the positions of the barrel and fusee are reversed and described as a 'reversed fusee'. Mudge was the first to use this arrangement in his marine timekeepers. With the conventional arrangement, viewed from the back of the watch, the fusee is to the left, anti-clockwise of the barrel. Winding the spring draws the chain from the outer edge of the barrel on to the outer edge of the fusee and the force on the fusee pivots is multiplied by the ratio of the smallest to largest diameter of the cone.

With the reversed fusee the barrel is to the left, anti-clockwise of the fusee, and the chain passes from the outer edge of the barrel to the inner edge of the fusee, between the pivots of the fusee and centre wheel. With this arrangement the force on the fusee pivots is proportional to the ratios of the cone.

For some obscure reason the obvious merit of this system was never generally taken up in England. It was used by Breguet in his early tourbillons but, in later manufacture, he reverted to the conventional arrangement. Fig. 201, reversed, and 201 fig. 147, conventional, illustrate the different positions of the fusee. 147

If the great wheel is fixed to the barrel and the fusee dispensed with the arrangement is called a 'going barrel'. The power at the great wheel decreases continuously during unwinding but this defect is kept to a minimum by the use of stop-work which 226 restricts the winding to the middle turns of the spring. Thus, if the watch requires four turns of the great wheel to maintain the action for thirty hours and the mainspring is capable of six turns, then it is 'set-up' one turn with the stop-work in the locked-down position. Winding the watch four turns brings the stop-work into the locked-up position and so the sixth turn is never wound. This system was used extensively by Breguet, who found it convenient to dispense with the fusee in his thin watches.

In fig. 254 can be seen an arrangement by Breguet using two going barrels without 254 fusee. This arrangement was used mainly in his marine chronometers and deck watches of which this is an example. There are examples of timekeepers by Breguet using two barrels and a double-grooved fusee and at least one example using four barrels without fusee but these were not, so far as the authors are aware, used in watches. By using two barrels a reduction in the thickness of the springs can be achieved and the number of turns in the barrel can be increased to eight; so that by setting up the mainsprings two turns and using the middle four turns, a more even torque results. By placing the barrels one each side of the centre pinion the couple relieves the pivots of the burden of resisting the spring torque and reduces the wear from this source. The reduction in arc during the last few hours of running is only a few degrees and amounts to no more than is often found in fusee watches caused by the friction of the thick spring during the early and middle turns of unwinding. Sometimes two barrel systems are fitted with geared arbors and are wound by a common square.

The necessity for keeping the spring torque within reasonable bounds is important if the rate is to be at all constant. It is well known that the verge escapement gains with an increase of power and loses with a decrease in power and the reasons for this are easily seen. The same species of error occurs in detached escapements, but the causes are more obscure. A complete analysis of the functioning of detached escapements is beyond the scope of this work but the following may help in an understanding of the problems of ensuring a close rate.

Consideration will be given to the lever escapement, but the chronometer is, to a lesser extent, affected in the same way.

The line of centres of the escapement is drawn through the centre of motion of the balance staff and the pallet staff. The unlocking takes place before this imaginary line is reached by the impulse pin and, depending on the angle of movement of the lever required to complete the unlocking, a certain proportion of the impulse will also take place before the centre line. If the escapement is set exactly in beat, the quiescent point of the spring will also coincide with this line.

Consider now what happens when the free oscillation of a balance and spring meets a disturbing force. If the force momentarily impedes the approach to the quiescent point the vibration will be slower. Conversely, if the force momentarily impulses the balance the vibration will be quicker. The reverse takes place after the quiescent point, so that a force impeding the balance will quicken the vibration and an impulse will slow the vibration. These forces occur continuously in escapements, and from this it can be seen that in the lever escapement the effect of unlocking is to slow the vibration. This is then quickened by the impulse before the quiescent point and again slowed by the impulse after the quiescent point is passed.

The resistance of the unlocking is brief and the loss in rate would be counterbalanced by the gain caused by the equally brief impulse before the centre line. The main part of the impulse is given after the centre line and this will cause a loss as in all double-impulse escapements. This error is confined to the escaping arc. During the supplementary arc the balance is free to continue the vibration in its natural period.

Any variation in the forces acting on the balance will cause a change in the total arc. Since the amplitude of the escaping arc is constant the variation in total arc can affect only the amplitude of the supplementary arc. The escaping arc then is a losing proportion of the supplementary arc. If the total arc is reduced, the escaping arc becomes a greater losing proportion of the supplementary arc and the watch will lose. If the total arc is increased the escaping arc will be a smaller losing proportion of the supplementary arc and the watch will gain.

It may be seen that the watch will lose in the short arcs. The arcs can be kept fairly constant with a fusee but small variations caused by the effects of change of temperature and condition of the oil, with consequent change in coil friction of the spring, are unavoidable. Since it is impossible to keep the arc constant other means must be used to make the escapement isochronous, irrespective of the arc.

John Harrison, in his prize-winning marine timekeeper, fitted a pin against which the balance spring banks during the winding vibration. If the arcs are small the spring rests against the pin for almost the whole of the vibration and so is effectively shortened by the distance of the pin from the stud. In the long arcs the increased angle of unwinding of the spring carries it away from the pin and the whole length of the spring is used.

This somewhat unscientific but very practical method of quickening the short arcs was re-discovered in the nineteenth century by the Swiss makers of lever watches without fusees. They fitted a regulator to the balance cock with two pins that could be moved along the spring by an index, enabling the user to regulate the watch without recourse to a skilled adjuster. To quicken the short arcs the spring was given a little free play between the pins and biased towards the inner pin.

Pierre Le Roy discovered that for every spring there was a certain length that made the escapement isochronous. He gave no details of his discovery but it is probable that he discovered the number of turns of spring that suited his marine timekeeper. It is a fact that springs, too long or too short, are difficult to time, but the number of turns of the spring has much greater effect on the adjustment for isochronism. An isochronous spring will not make the long and short arcs equal. What is required is a spring that is non-isochronous in the opposite sign to the escapement error. John Arnold believed that spiral springs were the source of isochronal errors because they were never concentric except when stationary. This would seem to indicate that he was unaware that the errors arose from the escapement and not the spring. He patented a helical spring in 1775 and was no doubt dismayed to find that this also worked eccentrically. Bending the final turn at each end of the spring into a progressively smaller radius cured the wobble and produced a concentrically-vibrating spring. Arnold's methods of adjusting his watches were secrets carefully preserved from his rivals and never recorded. No doubt he soon discovered that the principal advantage of the helical spring lay not in its concentricity but in the much increased radius of the collet. By manipulating the terminal curves he could bring pressure to bear upon the collet to influence the period of vibration of the balance, while at the same time the spring was kept completely free of any external influence. In altering the terminal curves to influence the collet the spring must inevitably work eccentrically to a degree that would depend upon the final shape of the curve. However, Arnold was no longer concerned with the concentricity of the spring although that was his original problem when devising it. He was now the fortunate possessor of the true knowledge that reveals itself only to those who follow the correct path of endeavour. That he fully understood his somewhat fortuitously-discovered science is evident from his inclusion of an extra quarter-turn in his springs to bring the collet attachment at 90 degrees to the stud. This has an accelerative effect on the collet in the short arcs which is progressively reduced as the arc increases and the collet turns away from the influence of the terminal curve. Arnold's discovery of the correct scientific application of the balance spring for isochronism was a singular scientific advance in horology and sets him apart from those who sought the same ends by mechanical ancillaries.

Breguet was probably influenced by Arnold's terminal curve when he raised the last turn of his flat springs and curved them in the same way to make the action of the spiral concentric. The Breguet spring is not so responsive to manipulation of the shape of the curve because the vibrating spring has insufficient leverage at the very small radius of the collet. In the majority of his watches Breguet included a regulator which could be used as an isochronal banking to quicken the short arcs. His precision watches were free sprung and, after 1800, had large-diameter, thin-rimmed balances. These could expand under centrifugal force in the long arcs to make the vibrations slower. Breguet may have devised his spring simply for its improved appearance, for he was certainly very much concerned with the aesthetic qualities of his mechanisms.

The rarely-seen flexible stud is no help in achieving isochronism but it is a pretty arrangement for keeping the vibrating spring concentric and relieving the balance pivots from side pressures. The watch illustrated (fig. 335) has two springs, both with flexible studs.

Lack of isochronism can be a most irritating source of error causing small, haphazard changes in daily rate that cannot be anticipated. If the force of impulse could be kept constant so that the balance amplitude was invariable, the escaping arc would remain a constant proportion of the total arc and the rate would be stable. This can be done with a constant force escapement or a remontoir, winding a secondary spring at regular intervals.

These complex mechanisms are sometimes found in clocks but very rarely in watches by reason of the limitations of space and the difficulties of construction.

The watch movement by Haley in the Clockmakers' Company collection is an example of a constant force escapement. The spiral spring impels a pivoted detent to impulse the balance at each oscillation and is then rewound by the toothed wheel driven through the conventional train and mainspring. The delicate mechanism is inclined to trip and is not suited to use in a watch.

The Mudge movement, also in the Clockmakers' Company collection, has a remontoir released at one-minute intervals to wind a secondary, spring-driven wheel, supplying power to the escape wheel. This is a far superior arrangement because the escapement functions in the conventional manner without concentrating the impulse forces on one component, as is the case with the constant-force escapement. The small variation in balance amplitude that occurs between windings is of no consequence because it occurs exactly the same number of times each day.

More recent developments in nickel steel alloys have produced materials which do not follow Hooke's law. When applied to an escapement especially devised to utilise the characteristics of the material the isochronal error can be eliminated naturally without necessity for skilled manipulation of the spring. One watch by George Daniels employs a double-chronometer escapement designed to utilise the short arc-gaining characteristic of the Ni-Span-C nickel-steel alloy. The escapement impulses the balance at each vibration via pallets on the balance axis and therefore no oil is required. The unlocking occurs in both directions on the centre line and the whole of the impulse is given after the centre line. This combination produces a losing rate but the angles of the impulse and draw are exactly matched to the acceleration of the balance spring as the arc decreases. By this means the rate remains constant irrespective of change of arc. Because the isochronism is a fixed, natural function of the integrated design no adjustments are required.

One of the most disturbing influences on the rate is change in temperature. A watch without temperature compensation will lose about 10 seconds per degree centigrade per day, and even if it is worn this might make a difference of one and a half to two minutes a day between summer and winter. Since the error lies mainly in the balance spring it was not unnatural that early attempts to equalise the hot and cold rates should have been directed towards varying its active length to effect a correction.

John Harrison was the first to use this system in the watch he designed and had made by John Jeffreys to test his ideas for the construction of his prize-winning chronometer. The watch, dated 1753, in the Clockmakers' Company collection,   146 includes among other features a laminated steel and brass strip fixed to the plate at one end, carrying a pin in the free end which touches the balance spring close to the stud. At a rise in temperature the strip will curl away from the stud and reduce the active length of the spring. The reverse occurs with a fall in temperature.

This system is unsatisfactory for a precision watch because the movement of the pin along the spring introduces further complex errors of isochronism. It was used at a later date by Mudge in the lever watch he made for Queen Charlotte in 1769. He fitted the pin to a pivoted regulator moved by the bi-metallic strip with change of temperature. The isochronal errors remained and the temperature correction was sometimes erratic due to friction at the pivots.

The superior principle of putting the compensation in the balance, thus leaving the balance spring undisturbed, was first used successfully by Pierre Le Roy. In 1765 he produced a marine chronometer using mercurial compensation to change the radius of gyration of the balance for change of temperature.

Such a balance is unsuitable for a watch but in 1773 John Arnold devised a balance with a spiral bi-metallic strip at the centre which, by means of a pivoted linkage, moved weights radially at the balance rim. Finding that the friction of the linkage made the radial movement of the weights uncertain he continued his experiments and in 1780 produced a balance with two compensation strips bent into a double loop   172 for compactness. Each carries a weight moved radially by temperature change to alter the radius of gyration to compensate. The thin rods carrying the weights pass through locating brackets in the balance rim. The small friction arising from this is almost eliminated by fine wires adjusted for length to hold the rods free of the bracket holes, as the compensation loops change form with change of temperature.

These were the first successful compensation balances to be used in watches. Combined with Arnold's pivoted detent escapement they could produce a formidable performance. (See the section on Performance of Early Watches, pp. 93–94.)

Although very successful these balances have certain deficiences that complicate their adjustment for temperature correction. The amount of movement of the weight depends upon the ratio of the thicknesses of the two metals. If the brass is too thick, the weight will move radially inwards too far and the watch gain in heat. If too thin the compensation will be insufficient and the watch will lose in heat. To correct this it would be necessary to change the weights to effect the desired change in the radius of gyration of the balance. Obviously the overall rate would be affected and need
 correction by changing the rating screws at the ends of the balance arms.

Such complications of adjustment would present no problems to a man of Arnold's calibre but would be irritating and time-consuming. Being a resourceful and inventive man he soon perceived the advantages of forming the bi-metallic strips into arcs of a circle to form a divided rim to the balance. This form of compensation balance was introduced in about 1775 and with it Arnold achieved a lighter and more rigid compensation, simplicity of adjustment by moving a weight to or from the fixed end of the rim without altering the weight of the balance, and complete freedom from friction in the compensation and uniformity of movement of the rim to eliminate change of poise with change of temperature.

Altogether this balance must be regarded as a triumph of inventive ingenuity by a man of rare ability whose work demonstrated true scientific progressiveness. Not only is it simple in design, but equally simple for an experienced mechanic to make. The late eighteenth century abounded with skilled men, who, once they had comprehended the principles involved in Arnold's work, began to produce vast numbers of comparatively inexpensive pocket timekeepers and chronometers. These timekeepers could maintain a rate of accuracy under changing conditions that, twenty-five years earlier, was obtainable only by using complex mechanisms, constructed and adjusted by exceptionally skilled men.

After Arnold's contribution to the precision watch any person with £50 to spend could buy a timekeeper with a rate in use in the pocket hitherto considered unobtainable. A detailed account of Arnold's work is contained in Dr. Vaudrey Mercer's book *John Arnold and Son*.

Arnold's method of securing the two halves of the rim to the cross-bar with screws was simplified and much improved by Earnshaw. He devised the ingenious method of immersing a disk of steel in molten brass to fuse the brass to the steel. The brass was then hammer-hardened and all surplus metal turned away, to leave a bi-metallic balance. Cutting through the rim at diametrically-opposite points and adding weights, adjustable by sliding around the rim, completed a very rigid and sensitive balance which is still used today by Thomas Mercer, chronometer makers of St. Albans, England.

The predictability of compensation afforded by a well-constructed balance revealed a flaw in the system known as 'middle temperature error'. This arises from a confliction of the effect of change of temperature on the period of vibration of the balance and spring. The change in elasticity of the spring, and therefore its period of vibration, is almost directly proportional to the change in temperature. The change in dimensions of the balance is also proportional to the temperature change but the period of vibration will vary as the square of the change in dimension. Much experimenting was done in the nineteenth century to try and influence the amount of movement of the rims to compare more precisely with the linear effects of the balance spring. This did not apply to watches, mainly because the scale of a watch is too small but also because it is kept at fairly close temperatures in the pocket and the effects are too slight to be of concern. The marine chronometer is adjusted to compensate for a wide range of temperatures, usually 40–95°F, and the middle error may be 2–3 seconds a day.

The solution to the problem was supplied in the late nineteenth century by Charles Edouard Guillaume's bi-metallic, brass and nickel-steel alloy balance. This governed the movement of the rims to suit the linear properties of the balance spring. Because no mechanical ancillaries were required the effects were stable and the balance could be made small enough for a watch. The Swiss makers especially used these balances in pocket chronometers with helical or spherical balance springs and pivoted detent escapements.

The final solution to compensation for temperature was applied, in the early twentieth century, to the balance spring, where it had begun in the mid-eighteenth century. Again Guillaume solved the problem with the invention of a complex nickel-iron alloy containing a variety of elements that rendered its elasticity almost insensitive to changes of temperature. Such small changes as might occur with different batches of wire were of no consequence in ordinary civil use where the variations of rate arising from daily use or abuse swamp any residual temperature error.

The mono-metallic balances with continuous rims were more rigid than the earlier, cut-compensation balances and had the advantage of freedom from centrifugal distortion.

When higher precision was required any residual error was corrected by the use of very small bi-metallic affixes to the balance rim.

A later improvement for fine adjustment was the uncut, bi-metallic 'ovalising' balance, the rims and arms made of materials with differing co-efficients of expansion. The Hamilton Watch Company of America developed this system to perfection for their marine chronometers and deck watches. By the use of Invar for the arms and 379 stainless steel for the rims the balances become slightly oval along the axis of the arms for a fall in temperature, the reverse occurring for a rise in temperature. By changing the positions of the screws in the rim the small residual temperature errors can be eliminated. Once the error in any batch of wire has been ascertained the balance can be made uniform to compensate and, in practice, no adjustment is required after manufacture.

Although, for precision watches, the compensation curb was abandoned with the introduction of the compensation balance, it was still used, especially by Continental makers, for watches of high quality where an exact compensation was not important.

Breguet's compensation curb carried only one pin, and the spring, passing between this and a fixed pin, was allowed more or less free play according to the change in temperature. His cylinder escapements were always fitted with this device and for this escapement it works very well. Earnshaw used a similar curb in some of his watches, 240 and these employed two curbs with pins at the free ends and the fixed ends screwed to a regulator rack. With change of temperature the pins opened or closed like pincers and altered the play of the spring to compensate. The lever watch by 167 Perigal uses this system and, within the limits of change to be found in wear, works well enough to give a good average rate over a long period.

Shock-proofing

Damage to the balance pivots is one of the most disastrous things that can happen to a precision watch. If the watch is dropped flat the pivot will spread at the tip and if
106 dropped on edge the pivot will either bend or raise a burr. Whatever damage occurs

it can be certain that the rate will be ruined. When it is considered that dropping a heavy watch a height of 5 cm, so that it lands on edge, will usually break the staff, it is surprising that the development of shock-proofing was so neglected. The danger of this happening is far greater than that of the fusee chain breaking, but to protect the escapement in the unlikely event of a chain failure a guard was often fitted. This was either a pin fixed in the plate close to the barrel or a curved plate separating the barrel from the escapement.

Only Breguet seems to have appreciated fully the advantages of protecting the pivots against damage. Not only does the rate suffer, but with it goes the maker's reputation, for it has always been universally understood by the owner that any deterioration in the going is entirely the fault of the maker. Breguet's 'parachute' was
176 a long steel spring screwed to the cock with the jewel hole and end stone set into the free end.

This system was originally devised by him to insulate the balance wheels of his perpetuelles from the vibrations of the platinum winding weight. The pointed pivots are supported in flexibly-mounted jewels with conical holes lightly sprung on to the points to eliminate all but rotational movement. The balance is then prevented from rattling in its bearings with the oscillations of the heavy weight. The pivots pass through holes in cocks at each end of the staff and the flexible jewels are mounted above. If the watch is dropped, the pivots will push aside the jewel and any damaging forces will be absorbed by the arbor of the staff against the holes in the cocks.

286c Later, pointed pivots were supported in this way at one end only and the jewel spring fitted with a screw to allow adjustment for complete freedom of the pivots.
258 Watches with conventional pivots needing through-holes and end stones had springs allowing flexure both vertically and horizontally, and these were sometimes provided at both ends of the staff. The high number of surviving Breguet watches and the excellent condition of their escapements is a testimony to the effectiveness of his invention.

The pointed pivot has the additional advantage of a constant radius of friction. This is useful in maintaining the balance arc constant in the vertical and horizontal positions. With conventional pivots the radius of friction changes from the tiny radius of the end of the pivot in the horizontal position to the full radius in the vertical position. This causes increased friction and results in a slower rate in the vertical positions. In his watches for civil use Breguet was more concerned with achieving a reliably close rate for long periods than an unnecessarily close rate for shorter periods. For this reason he used the pointed pivots in his high-grade lever watches in which the imprecise location of the balance pivots was outweighed by the uniformity of the useful rate under different conditions of use. No other maker exploited the obvious advantages of this method of shock-proof suspension which, by the conventional standards of precision watch construction, may have been considered rather crude. Pointed pivots, without flexible mountings, were not taken up again until the twentieth century and then by manufacturers of cheap alarm clocks. When fitted with agate cones their vitality and reliability is as indestructible as Breguet's watches.

Tourbillons

The difficulty of maintaining a steady rate in the pocket is due in the main to the constant change of position to which the watch is subjected. To bring the piece to time in the twelve-up, nine-up and three-up positions requires the most careful adjustment to the escapement, balance and balance spring. The adjustments are complex and inter-related. In the final stages they are tedious and costly in time and the effect is not permanent.

The solution to the problem was supplied by Breguet with his tourbillon or rotating 49 escapement. With the escapement mounted on a platform fixed to the fourth-wheel arbor, and the fourth-wheel fixed to the plate concentric to the arbor, the escape pinion is rotated as a consequence of the rotation of the platform rolling it around the fixed fourth-wheel. By this means positional errors were eliminated leaving only temperature and isochronal adjustments to be made. The rotation of the platform every 60 seconds called for a powerful mainspring to overcome the inertia of the platform, for it must be started from rest at every vibration of the balance. To reduce the effects of stopping and starting the platform Breguet introduced four-minute and six-minute periods of rotation.

The performance of Breguet's tourbillons is phenomenal. Every effect of error of poise in the balance and spring is reduced to an average rate for each revolution. This is particularly beneficial with a compensation balance in which change of temperature can cause change of poise.

The construction of the tourbillon requires the most careful and costly work-manship and for this reason production was limited. No other French maker made them and, among Breguet's non-French contemporaries, only Urban Jürgensen made them.

In his first tourbillon Breguet used Arnold's spring-detent escapement and, in the second, Earnshaw's. Both have a one-minute period of rotation. Later watches with a four-minute period of rotation have Peto's cross detent escapement which is locked 217 with the detent in tension. Presumably this was done to prevent damage to the delicate detent spring if the carriage was unskilfully handled.

A still later series with a four-minute carriage has Breguet's own échappement 214 naturel, while the final series has lever escapements with a one-minute period of 229 rotation.

It was not until the end of the nineteenth century that tourbillon watches were again made in series. Victor Kullberg and Nicole Nielson made them in England, with the latter continuing production into the present century. Nicole Nielson's work is usually to be found in watches by Frodsham, Dent and Smiths. In Switzerland Albert Pellaton 51 was the principal maker and his work is usually to be found in watches by Patek Philip. Many Swiss tourbillons were made in horological colleges mainly as a demonstration of individual skills. The Clerkenwell watchmaker, Sidney Better, made 52 a small series of tourbillon carriages in the 1930s for the Northern Goldsmith's Company but economic difficulties obliged him to cease after only a very few had been finished.

In Germany the celebrated firm of A. Lange and Son produced many fine tourbillons during the 1920s and 1930s. The watches were not made as a series but more as an exercise in technical excellence. The earliest were based on the half-plate karrusel ébauche but with the addition of the fusee and chain. Spring detent or pivoted detent escapements were fitted to one-minute carriages of delicate construction and high finish. Later examples, made under the direction of Alfred Helwig, a master craftsman of extraordinary accomplishment, contained a variety of innovations of fine workmanship to please the connoisseur. One particularly interesting feature is the so-called 'flying tourbillon', in which the carriage has no upper support. Both the bearings for the carriage are fitted to the front plate of the movement with the fourth pinion fitted to the carriage arbor between them. The carriage rotates, fully visible and apparently without means of support, above the front plate of the movement. Some have two going barrels driving in series and others have fusee and chain. Fig. 373 illustrates an example with fusee and differential-maintaining power visible on the back plate. All the watches are characterised by the highest-grade workmanship regardless of cost.

The construction of tourbillon watches is today undertaken in London by George Daniels. The watches employ one-minute carriages with Earnshaw spring detent escapements. Experiments with the proportions of the escapements have shown it possible to make them self-starting from the run-down condition and free of any tendency to set if subjected to violent movements during use.

Although a tourbillon can equalise all the vertical rates of a watch it cannot make the vertical and horizontal rates equal. If the balance arc varies the errors of poise will continue to be averaged but the rate can be affected by the escapement error. The difference in the vertical and horizontal rates is caused by the change in value of the pivot friction between the two positions. The cause of escapement error is discussed on p. 101. The two errors combine in a complicated way in the vertical positions because the increase in pivot friction has a braking effect on the balance, which both reduces and slows the total arc and produces a slower rate.

In order to separate the errors so that they could be individually dealt with, George Daniels constructed the watch seen in fig. 386 with a one-minute tourbillon wound at 15 second intervals by a remontoir mechanism. The remontoir spring, of Ni-Span-C alloy unaffected by temperature changes, is beneath the carriage and mounted on separate bearings to prevent frictional disturbances to the freedom of the carriage. Because the balance arc cannot change in the horizontal position, escapement error is eliminated. Positional errors are equalised by the tourbillon. The common vertical error caused by pivot friction is equalised with the horizontal rate by a simple cycloidal pin with fine screw adjustment. As with the horizontal arc, the vertical arc cannot change and the earlier, non-linear defects of the cycloidal pin cannot occur.

The success in trials of the tourbillon during the late nineteenth century attracted the attention of Bonniksen, who devised an extremely ingenious method of constructing a rotating escapement which eliminated the extra precision of construction that made the tourbillon so expensive. He called this mechanism a karrusel. Apart from the continuous rotation of the escapement it is in principle quite different from the tourbillon, as the rotating platform is not mounted upon an arbor of the train. The fourth-wheel arbor passes through the centre of the carriage which is located in a large plain bearing in the front plate. The top pivot hole of the fourth-wheel is cocked to the carriage, which can be rotated concentrically with this arbor. The escapement is mounted upon the platform above the fourth-wheel, which engages the escape pinion in the conventional manner. Excepting that the escapement is above the fourth-wheel, the going of the piece would be no different from any common watch. However, screwed to the underside of the platform is a toothed ring concentric with the platform and engaged by the third pinion. With the going of the watch the platform is rotated by the third pinion so that it completes a revolution every 52·5 minutes. Watches employing this device kept a very close rate and were extremely successful in trials.

A later variation of the karrusel includes a slight rearrangement of the train to give a thirty-five minute period of rotation and allows the use of a centre seconds hand. Still later examples have an upper pivot to the carriage to reduce the friction of the lower bearing. These are much rarer and to distinguish them from the earlier type Bonniksen described them as tourbillons. Almost all have lever escapements with spiral springs, with terminal curves sometimes free sprung.

In Germany towards the end of the century, the firm of A. Lange and Son produced a series of high-grade karrusels. These have gold lever escapements, and compensation balances with spiral springs with regulators. The finish of detail of their karrusels, in common with all the products of the company, is of the very highest grade.

It may be thought in some quarters that the day of the mechanical watch is over. This view has been broadcast particularly by those who believe that the electrically-driven watch is, or will be, superior. Some extravagant claims have been made for the rates obtainable from such watches but there is evidence that they have not been as stable as some makers had believed they would be. No doubt they will improve both in accuracy and reliability but they have a long way to go before they equal Breguet's proven claim that his perpetuelle watches will run for eight years without attention. A watch capable of a rate of accuracy that no civilised person would wish to govern his affairs by is of little use if the battery fails without warning, and batteries are notoriously suicidal. The authors are of the opinion that, at this stage in the development of the watch, the mechanical watch retains its lead both in long-term reliability and intellectual appeal.

Escapements

VERGE

Until the application of the balance spring no fundamental change took place in the watch from its earliest beginnings. So long as the balance formed part of the escapement and directly controlled the unwinding of the mainspring the timekeeping was at the mercy of every variation in power delivered to the balance. Its only moderating influence was its inertia, and to make use of this it had to be kept oscillating; and without the balance spring recoil was essential. Thus the verge escapement combined the two greatest escapement faults—restriction of balance freedom and excessive reversal of the train. When it is considered that the balance spends roughly a quarter of its working life winding up its driving spring through a hand-divided train with no pretensions to constant velocity ratios, it is easy to see that the verge escapement is severely tried as a timekeeper. That it is capable of maintaining a reasonable rate, when properly made, without the controlling influence of a spring is evident from its daily average. Only by the most careful adjustment of fusee curve to spring torque can its rate be brought even within the bounds of consideration as a check on the passing hours. With the relatively poor mainsprings available, and the wheel engine yet to be invented, it is not surprising that the watchmaker turned his efforts towards case forms and decoration. With the introduction of the balance spring and the steady improvement of the train and mainspring the verge began to keep a much closer rate, so that it was possible by experiment to make improvements to the escapement. Such experiments were directed to increasing the supplementary arc, and reducing the recoil by opening the pallet angles to between 100 and 115 degrees: an increase of some 20 degrees. It soon became apparent that if the recoil were reduced without reference to the total arc the rate began to deteriorate. Excessive end thrust to the crown-wheel pivot caused rapid wear and the watch was liable to set during winding or would not be inclined to start if wound after running down. With the pallets finally set at 100 degrees to 105 degrees the escapement continued in use for a further two hundred years and resisted all efforts at refinement. If jewelled and very carefully made it would, for a few months, show a fairly consistent rate; but with the thickening of the oil the rate deteriorated so that it was inferior to that of a plain unjewelled escapement which would, by virtue of its many compensating inefficiencies, keep an indifferently close rate for many years without any attention. The usual arrangement of the escapement between the watch plates is shown in fig. 1.

CYLINDER

The development of the cylinder escapement by George Graham introduced a new era of timekeeping. The absence of recoil and the inherent ability of the escapement to compensate to a large extent for variations in impulse due to its frictional rest, produced rates that were consistent enough to warrant the application of correction for temperature. Encouraged by these results makers devoted much labour to improving the escapement. Early examples were prone to self-destruction due to the heavy brass escape wheel cutting the impulse lips of the cylinder. By curving the impulse face of the wheel teeth its velocity was more constant, and the pressure on the lips of the cylinder was reduced during the latter half of the impulse; this helped considerably in reducing wear. With a reduction in velocity at the beginning of the impulse and the consequent increase at the end of impulse it was soon discovered that the increased velocity of the wheel at the moment of leaving the impulse lips was sufficient to cause pitting of the resting surface at the point where the wheel tooth dropped on to the cylinder. The introduction of the lighter steel wheel almost completely cured these faults, which were finally overcome by making the cylinder of ruby.

Once the difficulties of producing the escape wheel with its elevated impulse teeth had been overcome the cylinder watch was produced in large numbers both in England and on the Continent. The best proportions for the diameters of the wheel and cylinder were proposed by Breguet, who also used a higher number of vibrations for the balance and always, after about 1794, his ruby cylinder of unique design, in which the half section of ruby cylinder was suspended below the balance staff on a cranked extension of the balance seat. No doubt the object of the design was to simplify the manufacture of the ruby but, as with so many sound mechanical designs, it offered additional advantages. The thin crank to which the cylinder is attached allowed a considerable increase in arc before banking occurred and since the weight of the balance is taken on a separate staff the robustness is not impaired as would be the case if the more conventional cylinder were used. The teeth of the wheel do not require undercutting at the root to give banking clearance and so are stronger and very much 3 easier to produce. The oil cannot creep away along the cylinder, the recess into which the ruby fits being too great a hurdle to overcome. The balance wheel is attached to the staff midway between the pivots and so distributes its weight more evenly between the pivots and helps smooth away positional errors. The one disadvantage in the arrangement is the difficulty of placing sufficient oil in the bottom pivot hole. This blind hole is mounted in a steel tube attached to the plate by a thin arm against which the ruby-suspending crank banks if the arc is excessive. The tube is small enough to give adequate clearance to the inside of the ruby cylinder, and the pivot hole is correspondingly smaller, so there is very little space for oil. Oiling these holes requires a special oiler mounted in an uprighting tool so that the oiler cannot touch the inside of the tube. If any oil gets to the inside of the tube it will quickly run up the staff and ruin the rate of the watch. Fig. 2 shows the arrangement 2 of the Breguet ruby cylinder and pivot hole, in which the end of the pivot can be seen. Fig. 3 shows the form of the escape wheel teeth. Fig. 4 shows the arrange- 3, 4 ment of the English cylinder as produced by Graham. Note the form of the teeth and the excessive cutting away of the cylinder to give banking clearance.

Occasionally a cylinder escapement is found which has a flat escape wheel. These escapements do not perform very well since the arc of the balance must be kept small if the cylinder is not to bank against the root of the wheel tooth. The oil soon creeps away along the teeth and leaves the acting surfaces dry, resulting in rapid wear. This probably accounts for their almost complete disappearance.

VIRGULE

Since the virgule offers no advantages over the cylinder escapement it is difficult to see why it should have been made at all. It is quite incapable of retaining any oil at its working surfaces and soon sets if not oiled frequently. At first sight it would appear to have the advantage of being easier to construct so far as the wheel is concerned since, as can be seen from fig. 6, the impulse teeth are removed from the stalks of the wheel 6 and replaced with a single impulse curve on the balance staff. The construction of the staff, however, gave rise to fresh difficulties because it involved a certain amount of freehand work in filing and polishing the impulse curves to shape. For this reason, and the realisation that it would never perform so reliably as the cylinder, it was soon abandoned, although it did enjoy considerable popularity on the Continent for about twenty years.

To reduce the tendency to set, Beaumarchais introduced a further set of pins to the other side of the escape wheel and an additional impulse curve to the staff. Fig. 7 7 shows the form of this staff, which was extremely difficult to construct so that very few were made. The double virgule, as this escapement was called, did have the advantage that it was disinclined to set for want of oil but would, if neglected, wear its pins away.

DUPLEX

Although of French origin this escapement was not much employed on the Continent except by Breguet, who used it in some watches and small clocks. The principle of using the long teeth of the escape wheel for locking with reduced power and the shorter vertical teeth for impulsing proved most satisfactory and appealed to English makers who used it extensively. It is said that the French preferred the cylinder as this enabled them to produce a thinner movement. The Englishman took a more practical view of his watch as a timekeeper and the construction of English watches as a whole at this period shows that he was content to be conscious of the machinery in his pocket. The escapement does require the escape pinion and staff to be rather long if the effect of play in the bearings is not to affect the rate, which is extremely close when properly constructed. The frequent employment of this escapement in England led to great experience in its construction and proportions, so that by the middle of the nineteenth century, three-quarter-plate watches, especially those by McCabe, would run for years to within a few seconds a day.

In common with other single-impulse escapements it is inclined to set, but English watches with their high mainsprings supplying greater power could maintain a large balance arc, and so failure from this fault was almost eliminated. When constructed in this manner the most careful workmanship was essential. The long locking teeth need to pass through the passing slot as nearly at the centre of motion of the staff as is practicable if the side thrust is to be kept to a minimum; and since the diameter of the locking roller must be kept as small as possible to reduce the friction, the allowable tolerance in pitching the escapement is extremely small. The intersection of the impulse pallet with the vertical teeth is, of necessity, very shallow. The clearance required by the pallet on its return swing still further reduces the engagement of the teeth during impulse. Any variation in the depth of this engagement resulted in a variable impulse and a bad rate. It was in this department that the escapement eventually failed, for the power required to maintain a good rate caused considerable wear of the escape wheel teeth so that the escapement would trip. The difficulty of making a new wheel prompted many repairers to convert the escapement to the, by then, well-developed lever. Fig. 8 shows the arrangement of the escapement in its most common form. The passing slot for the locking teeth can be seen below the impulse pallet. Fig. 9 is of the original form of the escapement with two wheels, one for the impulse and the larger wheel for locking. This escapement is from the Breguet watch shown in fig. 212. Fig. 10 shows the fairly common Chinese duplex in which two oscillations of the balance are required to pass the double-locking tooth, the impulse being delivered at each alternate oscillation. These were used to obtain jump seconds. The extra power required to drive the balance through the second locking vibration was so great as to render the escapement practically useless as a timekeeper. The curious horned weights attached to the balance rim were, the authors are assured, intended to ward off the devil. It is not known whether they were any more effective than the escapement.

CHRONOMETER

The first recognisable chronometer escapement embodying the principle of impulse at alternate vibrations and complete freedom of the balance during the supplementary arc was made in 1765 by Pierre Le Roy for a marine timekeeper. His escape wheel gave impulse at the balance rim and was then locked free of the balance by a form of pivoted detent. His remarkable perception of the requirements of a portable precision timekeeper, combined with his inventive ability, assure his place in horological history as the man who first produced a successful single-impulse detached escapement.

The machine performed with remarkable accuracy even allowing for small changes of rate. By 1769 he had declared it impossible to improve, and retired from the scene. The design of his escapement precluded any useful reduction in size and it could not be used in a watch.

Arnold showed greater perception of the requirements of the escapement and included all the features that were fundamental to its principle of operation.

His escape wheel gave impulse to a pallet of smaller radius than the balance, to make better use of the inertia of the balance rim. His detent was pivoted in the vertical plane above the escape wheel which locked by means of pins set vertically in its rim. The horn of the detent, in the form of an inclined plane, was lifted vertically by a pin on the balance axis to release the escape wheel for the impulse. Although the friction of the unlocking was a fault in the design, Arnold showed his innate perception of the requirements of the escapement by fitting a separate, thin spring for the passing action. This fundamental advancement in the action was to remain a feature in all subsequent forms of the escapement. It is impossible to evaluate the usefulness of this escapement because the trials to which it was subjected were bedevilled by temperature compensation problems.

In 1772 Arnold abandoned experiments with the vertical detent and introduced a detent in the plane of the escape wheel. This he used with great success in a new series of marine chronometers. Confident that he had found the correct formula for an inexpensive, reliable and accurate timekeeper, he then produced a series of pocket watches employing this pivoted detent escapement. When properly used the watches could keep, and are still capable of keeping, a remarkable rate of time. (See Performance of Early Watches, pp. 93–94.)

The only reliable contemporary alternatives for pocket use were watches with verge or cylinder escapements and uncompensated balances. Such watches could show daily variations of a minute or more in use in the pocket and equally wide variations when static for summer and winter rates. It is a remarkable indication of Arnold's genius for scientific observation and development that within the space of a few years he had transformed the pocket watch from a bad timekeeper to a precision instrument capable of a uniform rate within a few seconds per day.

It was said earlier that the rate of a watch is dependent upon proper handling. The marine chronometer is kept static except in winding, when it is turned over once each day, whereas other watches are subjected to a variety of movements, some violent. Twisting or rotating the watch when winding, or placing the watch on a smooth surface, thus allowing the balance to rock the watch in the horizontal plane, can cause occasional setting of the escapement. Arnold's deep, domed consular cases seem to have been especially designed to encourage the balance to rock the watch and, if this does not sufficiently reduce the arc to stop the escapement, it will cause it to run very slow. A sudden twist of the watch, when carelessly pocketing it, for example, will almost certainly result in a set, if the motion coincides with the speed and rotation of the balance.

The cause of these inconveniences lies in Arnold's escapement proportions which are not ideally suited to pocket use. The ratio of impulse roller to escape wheel diameter is too high and gives an escaping arc of about 60 degrees. The total arc of the vibration is some 150 degrees in the vertical position and so the balance velocity is low and the arc is easily reduced to the escaping arc, when the watch will stop. The balance, although light in construction, is very large in diameter and requires a stiff balance spring. A strong mainspring is thus required to hold the arc to the 150 degrees via the small radius of impulse, and this produces heavy lockings at the escape-wheel teeth. Some later watches with smaller balances have a noticeably livelier action and are free from setting.

Variations in rate, with the deterioration of the very poor quality oil available in the eighteenth century, were an accepted part of the horologist's burden. The closer the rate the more noticeable the effects, and Arnold's rates were close enough to reveal any small defects in his escapement. When it is considered that a change in rate of one second in twenty-four hours represents an error of one in 86,400, and that it will take only a minute change in the action of the components to bring this about, the problems of the eighteenth-century horologist are seen to be formidable. It is natural that Arnold should suspect first the change in viscosity of the oil. Clearly oil will eventually deteriorate to a point where the natural action of the escapement will be impeded and the watch will stop. Leading up to this condition, many small and irregular variations are likely to occur and Arnold's detent is undoubtedly to blame for these. The heavy safety horn and the incorrect position of the banking screw causes the detent to bounce on return to the banking after the unlocking is completed. This is of no consequence while the oil is fresh because the detent is free to recover its correct position during the impulse and before locking occurs. However, long before the oil is sticky enough to prevent the escapement functioning it begins to hold the detent clear of the banking after the rebound. This results in the balance jewel meeting the passing spring at differing moments during the escaping arc causing small, continually-altering variations in rate. The passing spring is also at fault, and the excessive overhang from the resting pin allows flexure with change of unlocking forces created by change in the condition of the oil.

Arnold does not seem to have made any serious attempt to eradicate the faults in the escapement, but presumably believed the variations to be caused primarily by the necessity for oil at the detent pivots and abandoned the arrangement in favour of his spring detent. Examples of this pivoted detent movement are rare and collectors who own one can count themselves extremely fortunate in possessing one of the first successful attempts at producing a uniform precision pocket timekeeper. The escapement shown in fig. 11 is from watch No. 64 shown in fig. 172.

Louis Berthoud's pivoted detent escapement, designed many years later, is much superior in design and therefore more certain in its action. The relatively larger impulse radius reduces the risk of setting, for the escape wheel has an increased mechanical advantage; thus the power required to impulse the balance is smaller and the action of the whole escapement becomes lighter and more delicate. The detent is again relatively shorter and very much thinner and lighter in construction. Considerable

115

weight is removed by eliminating the safety dart and using a roller to carry the impulse pallet and supply the safety action. The excessive overhang of the passing spring is still to be seen, but the lighter lockings tend to reduce considerably the effects of this fault. Banking is effected by a screw rigidly mounted in a bracket screwed to the plate and meets the detent as nearly as possible at the centre of percussion. The inclination of the passing spring together with the short detent ensures a greater inclination of the detent during unlocking, and this makes the action more certain and in the case of the passing spring reduces the friction on the return vibration. The rate of a small marine chronometer in the possession of Cecil Clutton entirely fulfils the promise of the design.

Breguet, always seeking and often making improvements in escapements, devised the pivoted detent escapement shown in fig. 14 in which the detent, of diminutive proportions, performs the sole function of locking the escape wheel. The passing 'spring', in the form of a pivoted pawl, is mounted on the balance roller and returned by a spring screwed to the roller. The slight advantage gained by this system in reducing the weight of the detent is completely swamped by the effects of oil at the passing spring pivots and the friction of the return spring. As with all Breguet's escapements the proportions are admirable. Fig. 15 shows another of Breguet's escapements which is in principle an exact copy of Berthoud's and demonstrates in what high regard Breguet held this arrangement, and rightly so, for its proportions have never been improved upon. It represented the final word in pivoted detent escapements.

The escapement shown in fig. 16 is by Kendall. Its chief interest lies in the unusual design of the detent and clearly shows what can happen when an attempt is made to achieve an object without understanding fully the principles involved. The radius of the impulse pallet is far too small and is exceeded by the unlocking radius which is far too large. There is no passing spring but the effect of this is achieved by thinning the length of the detent (except for the tip) into a thin spring. The tip is formed to the shape of an incline on the underside. It can be seen in the illustration that the unlocking stone has moved the detent aside allowing a tooth to fall on to the impulse pallet. The angle of the tooth face is radial and causes a considerable change of velocity in the wheel as the two surfaces meet, and then transfers the impulse from the tip of the pallet to the tip of the tooth. On the return vibration the unlocking stone meets the incline at the tip of the detent and deflects it up and so passes underneath. The vibration caused by the detent recovering its shape is a most disturbing influence and causes the detent to move against its banking, which in any case is insufficiently rigid. The greatest variable factor in the escapement, however, is the depth of the unlocking stone with the detent during the passing action. This is governed entirely by the position of the watch affecting the end shake of the balance staff and detent, and must change many times a day if the watch is worn. The hook-like projection is a safety device and in the event of tripping returns the detent so that a tooth can safely lock. This escapement is from the watch shown in fig. 169. Fig. 13 is from the watch illustrated in fig. 225. Fig. 14 is from the watch illustrated in fig. 237. Fig. 15 is from the watch illustrated in fig. 171.

Although Berthoud or Earnshaw may have been first in the field with the spring detent it was Arnold who, quickly perceiving the advantages of the system, made the fullest use of it. He abandoned his pivoted detent and concentrated his efforts on the development and production of the spring detent. To protect the thin spring against buckling he arranged it so that the detent was held in tension by the escape wheel when locked. To accomplish the unlocking it is necessary to move the locking stone towards the centre of the escape wheel, which has vertical extensions to the impulse teeth, thus enabling the detent to pass over the rim of the wheel during unlocking. Fig. 17 shows the arrangement of the escapement and is taken from the watch shown in fig. 159. That it is a later development of Arnold's spring detent escapement is apparent from the short detent employed. In earlier models the detent was half as long again and extended to the edge of the plate. The longer the detent the more uncertain is the action between the passing spring and the unlocking stone, due to freedom at the balance pivots. Arnold seems not to have been very sensible of this for his early spring detents were far too long and he never did achieve a sufficient reduction in their length. To be certain of the action at this point it is necessary with a long detent to increase the depth of contact of the unlocking stone with the passing spring, and this results in an excessive displacement of the detent to effect the unlocking.

By direct comparison with the mechanical grace and style of his pivoted detent escapement the arrangement is cumbersome and unconvincing, but its performance was superior when used in marine timekeepers. As to its use in the pocket it suffered from small variations in rate due mainly to the long detent and the still too small impulse radius of the pallet, now mounted on a roller to supply the safety action.

The gradual development of the proportions of the components culminated in the shorter detent and a lighter escape wheel working with a larger impulse roller. In this final form it worked very reliably and showed great consistency of rate.

Earnshaw's spring detent escapement shown in fig. 18 taken from the watch shown in fig. 240 is the same basic arrangement as used in Arnold's pivoted detent, with the exception that the thin spring at the foot replaces the pivoted detent axis. The proportions are altogether superior to Arnold's in that the ratio of impulse radius to escape wheel diameter is lower and the detent is shorter and lighter in construction. It can be seen that the impulse face is so severely undercut that wear must result from the transfer of pressure from the tip of the impulse face to the tip of the wheel tooth. However, notwithstanding this fault (later corrected by Earnshaw) it is altogether a more attractive arrangement than Arnold's, being easier to manufacture and requiring less motive power so that a thinner watch could be produced. At its best and with a compensated balance, its rate was often superior to Arnold's and proved so successful that with only very minor alterations it was universally adopted. One such alteration was a still further increase in the relative size of the impulse roller. An example of this is seen in fig. 19 taken from the watch shown in fig. 313. Here the roller (made of sapphire in this instance) is half the diameter of the escape wheel. The detent length is still shorter than Earnshaw's but finally remained at this proportion, since excessive shortening merely increases the resistance to unlocking. Watches fitted with the escapement made to these proportions were unexcelled as timekeepers when properly adjusted.

The Peto cross detent is shown in fig. 20 taken from the watch in fig. 181. Its advantages are illusory. The cockling of the detent spring that this system was said by its inventor to overcome never existed, for had the spring been made thin enough for this to happen it would have been insufficiently strong to return the detent.

For the same reason the Peto detent cannot be made weaker, for such weight as would be saved by thinning the body of the detent would be more than counterbalanced by the weight of the curved horn. This further increased the inertia since it is formed at the most rapidly moving part of the detent. The friction between the passing spring and the horn of the detent is undesirable and, like the escapement as a whole, contributed nothing to the science of chronometer construction. It was, however, used occasionally by Breguet in his tourbillons where it is more suited to withstanding the effect of the kinetic energy of the rotating platform during the locking of the escape wheel.

An interesting variant of the escapement is the very rare James Ferguson Cole double rotary detached escapement. It is seen in fig. 21 taken from the watch shown in fig. 296. The impulse in this escapement is delivered in the normal way to the balance roller, whilst the unlocking roller is carried on a separate arbor pivoted between cocks mounted on the plate. The two are geared together in the ratio of 2:1 so that the angular velocity of the unlocking roller is half that of the balance. Thus the balance is enabled to make more than a complete revolution without tripping the escape wheel. The angle of the passing spring is ingenious for it allows the full length to flex during the passing vibration, but uses only the short rigid end for unlocking. This is essential, for with the reduced velocity of the roller the action of the stone with the passing spring must be very positive if the detent is to be released in time to catch the wheel tooth. The disruptive influence of the extra mobile completely swamped the advantage gained by the greater angular velocity of the balance, as also did the greater number of turns required in the balance spring, for such springs are difficult to make isochronous. The escapement is no exception to the rule that the fewer and more constant the mobiles, the better will be the rate. The authors have seen only two of these escapements and presumably Cole was not encouraged to continue with their manufacture in quantity.

21
296

'ÉCHAPPEMENT NATUREL'

To overcome the early defects of setting with the single-impulse chronometer escapement Breguet devised a double-impulse escapement which he described as 'échappement naturel'. The description applies to the manner of impulsing the balance by 'natural lift', without need of oil, in the same manner as the single-impulse escapement. He did this by gearing a second impulse wheel to the conventional escape wheel so that they locked alternately on a central pivoted detent. The balance engages the detent by the fork and roller arrangement of the lever escapement. As one wheel is unlocked to impulse, the detent is carried by the fork and roller to catch a tooth of the second wheel. Because the wheels are geared together, both are locked leaving the balance free to complete the supplementary arc. The escapement is ingenious and typical of Breguet's rare inventive gift for solving problems with complex mechanical solutions.

Breguet used the escapement in a series of tourbillon watches with remarkable results in long-term consistency of rate. For closeness of daily rate it could not compete with the single-impulse chronometer and Breguet eventually ceased its manufacture. Its principal defect arises from the gearing together of the escape wheels, which

214

allows only a half of the space between the escape wheel teeth to be used for the impulse at each vibration. This gives rise to a shallow intersection of the wheel and impulse roller which can be increased only by reducing the diameter of the roller. This was a defect of earlier chronometers and has the effect of increasing the escaping arc and magnifying the escapement error. Further faults arise from the necessary freedom of the gear teeth causing small variations of impulse and inconstant banking of the detent. In this respect the single-impulse escapement was superior because the spring of the detent ensured positive banking.

LEVER

Thomas Mudge devised the lever escapement in the 1750s and used it in a small clock undoubtedly intended as a contender for the award offered in 1715 for a means of determining the longitude. The date of completion of the clock is not known, but a survey of Mudge's work would indicate 1760 at the latest. At a later date he used the escapement again, in a different form, in a smaller clock. In 1799 his son published *A Description with plates of the Timekeepers Invented by the late Mr. Thomas Mudge*. The book is concerned principally with Mudge's marine chronometers but contains references to other pieces by Mudge with lever escapements.

The first watch with lever escapement was made by Mudge for George III who gave it to Queen Charlotte. It is hallmarked 1769 and, in correspondence with Count von Bruhl, Mudge's patron and friend, was referred to as 'The Queen's watch' or 'Queen Charlotte's watch'. It remains in the Royal Collection and is illustrated on pl. IIIA by gracious permission of Her Majesty the Queen.

The similarity of the wheel and pallets to Graham's clock escapement can be seen in fig. 55. Graham's escapement with circular locking faces was established in the mid-eighteenth century as the most usefully accurate escapement for regulator clocks. By substituting a forked lever and balance wheel for the crutch and pendulum of the clock, Mudge produced a portable double-impulse, detached escapement capable of maintaining a daily rate superior to any other then known and which was eventually to displace all other escapements for watches.

Mudge himself had little interest in the escapement and, refusing to spend more time on it, turned his attention exclusively to the improvement of his marine chronometers with constant force escapements. He declared his lever escapement to be better than any other escapement for pocket watches but doubted if any other maker would put himself to the trouble of making it properly. It can be seen in the illustration that, as conceived by Mudge, it would be difficult to construct. His work was usually complex, delicate and supremely well finished. No contemporary compares with Mudge for ingenuity and skill in executing his own ideas and only he could produce the daily rates of which his work was capable. It is not surprising then that Mudge believed the escapement would not succeed in other hands.

As constructed by Mudge the escapement has two principal faults. The first lies in the radial escape wheel teeth which, at the limit of locking, coincide with the flat locking faces of the pallets. This causes excessive oil adhesion between the faces of tooth and pallet and must result in rapid deterioration of the rate as the oil thickens, in addition to insecure locking of the exit pallet. It should be remembered that eighteenth-century oil was unstable and soon became sticky. The second fault is in the angle of the pallet arms which embrace a quarter of the wheel circumference. This gives rise to long impulse planes with high friction and long pallet arms with high inertia. Both faults rob the balance of energy which can only be compensated by a stronger mainspring. The long pallet arms are understandable especially if the escapement was derived from Graham's clock escapement. The radial tooth faces are unaccountable and at variance with accepted horological practice both then and now. Why Mudge should have fallen into such an obvious error cannot be explained and it is certainly at odds with his amply-demonstrated understanding of mechanical principles.

Mudge's opinion of his escapement and his want of faith in the ability of others to make it was not shared by his friend and patron Count von Bruhl. He saw clearly the merits of the escapement and set himself the task of having others made.

Mudge was prevailed upon to make a model for examination by the trade, and Josiah Emery, after some reluctance, for he was of the opinion that the escapement could not be made small enough for a watch, undertook the construction of a series of watches.

Emery arranged the escape wheel, pallets and balance in a straight line, which is surprising as this occupies more space than the angled arrangement used by Mudge. A comparison of the two escapements shows that Mudge, as in all his work, treated its construction as an exercise in supreme craftsmanship, while Emery was concerned with simplifying the construction to reduce the cost. The principles are similar in that both have jewelled pallets and a jewelled fork and roller action arranged above and below the plane of the lever. In Mudge's escapement the fork is jewelled and the roller cams are steel, whereas in Emery's the fork is steel and the roller pins have sapphire blocks let in at the contact points. Emery's escape wheel has eighteen teeth to Mudge's twenty but because Emery's pallets also embrace five teeth the angle of the pallet arms is relatively wider thus exaggerating the faults earlier described. The teeth are inclined forward to avoid the fault of radial contact and the exit pallet locking face is slightly inclined to ensure tangential locking at the moment of drop. This introduces a very small draw into the pallet face at the limit of locking. The entry pallet face is tangential at the limit of locking and this introduces a slight draw at the moment of unlocking. Clearly there will be a small recoiling of the escape wheel during the unlocking but this should not be confused with the later deliberate angulation of the locking faces to draw the pallet to the limit of locking. The recoil of Emery's wheel is proportional only to the angular displacement of the pallets during unlocking, and is about three degrees. The angle of draw required to impel the lever firmly to the banking is between twelve and fifteen degrees depending on the escaping angle but, not surprisingly, Emery did not see the advantage of this. By introducing the angulation to the locking face of the exit pallet Emery showed a clear under-standing of the principles of locking which, surprisingly, seem to have confused Mudge. Although Emery's lever is symmetrical and so does not need the balancing weight used by Mudge in his angled lever, its heavy proportions show that, like Mudge, he was not aware of the necessity for lightness in this component. The inertia of the lever is a principal factor in absorbing the energy of the balance.

Mudge's lever watch used a plain balance with two spiral balance springs; one compensated for temperature with a bi-metallic curb; the other was adjustable for meantime regulation with a cycloidal pin. Emery made his first lever watch in 1782, twelve years later, and so was able to benefit from the use of an Arnold compensation balance and helical spring. This should have put the advantage of rate with Emery but the rates of the two watches were probably equal and within five seconds per day. This is suggested by Mudge's comment in 1776 that one of Arnold's watches varied by not less than ten seconds per day, double the error he ever found in the Queen's watch. The rates of Emery watches worn by the authors are also of the order of five seconds per day. The result is most creditable to Mudge and underlines his complete mastery of the dubious art of tediously-difficult adjustment. Emery made about thirty lever watches between 1782 and 1795. Excepting for a change in the fork and roller action by the introduction of a cranked roller with pivoted journal, he made no changes to his basic design and no attempt to develop the proportions. His work is now rare and highly prized by discerning collectors. In addition to their aesthetic qualities his watches represent a service to both the inventor and horology in that they brought the escapement to the notice of later makers who fully exploited the qualities that only von Bruhl had understood.

During the last decade of the eighteenth century in England, other makers produced watches with lever escapements. Notable for their similarity to Emery's work are watches by Perigal and Pendleton. The Perigal escapement uses the right-angled lay-out of Mudge. This may have been influenced by the Queen's watch, for Perigal was the Royal watchmaker and would have had access to it. The detail finish of the escapement is similar to Emery's work as is the design of the movement. Pendleton's escapement uses a loop with two vertical pins to impulse the balance diametrically opposite the lever but the balance is identical to Emery's, and the design of the movement also. This principle of impulsing was used by Mudge in his first lever escapement. It is intended to reduce the friction of contact as the balance and lever rotate in the same direction during impulse. It is not a happy arrangement for the escaping angle must be increased to ensure a safe intersection of the components.

The escape wheel teeth of the Pendleton are slotted at the tips for oil retention. The teeth of the Emery and Perigal have crescents cut in the locking faces for the same purpose. None of these escapements has draw to keep the lever firmly against the bankings and consequently all suffer in use in the pocket from variations in rate. The many points of similarity in the details of these escapements would indicate that their makers were in close touch and perhaps assisted each other in their work.

The only other people known to be making lever escapements in England at the close of the eighteenth century were Grant and Leroux. Their work should be considered separately for they produced watches of quite extraordinary merit in their respective ways.

Grant's contribution to the development of the escapement was to demonstrate that lighter components and better impulse angles could improve the action and, thereby, the rate. It is not known when he made his first lever escapement but his earliest-known example is hallmarked 1795. The escapement has a brass wheel with upraised teeth lifting the pallets towards the centre. There is no obvious reason for this arrangement except that the lever can be brought to poise without a counterweight. The locking faces of the pallets are curves struck from the lever pivots to ensure that the escape wheel is quite still during the unlocking and does not absorb any balance wheel energy. The raised teeth of the escape wheel allow the use of a thin rim to reduce weight. The pallet arms and lever are reduced throughout their lengths to leave only the minimum metal required to secure the pallet stones and guard pin. The escapement has a most lively action and the acceleration of the balance from rest demonstrates the advantage of the light components. The escapement of a watch also by Grant shows the same light construction and curved locking faces passing above the rim of the escape wheel, but the pallet arms are lifted away from the wheel centre in the conventional manner of the anchor escapement. It can be seen also that the pallets now embrace only four teeth of the wheel. This small alteration increases the efficiency of the lifting action. The pallet arms are shorter with the obvious advantage of reduced inertia, but the greatest gain is in the reduced friction of the shorter impulse face of the pallet.

The advanced workmanship evident in Grant's watches played a significant part in improving their daily rates. These are however only marginally better than the rates of Emery and others. Grant was unaware of the absolute necessity of holding the lever firmly against the banking during the supplementary arc. His curved lockings allowed a greater angle of movement to the lever to ensure the free exit of the roller pin from the fork but could not prevent variation in re-engagement if the lever were disturbed. Grant seems rarely to have made two escapements alike and his work is always interesting. Fig. 27 shows an arrangement of his escapement using two balance wheels geared together. Each has a balance spring, but impulse is delivered to only one balance and is transferred by the toothed wheels to the slave balance. The watch is obviously experimental and is unjewelled. The fork and roller are typical of Grant as also are the upraised teeth of the escape wheel. The object of this curiosity was to keep the balance arc constant whilst the watch was moved about in wear. It performs its intended function so well that without actually starting the escapement by a touch of the balance it remains inactive in spite of the most vigorous agitation and, conversely, when vibrating, the balance amplitude is indifferent to the most energetic actions of the wearer.

The final important lever watch to be considered (fig. 162) is by Leroux. Fig. 28 illustrates the maker's quite remarkable comprehension of the requirements of the lever escapement. It can be seen that the escape wheel teeth have impulse inclines in the manner of the cylinder escapement. The pallet stones are very thin and have no impulse faces. The whole of the lift is imparted to the edge of the stone by the incline of the tooth. The locking is effected by the pointed tip of the tooth falling onto the face of the stone. This face is inclined to the tangent of the wheel tooth by an angle of 10 degrees so that when locked the wheel will draw the pallet towards its centre and ensure the lever is held firmly against the banking. This is the only eighteenth-century watch examined by the authors that has this angular draw to the locking faces. It is all the more remarkable because it is hallmarked 1785 and was therefore made during the Emery period; yet almost nothing is known of its maker. In addition to being the first watch with draw it is also the earliest-known watch with equal impulse and locking. Because the whole of the lift is given by the escape wheel teeth, the locking faces of the pallet stones, and therefore the lifted edges, can be struck from a circle whose centre is at the pallet staff. In the early nineteenth century, when the lever escapement was adopted by an increasing number of makers, much attention was given to the inequalities of the action of the two pallet stones. This arose mainly from incorrect proportions in the escapement preventing self-starting from rest. One successful solution was the pointed pallet escapement used by J. F. Cole and S. Mairet. Here, in the Leroux watch, is the pointed pallet escapement made nearly fifty years earlier. It is worth noting that the first lever escapement with equal locking and impulse was made, somewhat fortuitously, by Mudge when he converted a small verge clock to a lever. His escape wheel was vertical to avoid the use of a contrate wheel with a horizontal lever and balance so that the pallets engaged the teeth at a 90 degree angle. This rather ungainly arrangement was used later (*c.* 1800) by Grant in his chaff-cutter escapement.

Further points of great technical merit in Leroux's escapement are the very high impulse angle of the pallet edges achieved by embracing only two teeth of the escape wheel, and the short lever to transmit the impulse to the balance through a small escaping arc. The combination of high lift, short lever, equal locking, impulse with draw and the small escaping angle set Leroux's escapement apart from all others as an example of high efficiency with certainty of action. Such a combination of virtues was not to be found again in a lever escapement until *c.* 1820, with the manufacture of the English lever with draw. Even the illustrious Breguet, whose work will be discussed later, did not achieve Leroux's efficiency of design.

29   An interesting lever escapement is shown in fig. 29, from the watch shown in
184   fig. 184. Its arrangement is a composite of all the features discussed in the escapements illustrated. The date of the watch is not known for it came into the possession of Cecil Clutton as a movement and was restored and re-cased by George Daniels, but it may be presumed to date from about 1800. It can be seen that the circular termination of the lever is derived from Pendleton, whilst the pallet stones are curved like Grant's but with pointed acting tips to allow of the impulse from Leroux's escape wheel. The balance wheel, which can be attributed to Arnold, is controlled by two balance springs, one above and one below the arms, the lower one being affected by the regulator curb pins as by Mudge: a watchmaker's interpretation of the quotation, 'If you can't beat 'em—join 'em!' The maker is Taylor.

By the end of the eighteenth century the detached lever in England had virtually ceased to be made. Examples are rare enough to be highly prized by their owners. They are, with Arnold's early detent watches, the most exciting of English pre-nineteenth-century watches. If their makers were on the whole not sufficiently practised in the mechanical principles of their art this detracts nothing from their honest endeavours, and we are the richer for the interesting speculations they afford us in discussing their work.

The work of Litherland ensured the continued production of the lever watch in England but the escapement was no longer detached. The action of the unlocking and impulsing was affected by a toothed rack engaging a pinion on the balance staff.
30, 209 Fig. 30, taken from the watch shown in fig. 209, shows a typical arrangement. There are widely-spaced bankings and the balance can make a full turn in either direction, the locking faces being long enough to accommodate the increased movement of the pallets. The performance was no better than a good cylinder watch but was undoubtedly easier to produce than any previous escapement except the
31   verge, and was popular for cheap watches. An early variant is seen in fig. 31 in which the balance has been detached by replacing the rack with a simple roller. The stone impulse pin is fitted to a thinner projection of the roller so that the safety action takes place at the same elevation as the impulse. This device was the subject of a

patent by Massey and in its earliest form is seen in fig. 32 where the roller is made entirely of steel. It can be seen that both these escapements have curved locking faces with the consequent absence of draw and thus their makers ensured that the rate was little better than the rack lever. The use of the fifteen-tooth escape wheel necessitated an extra wheel in the train but it simplified the application of a seconds hand. The low lift-angle of the impulse faces and the well proportioned lay-out assured a good rate when eventually their makers overcame their conservatism and introduced the draw.

It is curious to note that with the lever once again detached London makers took a fresh interest in this escapement, and by the middle of the nineteenth century were producing it in quantity almost to the exclusion of all other escapements except the duplex and the chronometer, which were still used in watches of the highest precision. Exactly when the draw was reintroduced is open to conjecture, for no one actually claimed it as his invention, which is unusual in the history of watchmaking. Possibly anxious to create the illusion that the improvement in rate was due entirely to superior workmanship, the first users felt that they would not be doing themselves justice if they admitted to resorting to a 'dodge'. Examples of fifteen- and thirty-tooth Massey levers can be found with and without draw, but in the experience of the authors, watches with the all-steel roller never have draw. Allowing a little time for Massey to get into production with his patent issued in 1814, and presuming that the jewelled roller followed soon after, and with a further short period elapsing to account for the examples without draw, the year 1818 may be acceptable. Most certainly by 1820 it was accepted by the best makers that draw was essential for a steady rate. Its use without sensible alteration was continued into the twentieth century, although towards the end of the nineteenth century the club tooth escape wheel began to replace the pointed or ratchet tooth wheel.

One interesting and successful variant was that of Savage. Fig. 33 shows his arrangement in which it can be seen that the safety of the locking action during the supplementary arc is ensured by a pin acting upon the edge of the roller. The unlocking is accomplished by the two pins placed one each side of the impulse notch in the roller, into which the lever guard pin will pass during the unlocking, and the action is then transferred to the guard pin which impulses the balance. The difference in ratio of unlocking to impulsing lever length is small but significant, and results in a most lively action for a comparatively weak mainspring. It also offers the slight advantage that the friction of unlocking, by virtue of the wide notch in the lever, is disengaging, but this is an ancillary advantage and can in any case be achieved by correct proportioning of the escapement and impulse roller. The escapement was not much used, for it required the most careful pitching of the three pins, and any small errors of pitching soon rendered it less reliable than the conventional fork and roller. Variations in rate in these escapements can often be attributed to the widely spaced notch and roller pins occasioning overbanking in excessively long arcs.

A form of lever escapement popular in France for watches beating seconds was that devised by Pouzait. Fig. 39, taken from the watch shown in fig. 196, shows the arrangement. The balance diameter almost equals that of the back plate above which it oscillates, located in a skeleton cock. The escape wheel of thirty upraised teeth is at the centre of the movement and carries the sweep seconds hand. The impulse is delivered to the balance by the notch, seen between the spokes of the wheel, acting upon the large impulse pin. The safety action is effected by the pin on the exit side of the lever passing through a gap in the large safety ring.

Breguet's initial use of the lever escapement was in his first series of perpetuelles entered in the work-books in 1787. He had been planning the series throughout the 1780s and it is fitting that he should have wanted an escapement superior to the cylinder used in the previous series. His earliest-known watch with lever escapement is perpetuelle No. $3\frac{4}{86}$ made in 1786, and has a plain balance with compensation curb, long-since abandoned for the best work in England. If Breguet lacked experience with balances he certainly did not lag behind with his lever escapements. It is not known how he first became acquainted with the escapement. He visited England on more than one occasion and his inquisitive mind would soon have absorbed any useful ideas that he came across. He may have seen an example of Emery's work although only the principle of operation would have been attractive to him, his feeling for mechanical proportions being quite different from Emery's. However he came across the escapement, he lost no time in developing the idea according to his own brilliant concept of its merits.

34 Breguet's escapement was constructed with exceptionally light and delicate components so that the inertia of the lever was kept to a minimum. His escape wheel, of brass, is turned hollow on both sides to leave a thin rim; it has spokes with broad tips to the teeth which are slotted for oil retention. The pallets embrace only three teeth of the wheel to give a high-lift angle to the jewelled impulse faces. The locking faces are curves struck from the pallet centre with the jewels let into the very light pallet frame. The lever is exactly counterpoised to prevent disturbance of the lockings and is banked on the escape wheel arbor. A safety dart is formed on the tip of the lever and flanked by two vertical pins to engage the balance impulse nib at the centre line of the escapement. The escapement of No. 3 has a short lever so that the pallet axis is midway between the balance staff and the escape wheel arbor. All the later escapements have a longer lever. The shorter lever is better but requires a greater skill in manufacture because the tolerances in the roller and safety action are smaller for a given depth of locking and impulse. There is some evidence that Breguet was having difficulty in finding men skilled enough to make the escapements and the longer lever may have been a compromise towards a more certain action. It should be remembered that workmen on the Continent were not familiar with the escapement, being more used to the verge and cylinder which have a less complex action. In contradiction of this theory, Breguet never sought any shortening of the lever at a later date when he had an ample number of skilled men.

One effect of the long lever, which may have influenced Breguet to use it, is to reduce the force of unlocking. On the other hand the inertia is increased and the impulse angle of the balance is greater, the first fault robbing the balance of power and the second increasing the escaping error. It is surprising that Breguet did not see this as clearly as Leroux. By employing light compensation balances with helical springs with terminal curves, supported on pointed pivots with elastic suspension, Breguet completed the formula for an escapement that can run for many years to within a few seconds per day.

With the exception of a few miscellaneous watches Breguet did not use the lever escapement after 1800 but spent his time experimenting with a variety of detached, natural-lift escapements. In about 1810 he turned again to the lever and produced a new series which was used, almost without change, until the middle of the nineteenth 36, 37 century. The new escapement continued with the long lever and the counterpoise banking against the escape wheel arbor. To give divided lift, the escape wheel has club teeth which are pierced for oil retention. The piercing is in the form of a conical hole drilled from the back of the tooth to appear as a tiny hole at the locking corner. The pallet jewels are set in vertical slits in the pallet frame and have straight locking faces but are still without a positive draw. The entry stone is radial to the wheel when locked, while the exit is angled very slightly to ensure a safe and positive locking. This causes a slight recoil of the wheel during the unlocking of both stones but not sufficient to draw the lever back if the safety dart touches the roller. Clearly Breguet was reluctant to suffer the increase of power required to overcome the increased locking force of the draw. This is not surprising because Breguet did not regard his lever watches as precision timekeepers in the same category as his 'garde temps'. They were rather more a superior compromise between the inferior rate of the cylinder and the delicacy of the chronometer, and long-running reliability without need of servicing was a principal objective. On the whole his clients were more concerned with the appearance and novelty of their watches and a precise rate of going would play no part in their daily affairs. Even so, a properly-adjusted Breguet lever watch has a daily rate superior to many early chronometers.

As with the early levers the compensation balances are supported on pointed pivots 38 with parachute suspension. The balance springs are spiral with Breguet overcoils and regulators. With the exception of the proportions these escapements bear a close resemblance to the modern lever escapement, having a double roller, brass and steel compensation balance with moveable screws, and pallet stones set in vertical slits in the pallet frame. These escapements were produced in thousands in later years by the Swiss who did not fully comprehend them. Using a single roller with vertical guard pin for economy they increased the length of the lever still further to improve the safety action, resulting in the escapement having a poor action and becoming unreliable. It was probably this misunderstanding of the proportions of the escapement that prompted the Swiss to turn again to the cylinder escapement for mass production during the second half of the nineteenth century.

For his tourbillon watches Breguet used a right-angled lever escapement as seen in fig. 35. For this he was obliged to use a shorter lever due to limitations of space but 35 eased his anxiety over the unlocking forces by using curved lockings for the pallets. This escapement was made in 1810 and is the earliest single-roller escapement known to the authors.

During the second half of the nineteenth century the lever escapement was used increasingly by the best makers both in Switzerland and England. The proportions and angles settled into a near-standard pattern. The principal differences between the two separately developed schools lay eventually in the arrangement of the components and the shape of the escape wheel teeth. The English preferred the right-angled 24, 33 lay-out with the escape wheel, balance and pallets pitched in a triangle and with a brass escape wheel with pointed or ratchet-shaped teeth. The Swiss preferred the straight line lay-out with a steel escape wheel with club teeth. English makers used a 37 single roller with vertical guard pin acting at the edge, while the Swiss preferred a double roller with a smaller diameter for the safety action working with a horizontal dart.

126   Used with a short lever and a small escaping angle the single roller is quite suitable

for a pocket watch, which is not subjected to the violent movements of a watch worn on the wrist. The ratio of lever length to roller radius is about 4:1 giving an escaping angle of some 44 degrees. With a balance vibration of about 190 degrees the English makers felt it prudent to continue using the fusee to suppress isochronal errors. As development continued towards the end of the nineteenth century progressive makers increased the balance amplitude to about 260 degrees and reduced the escaping angle to about 40 degrees. This improved the daily rate but the reduction in lever to roller ratio made a change to a double roller necessary in order to prevent the guard pin jamming against the edge of the larger roller. An increase in the number of vibrations from 16,200 to 18,000 per hour further increased the velocity of the balance with consequent improvement in the stability of the rate. Quite why 16,200 should have been accepted as a standard vibration figure is not known. Emery, Grant and Perigal used 18,000 which is a satisfactory number for pocket watches and divides the seconds into fifths. Cummins, who was in the van of the resurgence of interest in the lever escapement in the nineteenth century, used 15,600 which is even more puzzling for this gives four and one third vibrations per second. English makers used very heavy balances for their watches, presumably to gain some sort of dominance over the latent errors of the escapement. The best makers formed their balance springs with terminal curves, sometimes two and even three turns above the spiral. When helical springs were used these also had terminal curves. Occasionally watches are found with 'duo-in-uno' springs in the form of a spiral rising into the helical form for the outer turns. The more fanciful springs have no magical properties and were probably the result of makers competing with each other to demonstrate the skill and quality of their work. They are in fact a disadvantage for they offer no means of controlling isochronal errors by applying bias to the collet through manipulation of the curves. Regulators were never used in high-grade watches and were regarded as a sign of inferior work. Thus it was that in the late nineteenth century the best English makers were so overwhelmed by the superlative quality of their work that they began to price themselves out of their own market. Aided by superb workmanship, a fusee to suppress the isochronal errors and by very thin balance pivots to reduce the friction of the heavy balance in the vertical positions, the watches were remarkably accurate timekeepers.

The Swiss had taken a different direction and had learned how to make simpler, cheaper lever watches which, while they could not equal the rates of the English watches, were more than adequate for civil use where fractions of a second are of little consequence.

The Swiss lever watches of the mid-nineteenth century were not of the same high quality as the English watches. They were made on a quantity-production basis although by no means mass-produced. Like the English they started with right-angled escapements with single rollers but their proportions were not so good and their timekeeping suffered to some extent. The levers were too long and the escapement angles too large.

By the 1860s the straight line escapement was in general use and had developed to the point where it could begin to rival the English escapements. Swiss watches had going barrels, club-toothed escape wheels and spiral springs with Breguet overcoils and regulators. The English eschewed regulators but the regulator was the key to the success of the Swiss going barrel watch. With it the isochronal error could be suppressed by adjusting the curb-pins to control the active length of the spring in the short arcs. This was a considerable advantage in reducing the cost of the watch and the adjustment did not need the skills of long years of practice.

By the 1880s, especially in Geneva, makers were producing watches with short levers, small escaping angles and double rollers for the safety action. These features improved the rates, and as the number of makers increased so prices dropped. Refinements in the manufacture by specialists of the balance springs, balances and escapement components ensured a steady improvement in the rates with less dependence upon adjustment after assembly. It was not long before it was generally recognised that a good Swiss lever watch was the equal of its English counterpart but far cheaper. The myth of the superiority of the Swiss watch was established and eventually, in the early twentieth century, caused the eclipse of the English lever and the dying-out of individual skills that had taken a century to develop. The finest of the Swiss levers display a wealth of technical development in the proportion and design of the escapements. They employed large, light compensation balances with spiral springs with terminal curves and regulators, small escaping angles, sometimes as low as 35 degrees for 11 degrees of lever angle, and steel escape wheels hollowed on both sides with club teeth bevelled on both edges to improve oil adhesion. As a result the balances accelerate rapidly from standstill and have a lively action for a relatively weak mainspring with minimal amplitude reduction in the vertical positions.

These qualities were further enhanced by increasing the angle of draw from 12 to 15 degrees enabling the lever counterpoise to be dispensed with and so reducing the inertia of the lever by half. The increase in draw causes a proportionately smaller loss of energy to the balance than the inertia of the counterpoise.

The development of the lever escapement by the Swiss makers has never ceased, so that, when used in a wrist watch, the best rates can equal those of the earlier pocket watches. The latest developments include escapements beating 28,000 and 36,000 vibrations per hour. Very fast beats have necessitated considerable re-designing of the escapement. The number of escape wheel teeth is increased from the conventional fifteen to twenty. The pallets, originally embracing three teeth of the wheel, now embrace four and the lifting angles are increased to ensure complete lifting during the very fast action. To prevent the action being too fast the escaping angle has been increased from 35 to 55 degrees for a total amplitude of 275 degrees. It is interesting to note that in the 1820s Breguet made a series of twenty-toothed lever escapements with the pallets embracing four teeth for an escaping angle of 56 degrees. Truly there is nothing new in horology! Such high-speed action has caused lubrication problems but these have been successfully overcome by new techniques with newly-developed solid lubricants. Plastics have not been ignored and at least one Swiss manufacturer makes an escapement wholly from moulded self-lubricating plastics. Bygone makers who could truly be described as artists would doubtless be horrified if they knew how their art had developed. Perhaps it would be some comfort to them if they knew it is now called micro-precision engineering!

Many millions of watches with lever escapements were made in America from the middle of the nineteenth century up to the middle of the twentieth century. The earliest escapements had single rollers and club-toothed escape wheels and so were a good compromise between the English and Swiss types. At one time the Howard Company used J. F. Coles's resilient escapement in which the pallets banked on the inclined flanks of the escape wheel teeth to absorb the shocks of over-banking. This must have been the only occasion when this curious aberration was taken seriously. Most American makers used the single-roller escapement until the end of the nineteenth century and the Waltham Company continued into the twentieth century producing wrist watches using this unsuitable escapement. Later American double roller escapements were inevitably influenced by contemporary Swiss practices, and while they have contributed little to the development of the escapement their performance is all that one would expect from a nation that knows no barriers to technical excellence.

ROBIN

The necessity for oiling the teeth of the lever escapement no doubt influenced Robin in his development of the escapement bearing his name. In its use of the lever with two locking stones and the impulse delivered to the balance at each alternate vibration via a pallet attached to the balance roller, it utilises the principles of both the lever and chronometer escapements. In this combination its inventor no doubt intended that it should possess the mechanical reliability of the lever with the superior rate of the chronometer. It failed in both respects, for with impulse in one direction only it was prone to setting and, unlike the lever escapement, in which the lever is helped to the banking by the action of the wheel tooth on the impulse face, the Robin lever relies upon the balance to perform this function for it. Thus, depending on the condition of the oil at the pivots and the balance amplitude, the clearance between the notch of the lever and the pin in the roller will, during the approach to unlocking, be a variable factor. Unless made with great precision the rate will be seriously affected by the pin actually striking the corner of the notch during entry. Figs. 40 and 41 taken from the watch shown in fig. 292 illustrate the escapement, in this instance of straight-line arrangement popular with Swiss constructors. The plates show the position of the lockings for alternate vibrations. The small drop to the entry stone in fig. 40 allows no impulse to the balance. The large drop to the exit stone seen in fig. 41 will allow of an impulse to the balance by the tooth nearest the balance staff acting upon the impulse pallet seen projecting from the roller beneath the unlocking pin. The banking is to the escape pinion by the ring formed in the lever. Constructed in this manner during the resurgence of popularity of the escapement during the second quarter of the nineteenth century it was very successful as a timekeeper if kept clean and oiled. The reason for this lies in the extreme angle of draw of the locking faces ensuring firm banking of the lever. This feature was not included by Robin and his contemporaries and their rates suffered badly for the want of it. Breguet made use of the escapement and fig. 42 shows his commonest arrangement in which he introduced a lifting face to the exit stone so that the balance, being impulsed in one direction by the lever and on its return vibration by the escape wheel, is less inclined to set. This modification necessitated the use of oil to the escape wheel which, being in contact with the balance roller, soon affected the rate. To overcome this fault he introduced a double escape wheel. Fig. 43 taken from the watch shown in fig. 210 illustrates the arrangement in which the upper wheel teeth impulse the balance.

With this arrangement its performance is as good as the lever, but the complex and costly construction prohibited its use in all but the most expensive of his productions in which, quite often, the high cost of the watch was brought about by unnecessarily complicated work.

DEBAUFRE

One of the first practical solutions to the problem of the recoil in the verge escapement was that of Debaufre. He moved the balance axis so that it was tangential to the face of the crown, or escape wheel, teeth. An additional crown wheel was mounted on the same arbor so that the two sets of teeth faced inwards and supplied impulse to the balance by inclined planes cut into a circular roller attached to the staff. During the supplementary arc the tooth-tip rested on the dead face of the roller close to the centre of motion. Fig. 44 shows the usual arrangement and is taken from the watch shown in fig. 195. It is said that the objection to this arrangement is that the impulse tends always to exert pressure on the end of the top pivot. This may have been objectionable if the staff pivots were unjewelled, as they may have been in the earliest examples, but is not in itself of serious consequence as can be seen from the going of marine chronometers which always run dial up with a heavy balance.

The prime cause of the varying rate associated with this escapement lies in the lack of constancy of the resting friction, which is continuously changing by reason of the freedom of the balance pivots in their holes. With the best workmanship its going was never better than a moderate cylinder watch but it was nevertheless produced spasmodically for over a hundred years by a variety of makers. A variant of this escapement which suffers from the same fault is seen in fig. 45, where a single wheel with radial teeth takes the place of the double crown wheel and impulses a double roller each with an inclined impulse face cut into its edge (Sully's escapement).

So that the escape wheel could run vertically between the plates with sufficient freedom for the wheel teeth, of necessity it had to be small in diameter and as this increased the ratio of resting friction variation was bad for the rate.

FASOLDT

An interesting and rarely seen variant of the lever escapement is that shown in fig. 46. The two locking stones engaging the large escape wheel and mounted on the lever serve the same purpose as those of the Robin, but the impulse is delivered to the balance by the smaller impulse wheel acting upon the stone seen engaged by a tooth of the wheel. Turning the balance anti-clockwise would withdraw the exit stone allowing the impulse to be delivered. The inertia losses in the escapement are high and the action is correspondingly weak. At the time of its production in the last years of the nineteenth century the lever escapement held a position of great supremacy for pocket timekeepers and this position was undisturbed by Fasoldt's escapement. The watches are beautifully made and were a brave if undistinguished attempt to introduce a little variety into escapements which, by that time, had inevitably settled into a pattern of dull uniformity.

J. F. COLE ROTARY DETENT

47, 308   Fig. 47 taken from the watch shown in fig. 308 shows a remarkable escapement by J. F. Cole that may well be the only example. It is a detached escapement in which an extra mobile is placed between the balance and escape wheel and serves both to unlock the wheel and supply the impulse to the balance. The teeth of the escape wheel rotate the smaller wheel by means of the vertical pins, but cannot advance by more than one tooth, for the velocity ratios are so arranged that the next tooth to engage will always butt its inclined tip to the approaching pin whilst the engaged pin is on the face of the preceding tooth. The unlocking stone of the balance when turning clockwise forces the small wheel to rotate as it meets one of the thin wire spokes projecting from the rim of the wheel, which it can do only by recoiling the escape wheel as the locked pin travels down the inclined tip of the tooth. Upon reaching the corner of the face of the tooth the impulse commences, for the small wheel has now been turned far enough to disengage the pin on the preceding tooth. At this moment the impulse pallet on the balance is presented to a tooth of the star wheel, seen mounted on the small wheel above the pins, and the impulse is delivered as in the duplex escapement. The small wheel continues to turn under the influence of the escape wheel until the next pin butts against the following escape wheel tooth. During the return vibration the unlocking stone causes the wheel to recoil as it brushes past the succeeding spoke. The action is then ready to commence again. This curious escapement, reflecting Cole's ingenuity at its best, has a most lively and convincing action. The recoiling of the wheel by the vertical pins causes wear to the corners of the escape wheel, so that tripping is provoked by variations in power affecting the velocity of the escape wheel. It was not found in wear to be very reliable, although its rate dial up is most precise. It is included for its extreme rarity and because the escapement cannot be seen in detail without dismantling the movement.

Repeaters

The application of repeating work to watches is credited to both Edward Barlow and Daniel Quare in about 1687. Barlow probably deserves the greater credit for he is reputed to have applied the essential rack mechanism to clocks at an earlier date. When Quare heard that Barlow had commissioned Tompion to make a watch including the rack-striking mechanism he quickly produced one to his own design and claimed priority of invention. Whatever the merits of either case it was Quare, with the backing of the Clockmakers' Company, who eventually received the credit by the decision of James II who tried both watches. His decision seems to have been made solely as a result of Quare's watch having one push piece while Barlow's required separate buttons for hours and quarters.

130   The essential components of the hour-striking mechanism consist of a snail of twelve steps rotating with the hour hand, a wheel of twelve ratchet-shaped teeth to raise the hammer one blow for each hour and a rack to turn the ratchet wheel by the number of teeth appropriate to the hour indicated by the snail.

A snail of four steps, rotating with the minute hand, is needed for the quarters; also a further rack to determine the number of quarter blows to be struck.

In the rest position the hour rack is held clear of the hour snail by the set-up of the repeating train spring. The quarter rack is held clear of the quarter snail by the hour rack.

In early watches the mechanism is wound by depressing the pendent. Later watches have a slide in the band of the case.

The ratchet wheel is fitted to the arbor of the repeating train spring and coupled to the hour rack either by gear teeth or a chain passing over a pulley. Depressing the pendent will pivot the hour rack into contact with the hour snail and turn the appropriate number of ratchet wheel teeth past the point of the hammer tail; because the spring arbor turns with the ratchet wheel, the spring will now be wound. Releasing the pendent will allow the spring to unwind and return the ratchet wheel. Each tooth of the wheel in passing will lift the hammer to strike one blow for each hour.

There are various methods of sounding the quarters. All require a quarter rack and the rack is always held clear of the quarter snail by the hour rack.

The earliest mechanisms used the quarter rack to lift the hammer tail and sound one blow for each quarter. Depressing the pendent pivoted the hour rack and turned the ratchet wheel as described. In this position the quarter rack is released by the hour rack to fall on to the quarter snail. Releasing the pendent will strike the hour as described. When the last blow is struck the hour rack will re-gather the quarter rack to lift it clear of the quarter snail. As it rises to the rest position, the hammer tail is lifted one blow for each quarter by teeth at the edge of the rack. A slight pause between the hours and quarters is arranged by suitable spacing of the quarter-rack lifting pin.

Tompion's predeliction for complicating simple mechanisms caused him to use two hammers to sound a double blow for each quarter. His quarter racks have two sets of teeth, one using the hour hammer after completion of the hour blows, and the other using the second, smaller hammer. The release of the small hammer is slightly delayed to produce the double blow, the first of which is louder than the second. The two sets of rack teeth are hinged to allow them to pass the hammer tail thus making the rack much thicker. To make additional room for it he moved the hour ratchet wheel from the front plate to a position between the plates and immediately above the repeating train spring.

Tompion's mechanisms are somewhat cumbersome both in design and execution and judging by the numbers seen with mutilated racks they must also have been unreliable in their action. His system of hinging the rack teeth causes the maximum possible friction in the passing action and requires a fierce return spring for the rack. This in turn necessitates a stronger repeating train spring which requires greater effort from the user in depressing the pendent and results in more wear to the mechanism in winding.

It was probably the considerable effort required to operate the mechanism fully that made necessary the all-or-nothing piece later applied to repeating watches. In principle this additional mechanism holds the hammer-tail pawl clear of the rack teeth until the pendent is fully depressed. If the pendent is not fully depressed the repeating train will unwind without lifting the hammers. In early watches this is achieved by a linkage coupling the hammer pawl to the hour rack tail. When the rack is fully down, the tail, in touching the appropriate step of the hour snail, will trip the linkage to release the hammer pawl. In later watches the quarter rack, in the rest position, holds back the hammer pawl. When the hour track is in contact with the hour snail, the linkage is tipped to allow the quarter rack to fall onto the quarter snail and so release the hammer pawl. It is sometimes said that Julien Le Roy was the inventor of this safeguard but this cannot be the case for it was employed in many Tompion watches in addition to other contemporary ones.

The use of a separate snail for the quarters allowed scope for additional steps to allow half-quarter or five-minute repeating. Whatever the number of steps the use of separate snails caused difficulty in ensuring, near the end of an hour, that the next hour was not struck before the quarter snail was in the correct position.

The first solution to this problem was used by Quare in a very early watch, No. 611, in the possession of the Ashmolean Museum, Oxford. A flirt piece is concentric with the highest step of the quarter snail. Projecting down from the flirt piece is a pin used to advance the hour snail by one step at each hour. As the minute hand approaches the hour, the pin contacts the star wheel of the hour snail, and the flirt piece is held behind the highest step of the quarter snail to allow the quarter rack to fall, if required, and strike the last quarter. As the quarter snail turns with the minute hand, the flirt carries the hour snail with it until the star wheel is suddenly advanced at the hour by a spring acting on the star wheel teeth. This sudden advancement instantaneously causes the following tooth of the star wheel to strike the flirt pin and advance the flirt to cover the lowest step of the quarter snail.

A later method, much favoured by English makers in the eighteenth century, relied upon correcting the position of the hour snail towards the end of the last quarter hour. The hour snail is driven by the cannon pinion and can revolve slightly on its carrying wheel which, with its star wheel, is held against a stop pin by a light spring. The quarter rack has an additional tail to correct the hour snail during the last quarter. It is only during the last quarter that the quarter rack can fall far enough for the correcting tail to touch the star wheel and turn it to ensure the correct step of the snail is presented to the hour rack. This system was used by Matthew Stogden who devised a most robust and reliable repeating mechanism based on Tompion's system. The principal difference lies in the quarter-rack mechanism. This is not used to raise the hammer tail but simply to switch off the quarters when a sufficient number of blows have been struck. Placed above the ratchet wheel for the hours is a further ratchet wheel for the quarters. The three teeth for sounding the quarters on the upper wheel are grouped together in a quarter of the circumference of the wheel. Depressing the pendent will turn the hour ratchet wheel until the number of teeth appropriate to the position of the hour snail have passed the hammer tail. Irrespective of the hour to be struck, all the teeth for the quarters will also pass the hammer tail. In unwinding, the hours will be struck as earlier described. The number of quarters to be struck will depend on the position of the quarter rack as determined by the quarter snail. If three quarters are to be struck then the rack will rest on the highest step of the snail and the hour rack will have further to travel before lifting the quarter rack to the rest position. During this travel the three teeth on the upper ratchet wheel will sound the quarters. If fewer than three blows are to be sounded the quarter rack will be on a lower step of the snail and will be gathered earlier. As the quarter snail is gathered, one of four teeth at its extremity will snatch the hammer tails out of the path of the ratchet wheel and no further blows will be struck as the mechanism unwinds. If the pendent is not fully depressed the all-or-nothing linkage will not allow the hammer tails to rise and no blows will be struck. This mechanism enjoyed great popularity with English makers throughout the eighteenth century. It can be used with one or two hammers and is easily adapted to half-quarter or five-minute repeating. In the mid-eighteenth century Mudge used it in minute-repeating form with an additional rack. It is probable he was the first to use this sub-division in a watch. Breguet used it in quarter- and minute-repeating form in his first series of perpetuelles and soon after developed it for ten-minute, half-ten-minute and five-minute repeating. His highest-grade repeaters produced in the early nineteenth century were a triumph of simplification of Stogden's system and completely reliable in their action. 145

For minute repeating the system is not very reliable in regular use. The engagement of the switching rack for the minutes is too shallow to allow for the effects of wear. The nineteenth-century Swiss system for minute repeating reverted to the earlier method of lifting the hammer tails by direct action of the quarter and minute racks and this is a more reliable method by reason of the slower and lighter action of the minute mechanism.

The earliest repeaters sounded on a bell in the back of the case and this system remained popular on the Continent until about 1780. Some English makers, seeking to make their large watches thinner, discarded the bell and arranged the hammers to tap on a block in the band of the case. Some of these, in common with bell repeaters, could be used silently by pressing a pulse piece in the band of the case.

In the last quarter of the eighteenth century some Continental watches had wire gongs. The earliest of these, particularly as devised by Breguet who was a frequent user of them, make a dull and unmusical sound. Later gongs, especially in Swiss watches, have a beautifully clear, bell-like note which is most gentle and pleasing. 212

All repeating mechanisms have a train of wheels to govern the speed of the striking. The earliest simply employ a sufficient number of wheels to create the necessary friction to slow the striking to the required speed. Later watches have a weighted pinion and manage with fewer wheels. The depth of the pinion with the last wheel in the train can be adjusted to control the speed. In about 1740 Julien Le Roy applied a tiny recoil escapement to the end of the train to reduce the number of wheels required. This system, much used by Breguet in his small thin repeaters, is very noisy and, combined with his pathetic gongs, makes some of his repeaters almost inaudible. In the late nineteenth century the Swiss introduced a silent centrifugal governor to control the speed and still further reduce the number of wheels. They seem however only to have been used in cheap watches produced in considerable quantity well into the twentieth century.

Clock-watches

A clock-watch is essentially a miniature striking clock sounding the hours in passing. It was inevitable that some of the earliest watches should have striking trains for they were derived directly from the spring-driven table clocks. The small striking clocks by Peter Henlein of the early sixteenth century are mentioned on p. 30.

An early example of a French clock-watch is that by Jacques de la Garde dated 58 1551. The earliest-known English examples date from about 1600.

The hours only are struck in early clock-watches which use the locking plate mechanism. An extra mainspring and train are required and this is released by a twelve-pointed star wheel attached to the hour hand beneath the dial. The star wheel lifts a catch to unlock the train which will run until re-locked by an extension of the catch falling into a notch on the locking plate. The locking plate has notches cut at increasing intervals at its edge to allow one extra blow for each release until, after a full turn, one o'clock is reached again. The locking plate is fitted to the back 106 plate of the watch where it is easily seen, and hour numerals at each notch allow a visual check when setting the hand. The hours are sounded on a bell in the back of the case. Watches of this type continued to be made especially by English makers until about 1700.

Later clock-watches employed the now well-developed rack-striking mechanism. This is much superior to the locking plate mechanism for the hands cannot become unsynchronised with the striking. A twelve-stepped snail is coupled to the hour hand and a toothed rack, released a few minutes before the hour, falls onto the appropriate step of the snail. When released by the minute hand precisely at the hour, the striking train gathers the rack one tooth for each blow of the hammer. When the last tooth is gathered the train is locked ready for the next release. This mechanism is easily adapted to sound the half-hours and quarter-hours by the addition of a four-stepped snail turning with the minute hand and a quarter rack. As with a striking clock the quarters are sounded before the hours. A further advantage of rack striking is that the racks can be released at will by pressing a button in the case, allowing the watch to be used as a repeater. The quarters are usually sounded one blow for the first, two for the second and three for the third. Some clock-watches will sound four blows for the last quarter in addition to striking the hour.

A late eighteenth-century Continental development of the clock-watch resulted in the grande and petite sonnerie mechanisms. When set to petite sonnerie the watch will sound the quarter and hours in passing, whilst when set to grande sonnerie it will sound both quarter and hour at each quarter.

Grande sonnerie watches are complex enough for most makers but Breguet deemed it necessary to add a minute repeating mechanism. His watches which have this additional complication sound the hours and quarters on two gongs, while depressing the pendant, and repeat hours, quarters and minutes on two additional gongs.

In the years following the French Revolution the Swiss produced many richly-decorated clock-watches for export to the East. Jacquet Droz produced fine enamelled examples with both going and striking trains automatically wound by pedometer 174 weights. These watches laid the foundation for the development in Switzerland of the very complicated watches that were, in the nineteenth century, to earn the Swiss their international reputation as master watchmakers.

135

With the advent of keyless-winding in the mid-nineteenth century the two springs of the clock-watch were wound by turning the winding crown opposite ways for each spring. These later watches were sometimes made in minute repeating form with an additional rack for the minutes also gathered by the clock-watch mechanism. By this means the user was spared the effort required to wind the repeater with a slide and needed only to touch lightly a button set in the winding crown. This is in fact a considerable advantage as anyone who has tried, in a drowsy condition, to work the later repeating slides with one hand will know. The ébauches for these watches represent the final development of the clock-watch. They were made in Switzerland and finished by individual workshops with their own escapements, cases and dials, etc.

Production of clock-watches virtually ceased in the first quarter of the twentieth century, probably due to their very high cost.

Self-winding watches

The first practical realisation of the self-winding watch is attributed to A. L. Perrelet in the 1770s. Earlier attempts are mentioned in eighteenth- and nineteenth-century writings but none survive and they are described as unsatisfactory and prone to rapid wear of the winding mechanism. Louis Recordon took out a patent for a self-winding watch in London in 1780 and it is known that he knew Perrelet and did business with him. A watch attributed to Perrelet and described in the *History of the Self-winding Watch* by Jaquet and Chapuis, has a winding weight pivoted at the centre of the movement. This would seem to be unsatisfactory for a pocket watch but the authors give a convincing account of its efficiency in winding.

Recordon's patent specifies a weight pivoted at the edge of the movement so that it oscillates while the wearer is walking. Both Recordon and Perrelet specified a fusee and chain, Perrelet maintaining the power with a differential in the fusee and Recordon with a Harrison maintaining ratchet. The Perrelet watch is extremely ingenious, for not only does it have a natural reduction in the winding ratio by means of the differential, it also winds through 360 degrees with the weight turning in either direction. The weights of these watches are locked on completion of the winding by the conventional pivoted steel piece raised by the fusee chain. Later watches by Recordon have a single going barrel.

In common with Recordon, Breguet also did business with Perrelet and it is probable that his interest in the self-winding watch stems from this association. The earliest of Breguet's self-winding watches, or perpetuelles as he called them, is No. $8\frac{10}{83}$, in the collection of the Clockmaker's Company. This also has a weight pivoted at the edge of the movement but in all other details is different from and quite superior to the Recordon watches.

As with all other aspects of Breguet's work the perpetuelles display his complete originality of thought and understanding of the requirements of a particular mechanism. His solution to the perfection of the system was no mere adaptation of a conventional watch but a radical and complete design that he never subsequently found necessary to improve. Because Breguet's perpetuelles were the foundation of his later successes and no other maker could improve upon them they are worth considering in detail.

170   His system employs two mainspring barrels wound together by a heavy platinum weight pivoted at the edge of the movement. The weight is supported mid-way between the upper and lower limits of its oscillation by a coiled spring. This spring is adjustable to ensure the weight is in the mid-position with one turn of winding of the mainsprings completed. This method of maintaining the equilibrium is virtually without friction and so the weight is responsive to every movement of the watch. When carried in the pocket the weight will oscillate vertically and, by means of a ratchet and pawl, turn the winding train to wind the mainsprings the four turns necessary to complete the winding. The ratio between the ratchet wheel and the barrel winding wheels is 24:1. Some twenty oscillations of the winding weight are required to complete one turn of the ratchet wheel. Thus the four turns of winding can be completed in about one mile of ordinary walking. The oscillation of the weight is assisted by two flexible springs at the edge of the case. When flexed to extreme these springs act as buffers to prevent damage through excessive exertions on the part of the wearer. When the mainsprings are fully wound a stop mechanism raises a bolt to lock the weight and prevent further winding. Each turn of the winding will allow fifteen hours' running and the completed four turns will give sixty hours' running as indicated by a hand and sector scale on the dial. Most common watches will achieve only eight or nine hours' running per turn of the mainspring but, with two springs, Breguet had ample power to raise the ratio by the addition of an intermediate wheel, increasing the running time. This was an important part of Breguet's design and ensured continuous running if the wearer was not very active or did not wear the watch every day. In the early examples there is no provision for manual winding for the watches were intended to run for eight years without cleaning and a winding hole would have allowed access to dirt falling into the movement. The escapements were lever and are described on pp. 125–26.

Breguet started manufacturing perpetuelles in the 1780s and said that both Marie Antoinette and the Duke of Orleans possessed examples in 1780. No. $2\frac{10}{82}$ is believed to be the earliest survivor and, in a nineteenth-century certificate, is described as a perpetuelle repeater. The earliest-known survivor is No. $8\frac{10}{83}$ which has a steel cylinder escapement and a plain balance without temperature compensation. A later example, No. $3\frac{4}{86}$, has a lever escapement and plain balance with compensation curb.

170   This watch is particularly interesting not only for its excellent lever escapement at such an early date but also for the English jewelling to the balance pivots. No. $2\frac{10}{82}$ has no jewelling. Although the art of jewelling was completely developed in England it is plain that this was not so on the Continent and Breguet was obliged to have his jewelling done outside his workshops.

The first thirty-one watches entered by Breguet in his 1787 sales books are all perpetuelles. The ébauches for these were made by Prudhomme in Switzerland. It would have been impossible to make all the ébauches in 1787, for this work involved making the plates, wheels, pinions and repeating work. The work must have been in hand for some years and the 1786 watch indicates how far advanced were Breguet's plans to produce the series. The ébauches cost an average of 300 francs each making a total of some 9,000 francs or about £2,300. This represented a considerable sum to Breguet and to help finance the work he entered into partnership with Xavier

Gide, a merchant horologist. This arrangement was not a happy one for Breguet, under pressure to produce increased quantities of commercial watches, was unable to complete the perpetuelles. With the dissolution of the partnership in 1791 Breguet finished and sold the first of the series. Two more were sold before he was forced to leave Paris in 1793, as a result of the effects of the Revolution, and many of the remainder were recorded as sold during the following six years. They were very expensive costing an average of 4,000 francs each.

Upon his return to Paris in 1795 Breguet turned his attention to simpler watches, probably because the perpetuelles were too expensive for post-revolutionary Parisians. Of the first series ten examples are known to survive. Watches finished during the Revolution have long pendents and white enamel dials while those finished after the Revolution have short pendents and engine-turned silver dials. Col. pls. VA and XVIA represent the two types. The escapements are lever with compensation balances with helical springs with terminal curves and regulators. At the visible edge of the movement is engraved 'Inventé et Perfectionné par Breguet à Paris'. Strictly speaking Breguet did not invent the perpetuelle but he most certainly re-invented it and staked his future on his expectations of its success. His confidence was completely justified and even today their performance in wear is all that the owner could wish.

With his introduction of thin watches in the early nineteenth century Breguet had made his perpetuelles appear old-fashioned and thick and he himself described them as 'perpetuelle ancienne'. He produced a new and thinner series in which the weight was below the top plate of the movement. For these the watch train and escapement are compressed to one side of the movement to make room for the weight. The stopwork is fitted to a separate wheel to reduce the height of the barrels. They included half-quarter repeating and late type lever escapements. A still later and sturdier series is illustrated in fig. 265 by the watch purchased by the first Duke of Wellington and is more suitable for use on horseback. Few other contemporary makers interested themselves in the perpetuelle and none achieved Breguet's admittedly expensive technical perfection. One notable maker was Jacquet Droz who produced a series of ornate clock-watches with both going and striking trains wound by oscillating weights.

A later attempt to commercialise the self-winding principle was made by the Viennese maker A. von Loehr. In 1878 he patented a rather crude method similar in design to the earlier Recordon, using one going barrel. The watches run for fifty hours when fully wound but the gear ratio of the simple winding train is too low and the wearer must keep fairly fit to be able to put himself to the necessary exertion to ensure complete reliability.

In the 1920s Messrs Le Roy of Paris produced a series of very thin self-winding pocket watches and some wrist watches. Both types employed a weight pivoted at the edge of the movement, which is unsuited to a wrist watch.

The self-winding mechanism for wrist watches was not taken seriously until 1924 when John Harwood patented his self-winding wrist watch. Like Breguet, Harwood saw the merit of keeping the mainspring almost continuously fully-wound to maintain a more constant power, and the elimination of manual winding to make the case dust- and moisture-resistant. Fig. 375 illustrates his system, employing a centrally-pivoted weight winding for part of one direction of rotation. There is no stop-work to lock the weight when winding is completed. Instead a friction clutch is used to turn the winding train. When the spring is fully wound the clutch allows the weight to oscillate without turning the winding train. This system was not wholly successful and could cause the watch to gain if the clutch were too fierce and tugged at the wound spring. There is no handsetting crown as with conventional wrist watches. The hands are set by rotating the knurled bezel after which the bezel is turned in the reverse direction until a red spot appears in a small hole in the dial. This disengages the mechanism from the internal teeth of the bezel. One of these watches is known to have been in use for twenty years during which time it has been cleaned only three times and has never failed its owner. The watches were not a great success with the public, partly because the slipping clutch could be unreliable, but principally because people found it difficult to accept a watch without a winding crown.

During the past thirty years Swiss makers have perfected the self-winding wrist watch so that it is completely reliable even though the wearer may live a most sedentary life. In their winding mechanism they are almost indistinguishable from A. L. Perrelet's work in that they use a centrally-pivoted weight winding in both directions of rotation through 360 degrees. Breguet described Perrelet as a good and gifted man. It is indeed a gifted man who could so completely anticipate the conclusion of two hundred years of development of the self-winding mechanism.

Stop-watches and chronographs

The centre seconds hand for watches coincided with the introduction of the cylinder escapement in the second quarter of the eighteenth century. It was probably intended to boost the accuracy of the new watches and no doubt the owners were soon disillusioned by its daily indications. To assist in the necessary daily correction of the hand a stop lever was added so that the balance could be stopped and re-started to coincide with a superior timekeeper. The early stop lever was excessively simple and merely pushed a piece of wire into the wheel teeth to stop the train. While this method would be suitable for a verge escapement, which is self-starting, it is quite unsuited to a cylinder escapement which needs to have the balance started in motion by a shake. Some later examples overcame this deficiency by extending the wire to push the cylinder banking pin when the lever was operated. These can be used for measuring short intervals of time or competitive events but not without the inconvenience of first opening the bezel to reach the stop lever.

The use of a centre seconds hand that can be stopped and re-started without stopping the balance is attributed to Pouzait. He used a separate train and spring for the seconds hand which advanced by whole seconds. Fitted to the end of the escape wheel pinion is a small castellated nut which allows a fly in the seconds train to pass through the castellations as the nut revolves. With a balance beating 18,000 vibrations an hour, and fifteen teeth in the escape wheel, six castellations are needed to allow sixty revolutions of the fly in each minute. The hand is stopped by moving a lever to hold back the fly and prevent its rotation. The seconds are indicated on a subsidiary dial. This mechanism was at first too complicated for a watch and occupied too much space. By the middle of the nineteenth century however it was

taken up by the Swiss makers who produced many fine examples. These have a 322
modified release mechanism for the fly in the form of a six-pointed star fitted to the
escape pinion. This reduced the height of the mechanism but with the slight disadvantage of adding the power of the seconds train to the escape wheel with the
seconds hand running.

Breguet, doubtless attracted by the dignified advance of the hand marking whole
seconds, devised his own form but without the additional train and spring. He too used
the six-pointed star fitted to the escape pinion. This engages a sixty-ratchet-toothed
wheel, which is fitted to a thin arbor that passes through the centre pinion to carry a
centre seconds hand. The point of the star needs to advance the seconds wheel through 175
only a third of the distance between two ratchet teeth, the remainder being advanced by
a jump spring. The action is rapid and the hand appears to make a single advance
for each second. The ratchet wheel is very thin and can be flexed above the plane
of the star to prevent engagement and stop the hand. This system seems to have been
used only by Breguet, probably because of the difficulties of making the delicate
components. It is worth noting that he was probably the first to drill through the
centre pinion to drive the seconds hands from the back of the movement. Other
makers put the fourth wheel at the centre of the movement and indicated hours and
minutes on a subsidiary dial.

Later watches marking whole seconds by a centre hand have escapements especially
designed for the purpose. Pouzait, for example, used a crude·lever escapement with 196
a balance vibrating once per second. This is a most unsatisfactory method, for every
movement of the watch affects the period of vibration, with disastrous results for the
timekeeping. The Chinese duplex escapement has a balance vibrating four times per 10
second, but two are passing vibrations when the wheel does not advance and one of
the return vibrations simply transfers the locking for a very small, almost indiscernible,
advance of the wheel. The fourth vibration is for the impulse and advances the hand.
It need hardly be mentioned that these watches were not intended to be timekeepers.

None of the foregoing types are suitable for measuring the time of events and maybe
regarded as more amusing than useful.

Louis Fatton, an eminent pupil of Breguet, patented an ingenious seconds timer 233
with a hand that advanced continuously by fifths of a second and left an ink spot on
the dial when a button in the case was pressed. If necessary the watch could be
stopped with an additional button to enable the timing to start from zero. The
tip of the hand is in the form of an ink reservoir with a spring-mounted pen
immersed in the ink. When the button is pressed a mechanism in the watch is
released to draw the pen down to touch the dial momentarily. The action is
instantaneous and does not affect the going of the watch. With this system a series
of short intervals can be timed in rapid succession. Breguet was much impressed by
this mechanism and produced many examples. One in particular is worth mentioning,
for the radius of the pen is decreased for each of five revolutions to give a separate
circle of marks for each of six minutes.

At the same time Breguet made a small series of double centre seconds watches.
One gold hand is fitted to the fourth pinion and the second, of blued steel, to a clutch-
driven arbor passing through the hollow fourth pinion. The steel hand can be stopped
by pressing an additional button in the case, which arrests the balance. By this means
140   two intervals can be observed during a single event. The two hands can only be brought

together again by starting the steel hand at the moment it passes above the gold hand. Neither hand can be returned to zero.

The second half of the nineteenth century saw an increasing demand in industry and sport for a more convenient means of measuring time. The principal inconvenience of being unable to set the hand to zero was overcome by the use of a cam fitted to the seconds arbor and set to zero by a pivoted lever. In early examples the wheel for the seconds hand turned on a pipe surrounding the cannon pinion as in the earliest English centre seconds watches. In later watches the so-called chronograph mechanism is fitted to the back plate of the movement and the seconds wheel arbor passes through the hollow centre pinion, as earlier advocated by Breguet. The drive for the seconds is via an intermediate wheel driven by and concentrically pivoted with the fourth wheel. Pressing a button in the case pivots the intermediate wheel to engage the seconds wheel and start the hand. The hand is stopped by a second press of the button which disengages the intermediate wheel. A third press returns the hand to zero and brings the setting lever onto the cam. The seconds train wheels are cut with very fine triangular-shaped teeth to reduce disturbance of the seconds hand during engagement. The various levers required to start, stop and zero are raised and lowered by a castellated nut turned one tooth for each press of the button. An additional concentric hand for split seconds was often added by using a hollow seconds arbor with the additional arbor passing through to carry the extra hand. The two arbors are coupled by a spring fitted to the seconds wheel and a cam fitted to the split seconds arbor. In seeking the lowest part of the cam the spring will carry the cam with the revolving seconds arbor. Stopping the split seconds hand, by pressing a button to apply a brake, causes the spring to ride over the periphery of the cam as the principal seconds hand continues to turn. Stopping and zeroing the principal seconds hand will not disturb the position of the stationary split-seconds hand. A second press on the split-seconds button will release the brake to allow the spring of the principal seconds to seek the lowest part of the cam and zero the split seconds.

Although the teeth of the seconds train are very fine it is impossible to prevent occasional butting of the flanks during engagement. If the teeth of the intermediate wheel are not aligned exactly with the spaces of the seconds wheel, the hand will jump forward or backward depending on the amount of misalignment. Adjusting the depths of the wheels to turn the butting into idle movement of the intermediate wheel will only cause delay in the starting of the hand. In a well-constructed and finely-adjusted chronograph mechanism the jumping of the hand is occasional and represents about one tenth of a second for a watch beating 18,000 vibrations per hour. Nevertheless it is a fault and cannot be eliminated. Occasional attempts have been made to engage the drive for the hand by a frictional clutch to eliminate the jumping caused by engaging moving wheel teeth. This system gives instantaneous and positive starting to the seconds hand but the relatively strong spring required to maintain the grip of the clutch against the inertia of the seconds hand causes excessive friction at the pivots. For measuring intervals in excess of one minute, chronographs are usually fitted with a minute recording hand automatically zeroed with the principal seconds hand. For long periods, such as twenty-four-hour races, the time-of-day dial can be used.

Stop-watches are used for timing short intervals. As with the chronograph the hand can be returned to zero but the watch is stopped at the same moment. To ensure instant re-starting the balance is given a twist by a spring lever. It might be supposed that this system would ensure more precise starting of the hand because it is not necessary to engage moving wheels as with the chronograph. In fact the error can be greater if the watch is stopped with the balance turning in the direction in which it will be re-started. If, for example, the balance is turning clockwise and is stopped after impulse is completed, the hand will not advance if the balance is re-started in the same direction. It will be one fifth of a second later before the hand can advance during the return impulse to the balance, and the final measured interval will be too long.

Modern requirements for measuring elapsed time are very demanding for events are often measured in hundredths of a second. For this reason electronic chronographs have developed rapidly and have completely outdated the conventional mechanical watch. The newest examples can measure in hundredths of a second for separate short intervals, at the same time as they record the overall length of the event to within a second or two in a day.

The authors rarely have occasion to measure the passage of time in fractions of a second but for those who are so obsessed with their day-to-day progress that they must measure it in hundredths of a second, they willingly concede the inhuman merit of the electronic chronograph. The user should however be sure to have a spare battery to hand if the slight defects of the mechanical chronograph are not to be finally proven superior to total failure of the electronics.

Monochrome plates

1   Verge.

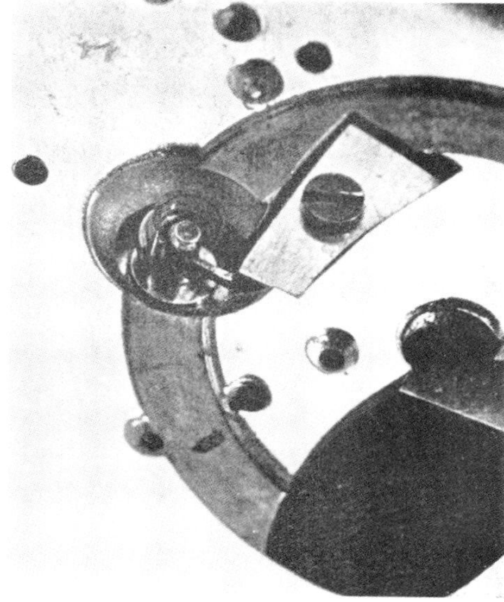

2   Breguet ruby cylinder.

3   Breguet ruby cylinder.

4   Graham cylinder.

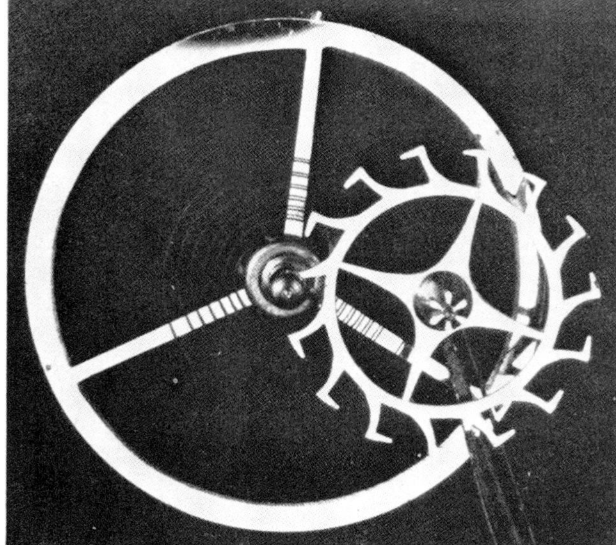

5   Flat wheel cylinder.

6   Virgule.

7   Double virgule.

8   Duplex.

9   Double-wheel duplex.

10   Chinese duplex.

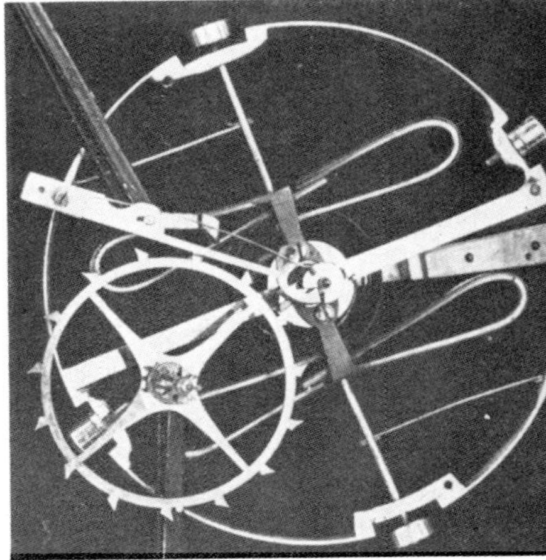

11   Arnold pivoted detent, see 172.

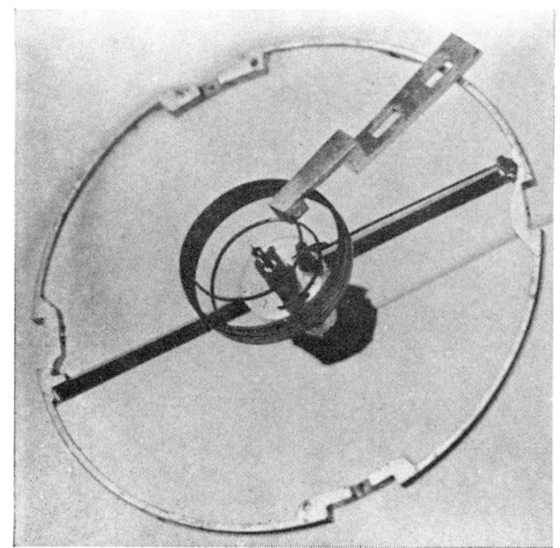

12   Arnold terminal curves, see 172.

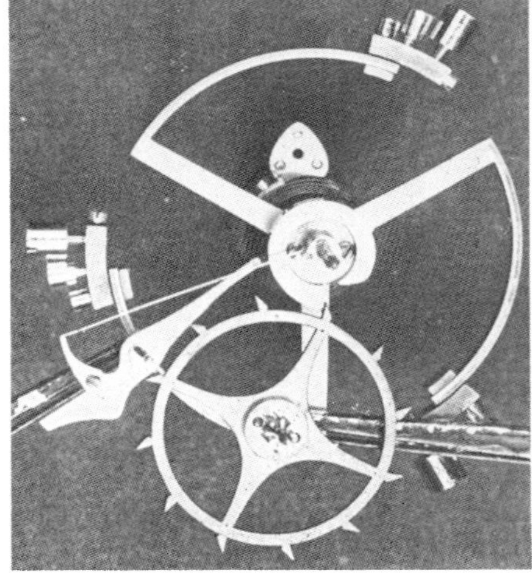

13   Berthoud pivoted detent, see 225.

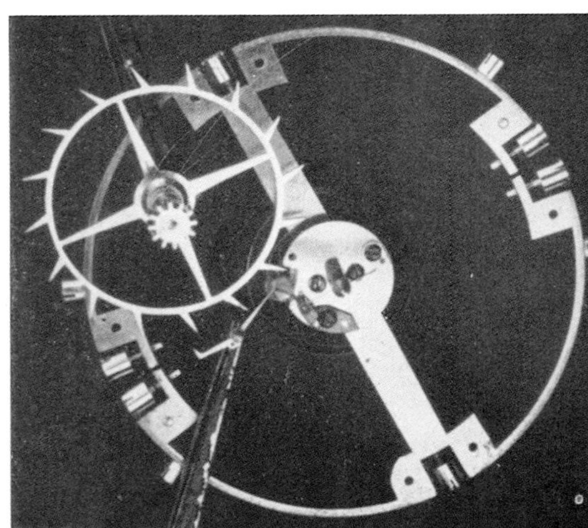

14   Breguet pivoted detent, see 237.

Breguet pivoted detent, see 171.

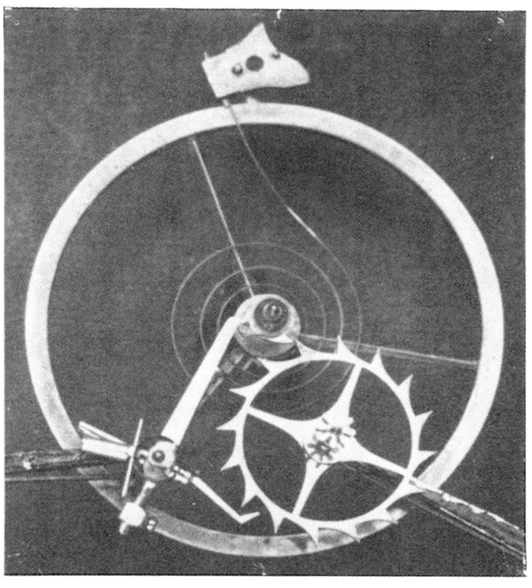

16   Kendal pivoted detent, see 169.

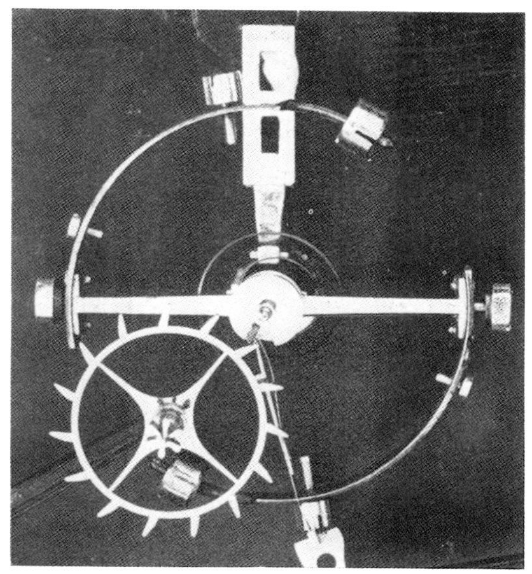

17   Arnold spring detent, see 159.

18   Earnshaw spring detent, see 240.

19   Frodsham spring detent, see 313.

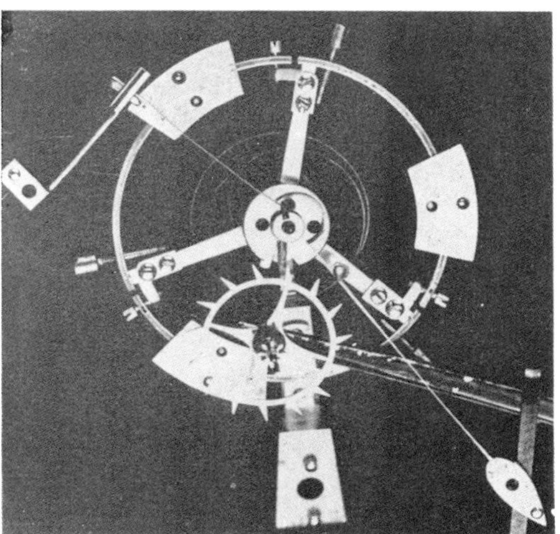

20   Peto cross detent, see 181.

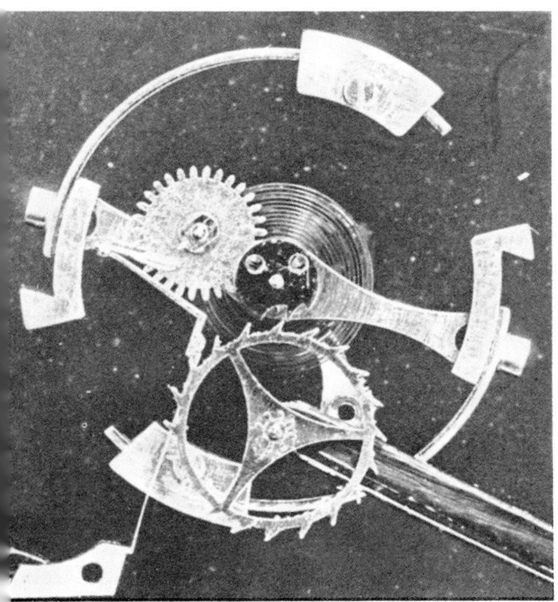

21   J. F. Cole double-rotary detached, see 196.

22   Emery lever, see 180.

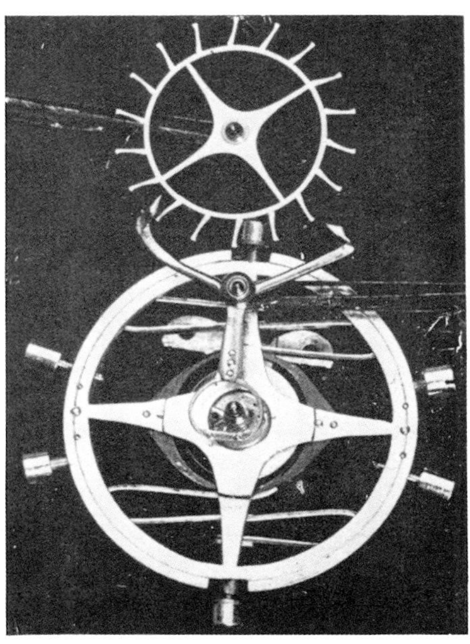

23   Pendleton lever, see 173.

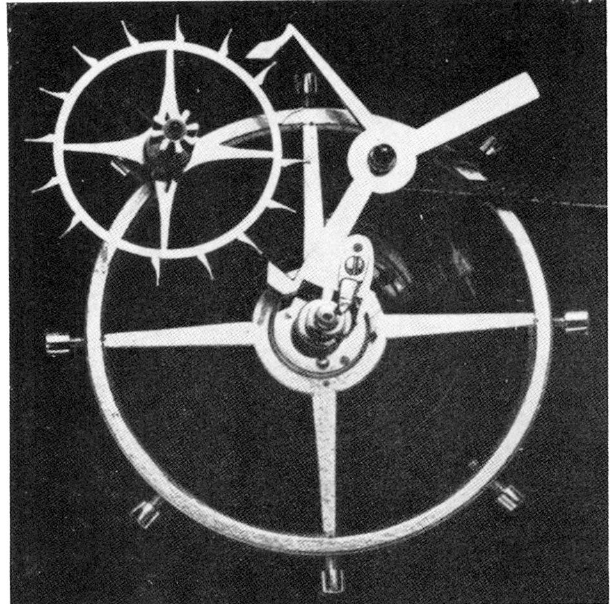

24   Perigal lever, see 167.

25   Grant lever, see 187.

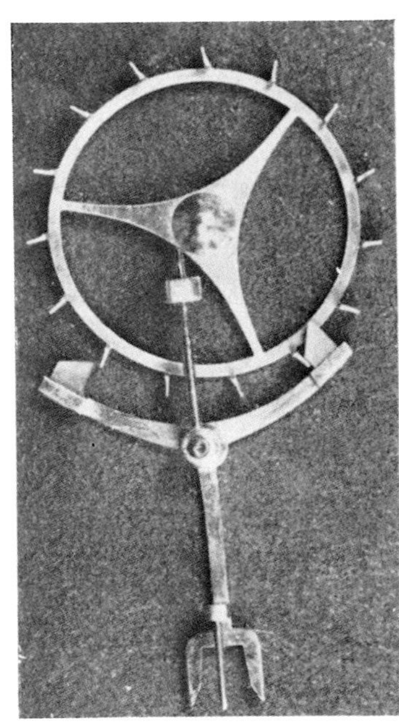

26   Grant lever, see 200.

27   Grant double-balance
wheel lever.

28   Leroux lever, see 162.

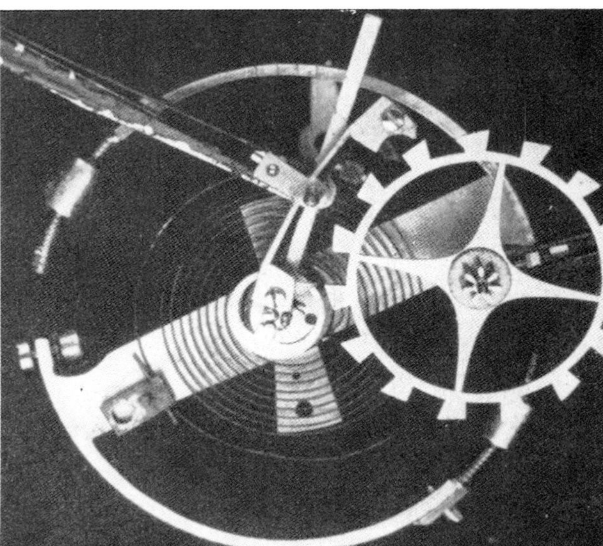

29   Taylor lever, see 184.

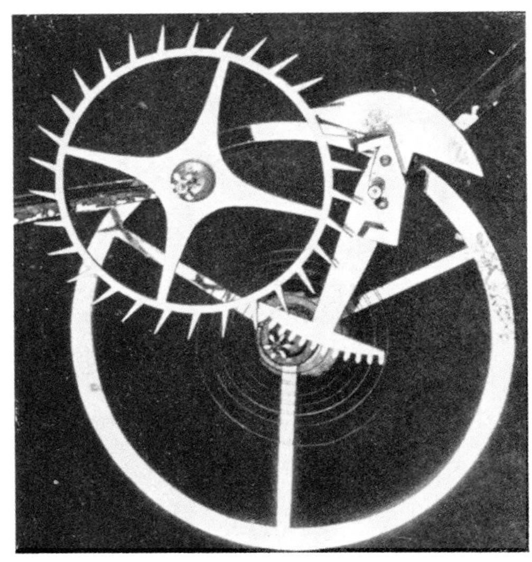

30   Rack lever, see 278.

31   Massey lever.

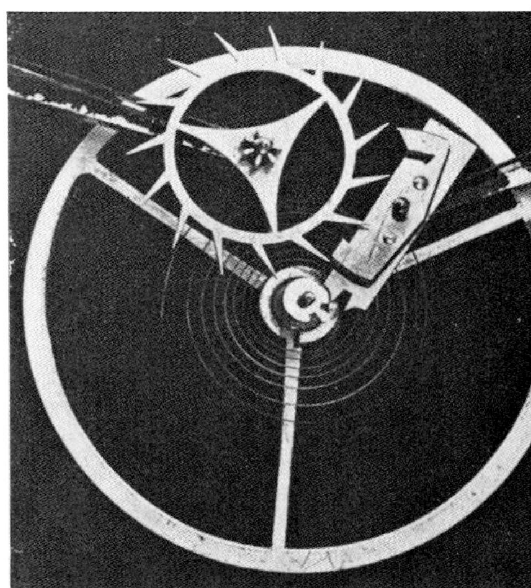

32   Massey lever.

33 Savage two-pin lever.

34 Early Breguet lever, see col. pl. XVIᴀ.

35 Breguet tourbillon lever, see 229.

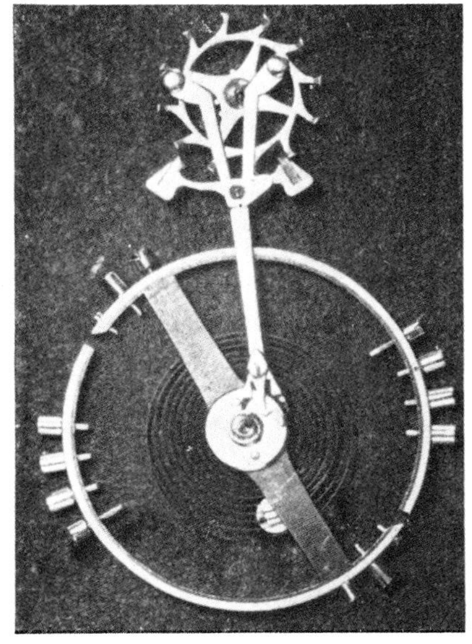

36 Breguet late lever, see col. pl. Vᴄ.

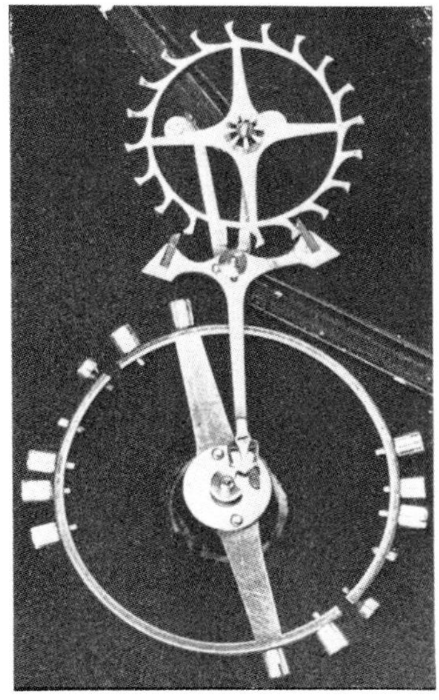

37 Breguet late lever, see 267.

38 Breguet pointed pivots.

39 Pouzait lever, see 196.

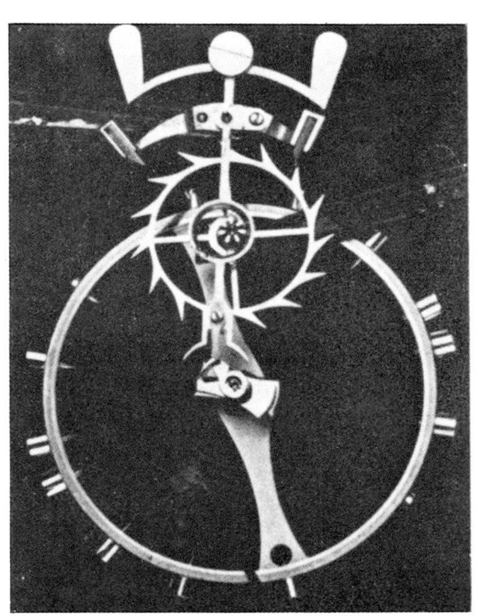

40 Late Robin, see 292.

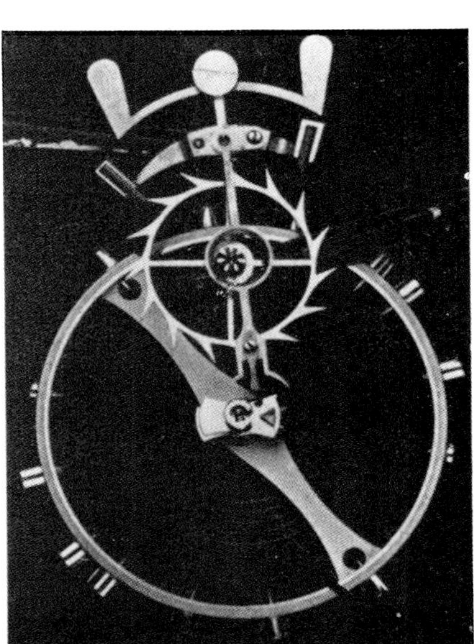

41 Late Robin, see 292.

42   Breguet Robin.

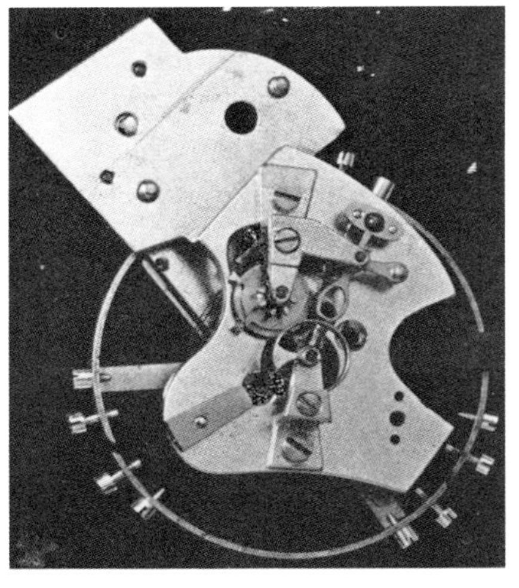

43   Double-wheel Robin, see 210.

44   Debaufre, see 195.

45   Sully, see 205.

46   Fasoldt, see 314.

47   J. F. Cole rotary detent, see 308.

48   Daniels' double-impulse chronometer,
see 387.

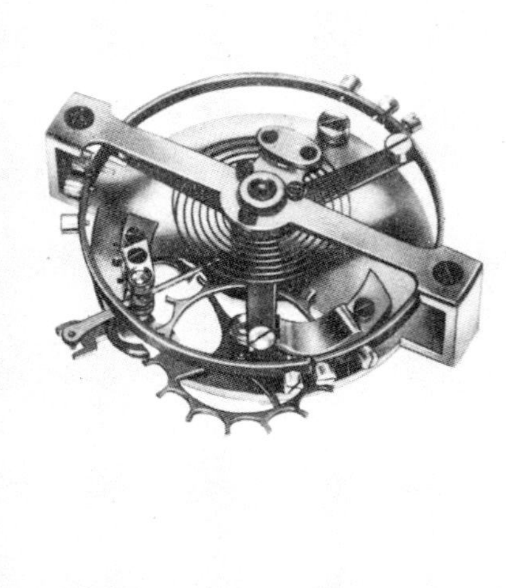

49   Breguet one-minute tourbillon
carriage with lever escapement, see 229.

50   Thomas Cummins balance wheel with
his form of Massey roller, see 266.

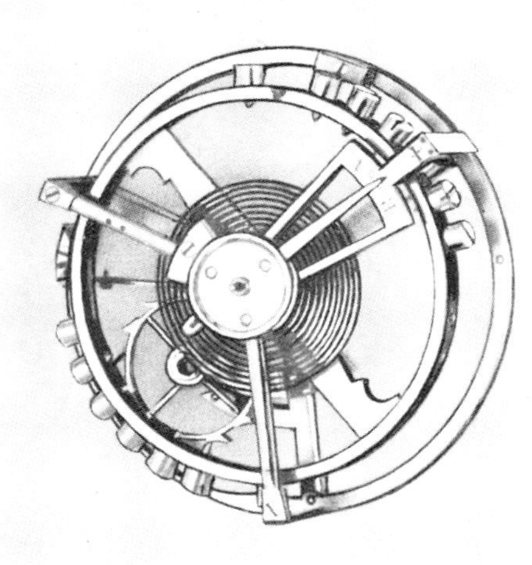

51   Albert Pellaton one-minute tourbillon carriage with spring detent escapement, see 343.

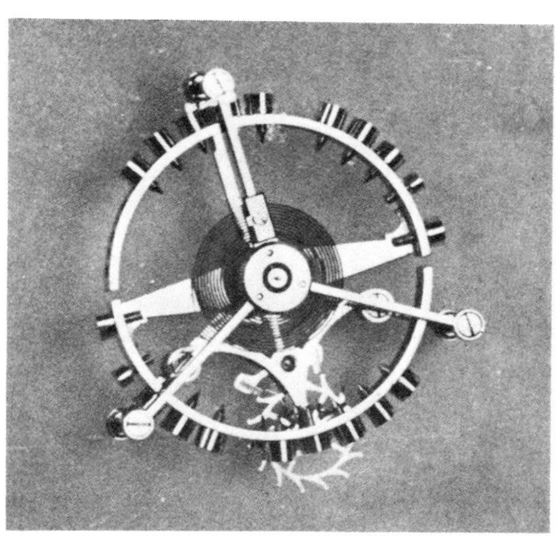

52   Sidney Better one-minute tourbillon carriage with lever escapement, see 376.

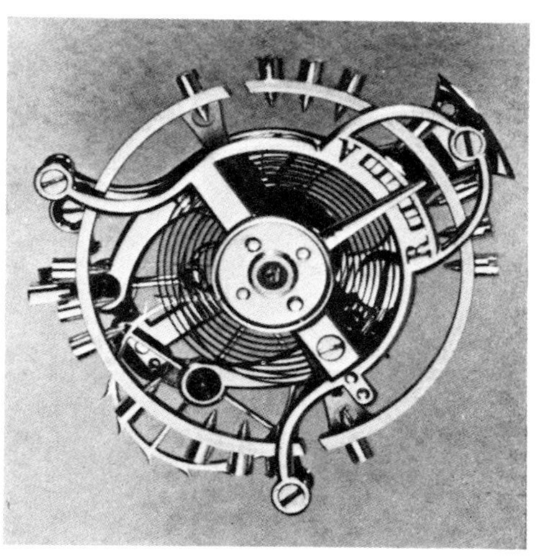

53   Girard-Perregaux one-minute tourbillon carriage with pivoted detent escapement.

54a   Daniels' tourbillon carriage, see 385.

54b   Daniels' tourbillon with remontoir, see 386.

55   Thomas Mudge's first lever escapement from his timekeeper of 1754 with his original system of engaging the balance diametrically opposite the lever staff. This is the Mudge marine clock in the British Museum, London.

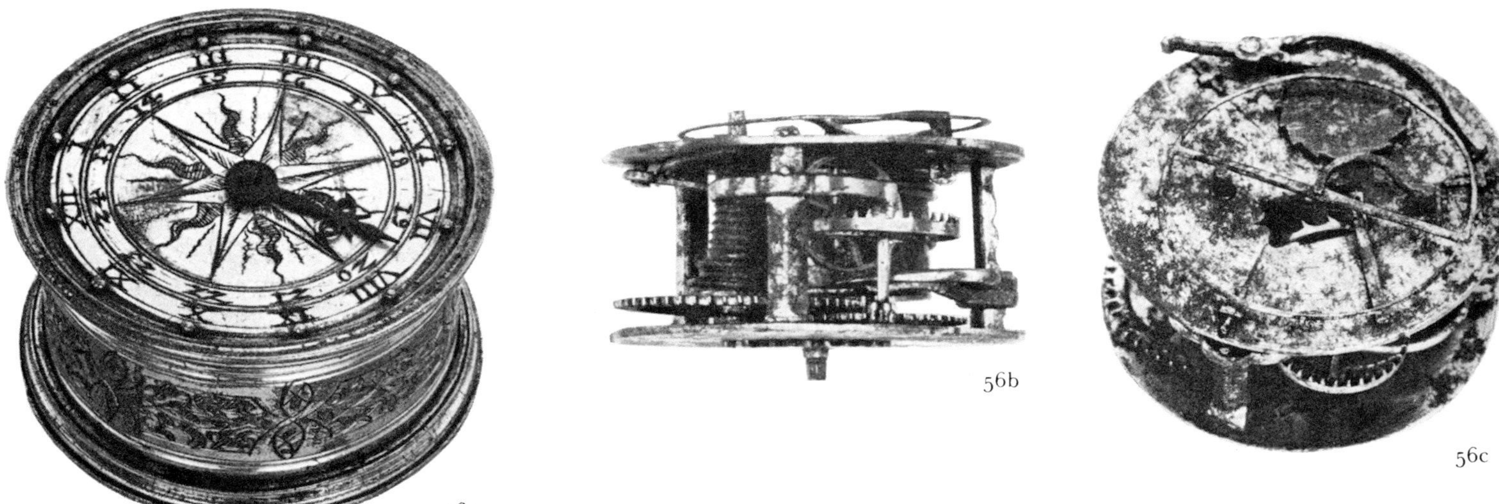

56a-c ANONYMOUS, probably German. Probably second quarter of the sixteenth century. Verge escapement without balance spring. Large, thin-rimmed iron balance wheel with thin flexible S-shaped balance cock. Iron train with long thin fusee and gut. Metal-gilt engraved drum-shaped case with 24-hour dial and iron hand, numbers 1 to 12 in Roman numerals, 13 to 24 in Arabic numerals. Touch pieces at the hours. The 2's are written as Z's. It may be that these early watches with fusees show Italian influence, unlike those with the more typically German stackfreed.

57a-b ANONYMOUS, probably German. Second quarter of the sixteenth century. Verge escapement without balance spring. Ratchet set-up regulator. Iron train. Long narrow fusee with gut. Drum-shaped metal-gilt and engraved case and dial with iron hand. 24-hour dial, numbers 1 to 12 in Roman numerals and 13 to 25 in Arabic numerals, the 2's as Z's, which is an indication of early date. The skeleton lay-out of the movement is similar to the large drum-shaped clock by Jacob Czech, dated 1525, belonging to the Society of Antiquaries of London and may therefore be of South German origin.

58a-c JACQUES DE LA GARDE, Loire, France. 1551. Verge escapement without balance spring. Set-up regulator. Fusee and gut. Hour-striking clock-watch. Case and dial engraved metal-gilt. This is the earliest dated watch. As opposed to all the German watches, the train is entirely of brass. Note the long narrow fusee typical of early French watches. It is believed that the winding key is original. The watch also has its original travelling case.

59a-e ANONYMOUS, German. Probably second quarter of the sixteenth century. Watch with alarum attachment. Verge escapement without balance spring. Bristle regulator. The train of both watch and alarum are of iron throughout. Stackfreed, unenclosed mainspring. The drum-shaped case and dial are of engraved brass-gilt. The dial has touch pieces at each hour and a single iron hand. The alarum has three spring feet by which it is attached to a rim on the top of the case of the watch. Depending from the alarum is the detent and the alarum is rotated until the detent is over the hour at which the alarum is to go off. The hand of the watch trips the detent as it reaches it.

60a-c ANONYMOUS, German. Third quarter of the sixteenth century. Verge escapement without balance spring. Bristle regulator. Stackfreed with uncased mainspring. Iron train with alarum. Brass-gilt case and dial. 24-hour dial, numbers 1 to 12 in Roman numerals, 13 to 24 in Arabic numerals. Movable alarum-setting dial in the centre. Single iron hand. The case pierced and engraved with apertures in the lid through which to see the hours. Note the S-shaped balance cock, making for easy adjustment of the depth of the escapement. The drum-shaped case indicates a date not later than the third quarter of the sixteenth century. Note also that 2's are engraved as Z's, indicative of an early date.

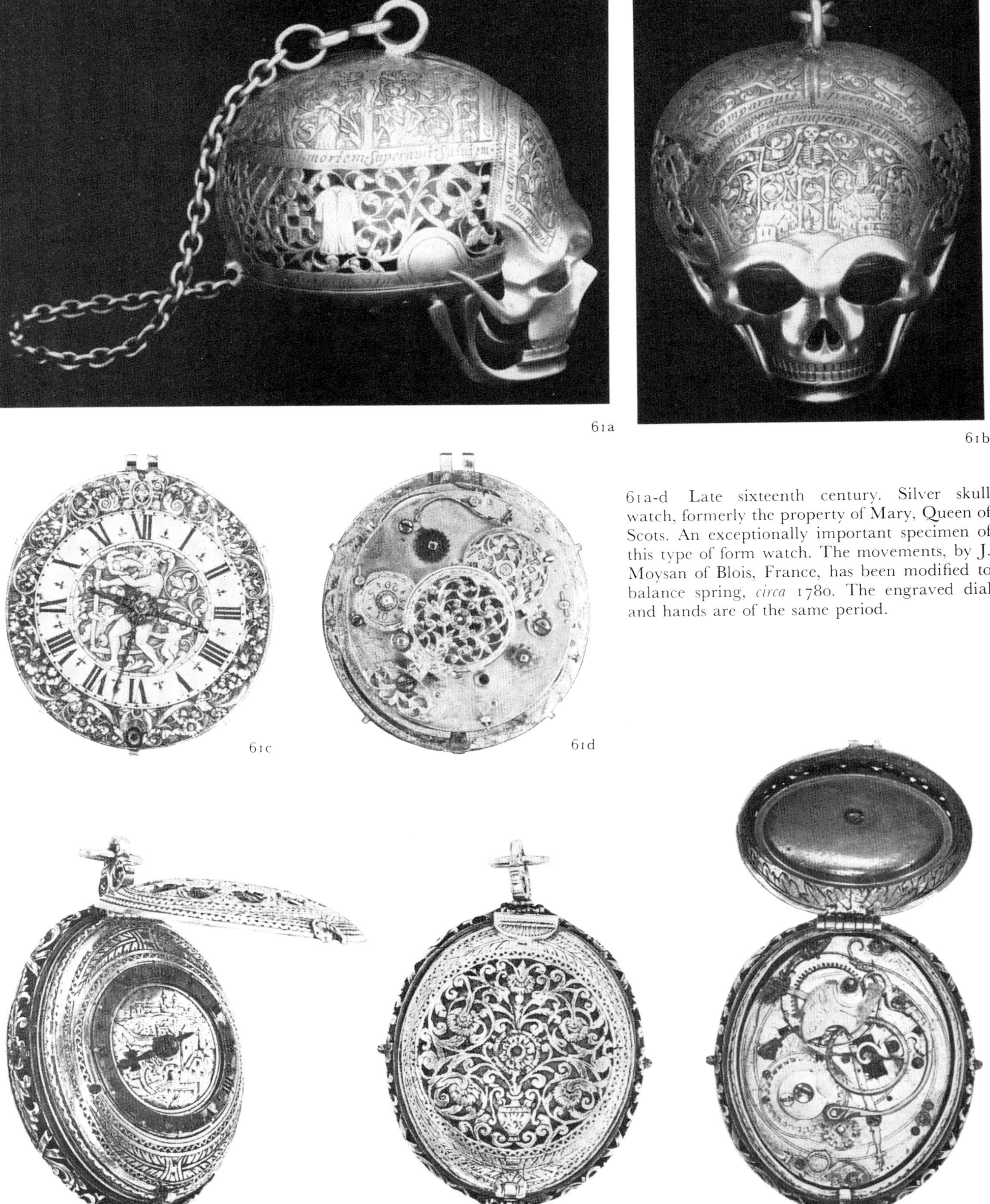

61a-d Late sixteenth century. Silver skull watch, formerly the property of Mary, Queen of Scots. An exceptionally important specimen of this type of form watch. The movements, by J. Moysan of Blois, France, has been modified to balance spring, *circa* 1780. The engraved dial and hands are of the same period.

62a-c ANONYMOUS, but bearing the initials A. R. in a shield; German. End of the sixteenth century. Verge escapement with bristle regulator. Stackfreed. Hour-striking clock-watch. The case which is engraved and pierced and the dial which has touch pieces at the hours are of metal-gilt. This is a typical watch of the end of the sixteenth century, at which time curved sides were just beginning to come into fashion. The hinged cover over the dial has apertures pierced in it for reading the hours.

63a-b P. CHAPPELLE, probably France. *Circa* 1600. Verge escapement without balance spring. Ratchet set-up regulator. Fusee and gut. The case and dial are of plain brass gilt. Single blued steel hand. This is an unknown maker, but from the style of the signature and the floreate design of the balance cock it is almost certainly French in origin. The movement does not hinge into the case but is held in by two spring clips disengaged by knobs emerging through the dial at 9 o'clock and 3 o'clock.

63b

63a

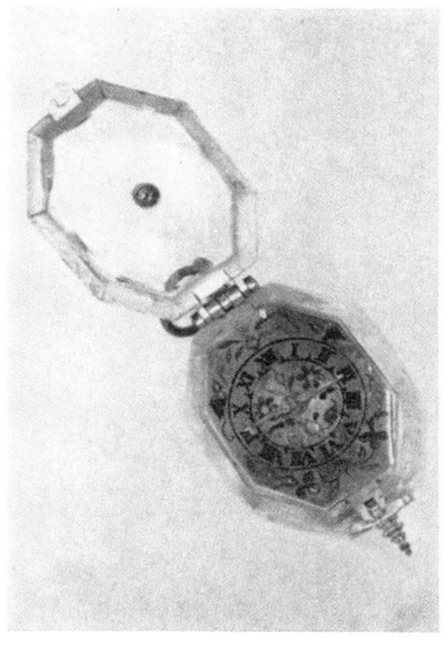

64a-b ANONYMOUS, German. *Circa* 1600. Verge escapement without spring. Stackfreed with bristle regulator. Champlevé silver and enamel dial with single gilt hand. Octagonal crystal case. This is an example of a very rare form of rock crystal case in which the movement is hinged direct to it, and not with a metal rim as is usual.

64a

64b

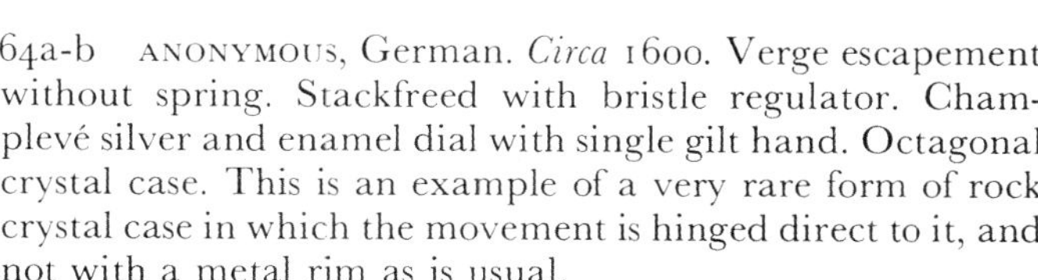

65a-c JACQUES BULKE, London, England. *Circa* 1600. Verge escapement without balance spring. Ratchet set-up regulator. Fusee and gut. Hour-striking clock-watch. The case, which is flat-sided and pierced, and the dial are of engraved gilt-metal. Single steel hand. Jacques Bulke is one of the earliest recorded British watchmakers and this must be one of the earliest surviving British watches. A characteristic English feature is the engraved decorative border round the edge of the watch plate, but in its brass train with fusee and almost all other mechanical and decorative features the watch shows strong French influence.

65a

65b

65c

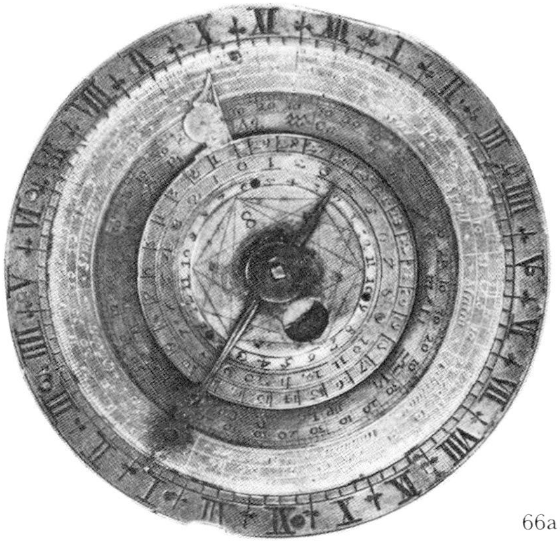

66a

66b

66c

66d

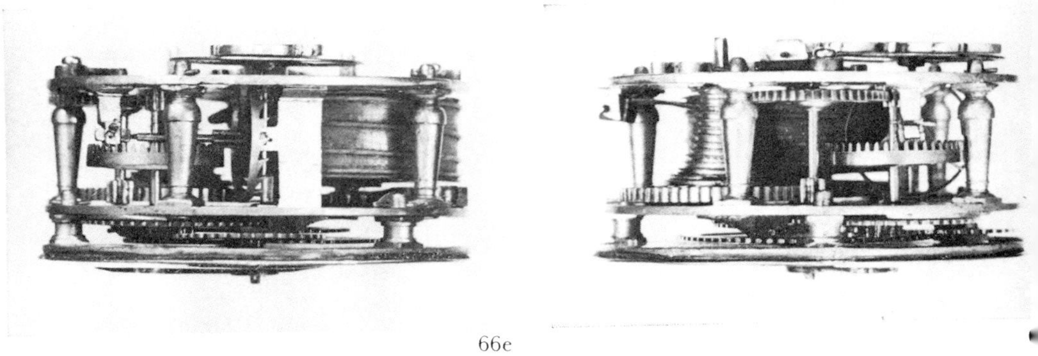

66e

66a-f   M. VALLIN, England, probably London. *Circa* 1615. Verge escapement. Plain balance. No balance spring. Fusee and gut. Gilt-metal and silver twenty-four hour dial with concentric circles for calendar, date of month, signs and degrees of zodiac, age of moon. Aperture for phases of moon. Single steel hand. Gilt-metal case with case band. The pinned-on cock of floreate pierced design and engraved decorative border of backplate are typical of early seventeenth-century English work. An unexplained feature of this watch is the signature 'M. Vallin' as the only and famous maker of this name in the early seventeenth century is N. Vallin.

67a

67b

67c

67a-c   C. CAMEEL, Strasbourg, Germany. *Circa* 1625. Verge escapement without balance spring. Ratchet set-up regulator. Fusee and gut. Plain silver dial and blued steel hand. Silver case. This is one of the more elaborate form watches of the early seventeenth century. By hinging back the legs and stomach of the bird the dial is revealed and the movement will also hinge through the same aperture.

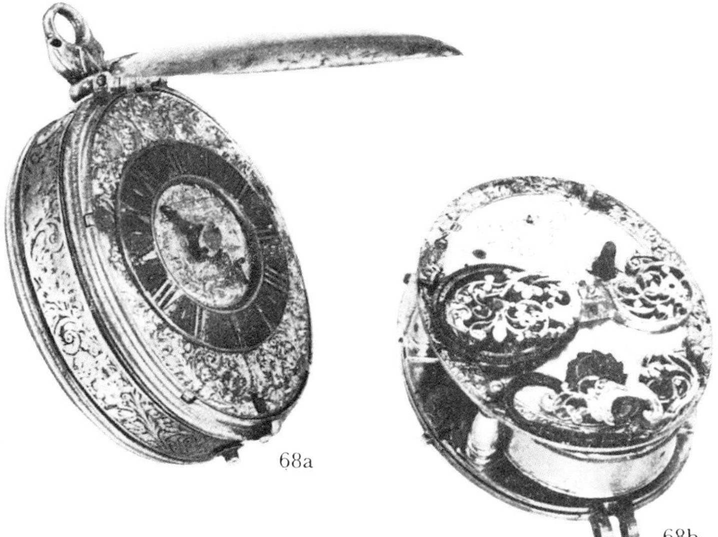

68a

68b

68a-b   ROBERT GRINKIN, London, England. First quarter of the seventeenth century. Verge escapement without balance spring. Ratchet set-up regulator. Fusee and gut. The case and dial plate of engraved gilt-metal, chapter ring of silver, blued steel hand. Note engraved border to watch plate, typical of early English work. Foliate design of balance cock showing French influence, pegged and pinned to the watch plate.

69a-b   ROBERT GRINKIN, London, England. First quarter of the seventeenth century. Verge escapement without balance spring. Ratchet set-up regulator. Fusee and gut. Engraved silver dial with single blued steel hand. Engraved silver oval case with solid dial cover. Note engraved border round the movement plate, typical of English work of this period. Balance cock fitted to a peg on the watch plate and pinned.

69a

69b

70a

70b

70c

70a-c   EDMUND BULL, London, England. *Circa* 1620. Verge escapement without balance spring. Worm and wheel set-up regulator. Fusee and gut. Silver engraved dial with single blued steel hand. Silver case of unusual design, being flat-sided and scalloped, reminiscent of the crystal cases of the period. The watch at present has a glass over the dial which is apparently original but may at first have been crystal held in by the earliest method of four tags bent over the edge of the glass. The signature is in a border round the edge of the dial, similar to the English engraved borders which by this time were beginning to go out of fashion.

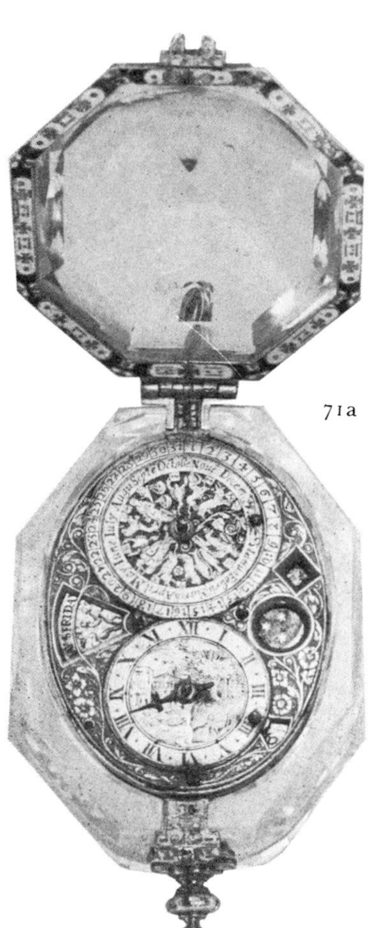

71a-c   SIMON BARTRAM, unrecorded, England. First quarter of the seventeenth century. Verge escapement without spring. Dial gilt with silver chapter rings for hours, month, date. Apertures for moon phases, day of the week and zodiac sign. Single steel hand. Rock crystal case with enamel rim to front cover, attached to back without metal rim. Note decorated edge to top plate and balance cock pinned to stud.

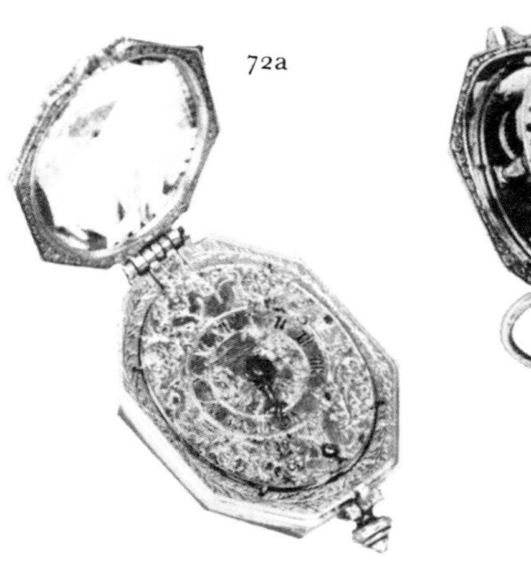

72a-b   DAVID RAMSAY SCOTUS [*sic*], England. First quarter of the seventeenth century. Verge escapement without spring. Ratchet-type set-up regulator. Fusee and gut. Engraved silver dial with single steel hand. Octagonal rock crystal case; the movement hinged to metal rims into which the crystal is fitted.

73a-b   DAVID RAMSAY, London, England. *Circa* 1625. Verge escapement. Fusee and chain. Engraved silver dial and steel hand. A typical crystal case with metal rims on to which the movement and cover are hinged.

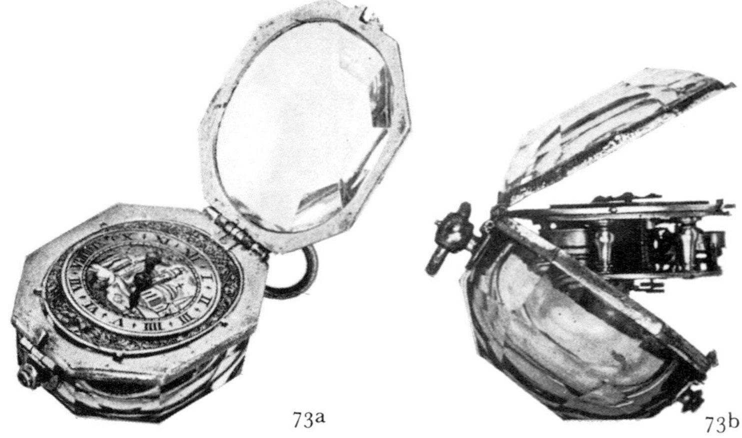

74a-c GERARDUS BAUER, Amsterdam, Holland. *Circa* 1625. Verge escapement without balance spring. Ratchet set-up regulator. Fusee and gut. Engraved silver dial and blued steel hand. Silver bud-shaped case chiselled in a chequer-board pattern.

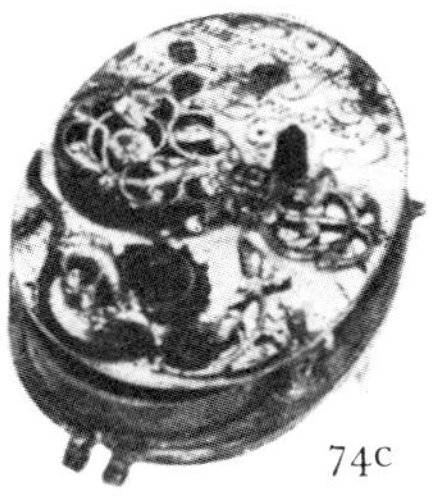

74a 74b 74c

75a-c DAVID RAMSAY, London, England. *Circa* 1625. Verge escapement without balance spring. Worm and wheel set-up regulator. Enamel dial painted in white with coloured flowers. Silver chapter ring. Blued steel hand. Faceted rock crystal case held in metal rims. Outer silver carrying case.

75a 75b 75c

76a 76b

76a-b JOHN SNOW, London, England. *Circa* 1630. Verge escapement without balance spring. Worm and wheel set-up regulator. Fusee and gut. Plain silver dial and blued steel hand. Plain silver case with crystal. This is probably original as it represents a very early method of fixing the crystal, which is held by a rim with tags which are turned alternately under the crystal and under the rim of the watch.

77a 77b

77a-b P. GREBAUVAL, Rouen, France. Second quarter of the seventeenth century. Verge escapement without balance spring. Decorative Continental-type ratchet set-up regulator. Fusee and gut. Engraved silver dial with single blued steel hand. Oval-shaped, engraved gilt case with solid front and back covers. The deep engraving is typical of French work as opposed to English engraving which was usually more shallow. The cover of the movement has on its inside a sundial and compass.

78a-d   DAVID RAMSAY SCOTTES [*sic*], England. *Circa* 1625. Signed. Verge escapement without balance spring. Ratchet set-up regulator. Fusee and gut. Balance cock pegged and pinned. The case and dial, which are in the form of a six-pointed star, are of silver and engraved all over. The single hand is of gold.

79a-c   JAMES VAUTROLLIER, London, England. *Circa* 1630. Verge escapement without balance spring. Ratchet set-up regulator. Fusee and gut. Engraved silver dial. Gilt hand. Engraved silver case. Note the circular window: from the engraving of the case it is evident that there always was such an aperture which was probably originally filled with crystal. The present fixing of the bezel is probably later. The decoration of the movement with engraved band round the edge is typically English. Note also the outer travelling case and original key.

80a-c   A. SENEBIER, Geneva, Switzerland. *Circa* 1630. Verge escapement without balance spring. Ratchet-type set-up with unusually large decorative click. Fusee and gut. Engraved gilt dial with silver chapter ring and steel hand. Eight-lobed crystal case with gilt engraved rims. The watch also has an outer case of wood covered with fish skin. Note the unusual rectangular-shaped cock foot, pegged and pinned to the watch plate.

81a-b   JOHN MICASIUS, London, England. *Circa* 1630. Verge escapement without balance spring. Ratchet set-up regulator. Fusee and gut. Engraved gilt dial with silver chapter ring and steel hand. Octagonal crystal case with gilt rims.

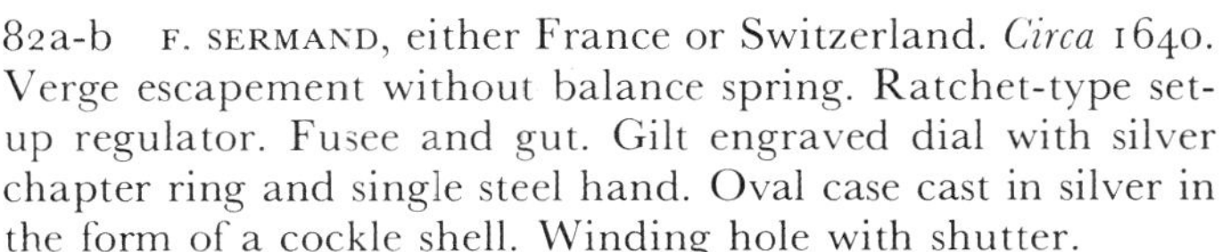

82a-b   F. SERMAND, either France or Switzerland. *Circa* 1640. Verge escapement without balance spring. Ratchet-type set-up regulator. Fusee and gut. Gilt engraved dial with silver chapter ring and single steel hand. Oval case cast in silver in the form of a cockle shell. Winding hole with shutter.

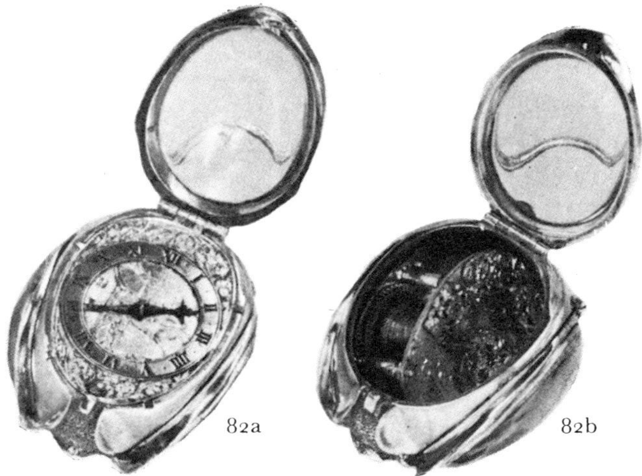

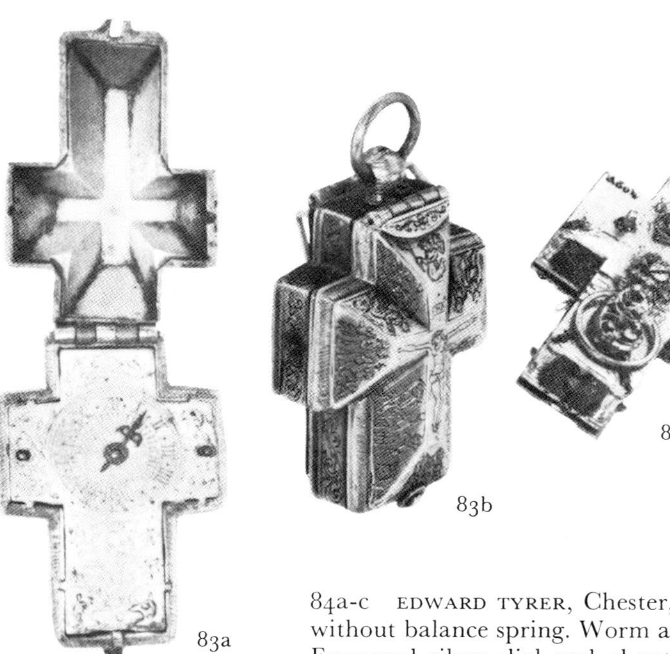

83a-c GEORGE COIQUE, France. Second quarter of the seventeenth century. Verge escapement. Plain balance wheel probably slightly later, without balance spring. Ratchet-type set-up regulator. Fusee and gut. Engraved silver case with gilt dial plate and silver chapter ring. Single blued steel hand, possibly of later date. A typical form watch of the type popular in the second quarter of the seventeenth century.

83c

83b

83a

84a-c EDWARD TYRER, Chester, England. *Circa* 1630. Verge escapement without balance spring. Worm and wheel set-up regulator. Fusee and gut. Engraved silver dial and chapter ring. Single blued steel hand. Single silver case of oval shape with hinged cover to dial. The movement hinges out of the case. Note the engraved band of decoration round the movement plate typical of English practice at this time; the balance cock is pegged and pinned to the plate. This is a fairly early specimen of the English 'Puritan' watch.

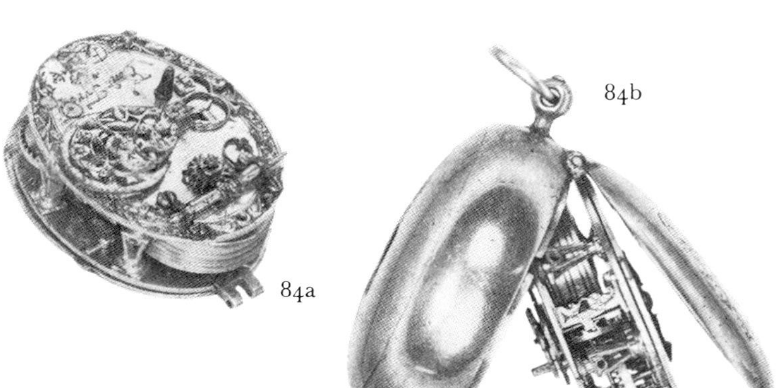

84b

84a

84c

85a-c BENJAMIN HILL, London, England. Mid-seventeenth century. Verge escapement without balance spring. Worm and wheel set-up regulator. Balance cock attached to plate by screw. Fusee and gut. Silver dial with subsidiary dials for calendar and astronomical work. Blued steel hands. Split bezel for glass. Pair case, the outer case covered in leather decorated with silver pinwork.

85a

85b

85c

86a 86b 86c

86a-c EDWARD EAST, London, England. Second quarter of the seventeenth century. Verge escapement without balance spring. Worm and wheel set-up type regulator. Fusee and gut. Engraved silver dial with single blued steel hand. Silvered pair case, the outer case decorated with leather and silver pinwork. Note cock with foliate engraving and piercing and irregular-shaped foot typical of this period, with a late example of fitting the cock to a peg on the watch plate and fixing with a pin. The thinness of the movement is typical of the half-century before the introduction of the balance spring.

87a 87b 87c

87a-c DANIEL FLETCHER, London, England. Mid-seventeenth century. Verge escapement without balance spring. Worm and wheel set-up type regulator. Fusee and gut. Engraved silver dial, the centre part pierced over a gilt background. Single blued steel hour hand; fixed calendar chapter ring with moving annular ring between hour and date chapter rings, with gilt pointer for the date, seen in the illustration pointing to the fourteenth day. Silver pair case, the outer case covered in black leather decorated with silver pinwork. Note irregular shaped cock foot and an early example of fixing by a screw. Foliate-type decoration of the cock plate typical of mid-seventeenth century English design. This is a very early example of what was to become the typical English pair case.

88a-b SAM ASPINWALL, England, Early seventeenth century. Verge escapement without balance spring. Ratchet-type set-up regulator. Fusee and gut. Plain silver dial with touch pieces on gilt background; silver blued steel hand. Cast silver oval case. Note the typical balance cock, pegged and pinned.

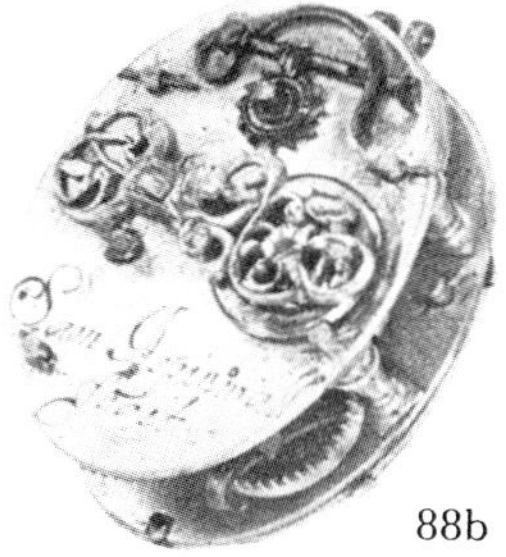

88a 88b

89a

89b

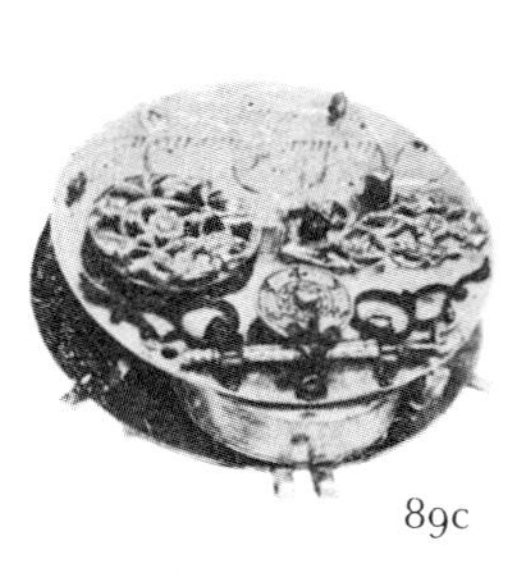

89c

89a-c NICOLAUS RUGENDAS, Augsburg, Germany. 1650. Verge escapement without balance spring. Worm and wheel set-up regulator. Fusee and gut. The case and dial, except for the chapter ring which is of white enamel, are chiselled radially to produce an effect which, 150 years later, was commonly produced by engine-turning, and are covered by a translucent rust-coloured enamel. Single gilt hand.

90a-c GOULLONS, Paris, France. *Circa* 1650. Verge escapement without balance spring. Worm and wheel set-up regulator. Fusee and gut. Enamel chapter ring with steel hand. Gold case enamelled with white flowers and foliage in relief with black markings.

90a

90b

90c

91a

91b

91c

91a-c BENJAMIN HILL, London, England. *Circa* 1650. Verge escapement without balance spring. Worm and wheel set-up regulator. Fusee and gut. Silver dial with matt centre and single steel hand. Silver cast and chiselled case in the form of a pomegranate. Note unusually long oval-shaped cock foot attached to watch plate by a screw.

92a-c ESTIENNE HUBERT, Rouen, France. *Circa* 1660. Verge escapement without balance spring. Worm and wheel set-up regulator. Engraved silver dial and steel hand. Chiselled and engraved silver case. Revolving slide to cover winding hole in back of the single case. A typical French watch of the late pre-balance-spring era. It also has an outer case of leather with silver pinwork.

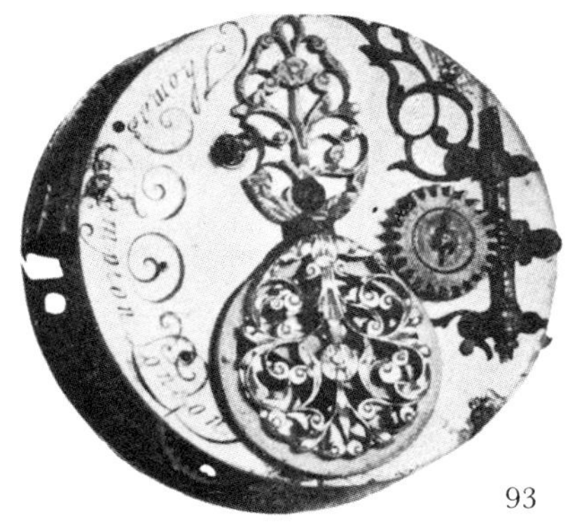

93 THOMAS TOMPION, London, England. *Circa* 1675. Verge escapement without balance spring. Worm and wheel set-up regulator (part of the decorative supporting hinges missing). Fusee and gut. This is a movement only and has no dial or case. It is the only known surviving specimen of a Tompion watch movement of the pre-balance spring era. Note the early type of cock with irregular edge to cock foot and plate, the cock screwed to the plate. The train is of high quality with no pinion of less than six leaves. Three-wheel train. 15-hour going period.

94a-c SEIGNIOR, London, England. *Circa* 1670. Verge escapement without balance spring. Worm and wheel set-up regulator. Fusee and chain. Matt gold champlevé dial. Single steel hand. Gold inner case, outer case covered in black leather decorated with gold pinwork. This almost mint watch is typical of the period just before the introduction of the balance spring.

95a

95b

95c

95a-c   IGNATIUS HUGGEFORD, London, England. *Circa* 1670. Verge escapement without balance spring. Worm and wheel set-up regulator. Fusee and chain. Silver dial, the hour numerals on heart-shaped plaques with a single steel hand surrounded by a revolving silver calendar ring. Silver pair case, the outer case covered in tortoiseshell decorated with silver pinwork. In 1704 this watch was used to oppose Facio de Duillier's application for a prolongation of his patent for jewelling. It was produced by the Clockmakers' Company to show that he had been anticipated in his invention, but in the nineteenth century it was found on examination that what appears to be a ruby end stone to the balance staff is purely decorative and does not operate as an end stone. The watch is additionally unusual in that the balance cock is made of blued steel.

96a

96b

96c

96a-c   MARTINOT, Paris, France. *Circa* 1670. Verge escapement without balance spring. Worm and wheel set-up regulator. Fusee and gut. Enamel dial with single steel hand. This is a very early example of a one-piece enamel dial. Gold pair case. The outer case is of gold filigree.

97a

97b

97c

97a-c   JEREMIE GREGORY, London, England. *Circa* 1670. Verge escapement without balance spring. Worm and wheel set-up regulator. Fusee and gut. Alarum. Silver dial and engraved centre and rotating alarum set dial. Single steel hand, pierced and engraved silver case and plain silver outer case probably of later date.

98a   98b   98c

98a-c   HENRY ARLAUD, London, England. *Circa* 1665. Verge escapement without balance spring. Fusee and gut. Gilt dial with silver chapter rings for age of moon hours and date. Aperture for moon phases, day of the week and zodiac sign. Silver case engraved with floral pattern and the words 'Richard Baillie at the Aibay'.

99a   99b   99c

99a-c   DANIEL CARRE, probably England. *Circa* 1670. Verge escapement. Plain steel balance, later converted to balance spring with later regulator. Fusee and gut. Engraved gilt dial with subsidiary dials for phases of the moon, month, day of the month, day of the week and hours. Pair case, the outer case covered in oxydised iron decorated with silver pinwork. Note the presence on the back plate of the worm and wheel set-up regulator belonging to the pre-balance spring period. The day of the month indicator is in French but the style of the watch is English, suggesting that it was made for the export market.

100a   100b   100c

100a-c   THOMAS TOMPION. London, England. *Circa* 1675-80. Verge escapement. Fusee and chain. Later enamel dial and steel beetle and poker hands. Silver pair case, the outer case decorated with tortoiseshell and silver inlay. This is a very early balance spring watch. It is unnumbered, indicating a date earlier than 1680 when Tompion started numbering his watches. Also, the foot of the balance cock has an irregular rim and there is no mask engraved at the juncture of the table and foot of the cock. All these are signs of the very early balance spring period.

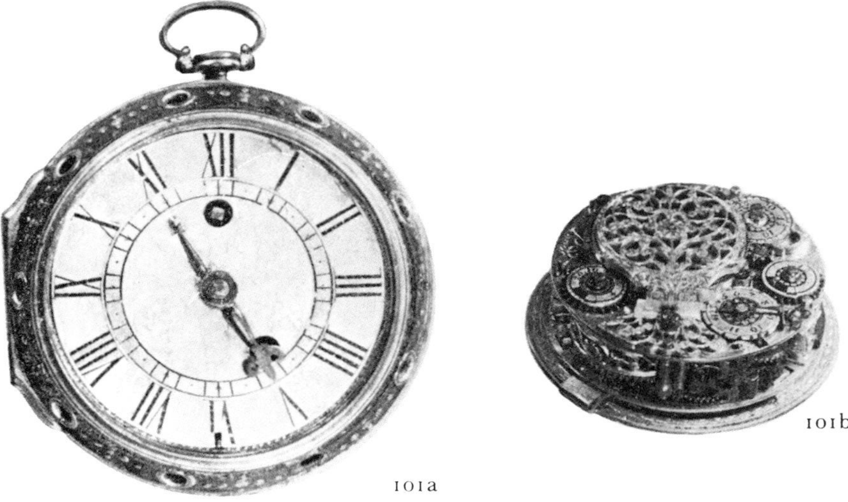

101a

101a-b THOMAS TOMPION, London, England. *Circa* 1676. Verge escapement. Steel balance wheel, spiral steel balance spring and regulator. Resting barrel. Clock-watch. The indicator dials on the watch plate are for regulator, stop-work, locking plate, up-and-down. Gold dial with engraved centre, single steel hand. Gilt outer case covered in leather. This is one of the earliest surviving Tompion watches and has no number. It shows that with the invention of the balance spring Tompion at first thought that the fusee was no longer necessary. Possibly with a single hand the resting barrel (similar in effect to a modern going barrel) may have been almost adequate. But with the introduction of a minute hand the necessity for continuing the fusee immediately became apparent.

101b

102a

102b

102c

102a-c HENRICUS JONES, London, England. 1675-80. Verge escapement. Fusee and chain. Silver dial with oval chapter ring for minute divisions, and expanding hand so that the pointer is always pointing to the outer edge of the oval-shaped chapter ring. Silver pair case, the outer case decorated with leather and silver pinwork. This must be one of the earliest surviving watches of the balance spring period. The regulator is of the Barrow type.

103a-b JONATHAN GROUNDS, London, England. *Circa* 1680. Verge escapement. Barrow-type regulator. Fusee and chain. Champlevé enamel dial with blued steel hands. Silver pair case. Note the very early type of balance cock without a mask at the juncture of the table and foot. The Barrow regulator is also indicative of the very early balance spring period. The signature on the dial is spelt Grounds, but that on the movement is spelt Growndes.

103a

103b

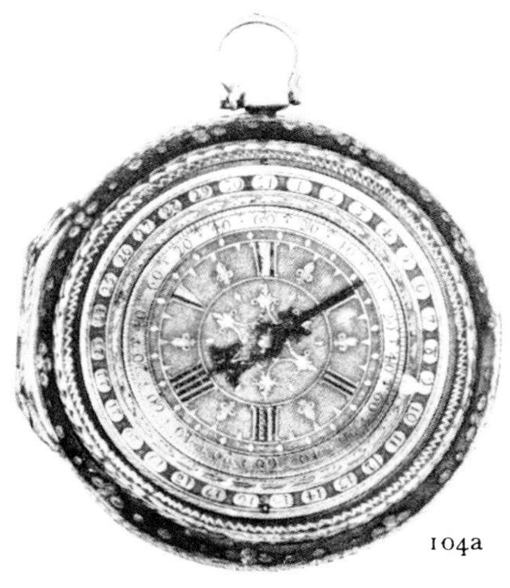

104a

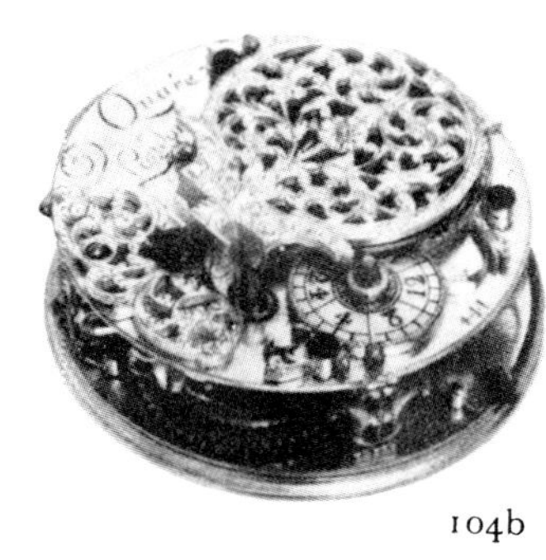

104b

104a-b DANIEL QUARE NO. 441. London, England. *Circa* 1680. Verge escapement. Spiral steel balance spring and regulator. Fusee and chain. Silver dial. An unusual combination of the six-hour dial with an outer calendar ring. Blued steel hand. Pair case, the outer case of green leather with gold pins which, judging from the curved ends of the hinge, is probably a little later than the inner case and movement. The watch is undoubtedly of very early balance-spring date, the balance cock which is of silver having an irregular-shaped foot and no mask engraved at the junction of the foot and plate. See also 108 for another Quare six-hour dial, Quare's watch No. 699.

105a

105b

105c

105a-c JOSEPH WINDMILLS, London, England. *Circa* 1680. Verge escapement. Fusee and chain. Champlevé silver dial with blued steel hand. Silver pair case, the outer case decorated with tortoiseshell inlaid with silver. Note irregular-shaped foot to the balance cock, typical of the earliest balance spring period.

106a

106b

106c

106a-c THOMAS TOMPION. No number. London, England. *Circa* 1680. Verge escapement. Plain steel balance. Spiral steel balance spring. Fusee and chain. Silver champlevé dial with central alarum disk and outer ring for dates of the month. Steel hands. Hour-striking clock-watch and alarum on a bell. Outer case of pierced silver. Inner case silver with engraved back and pierced band. The fact that this is an un-numbered balance spring watch indicates a date between 1675 and 1680. As well as the Tompion regulator dial the worm and wheel set-up regulator fitted to the top plate and normal to pre-balance spring watches, has been retained, indicating a date not long after 1675.

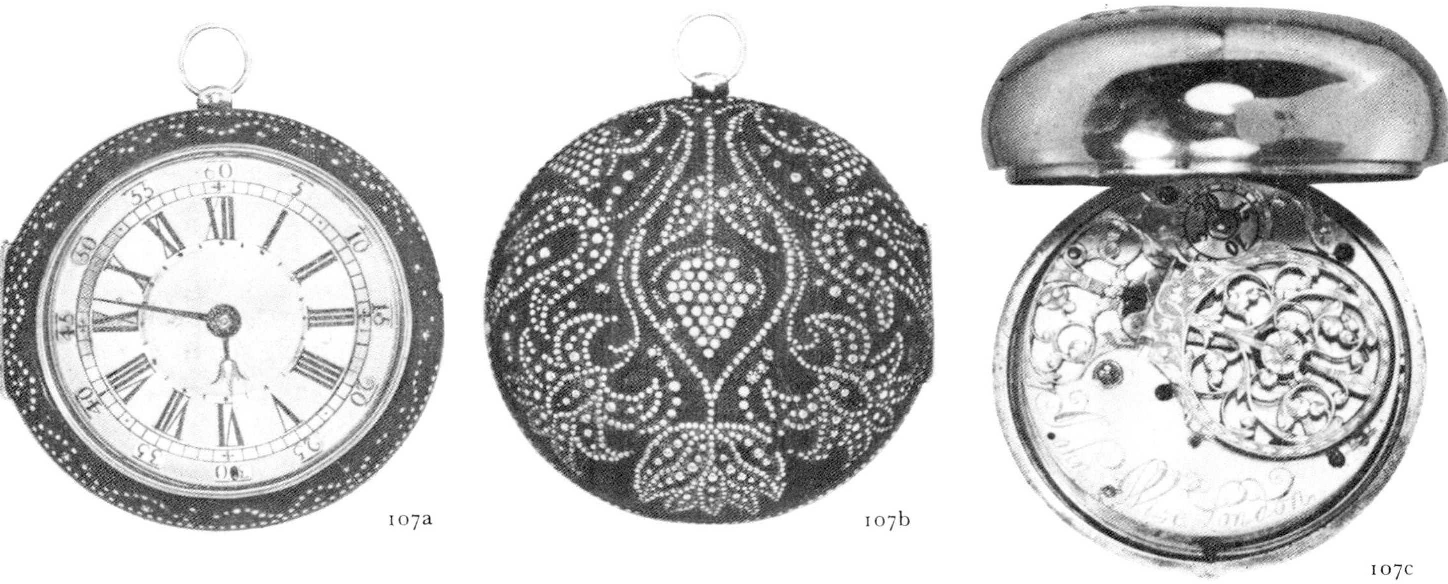

107a-c   JOHN WISE, London, England. *Circa* 1680. Verge escapement. Plain balance. Spiral steel balance spring of two turns. Fusee and chain. Silver champlevé dial. Blued steel tulip and needle hands. Silver inner case. Outer case leather with silver pin-work decoration. All the details of this watch, especially the unsigned dial centre, show it to be among the earliest balance spring watches.

108a-b   DANIEL QUARE NO. 699. London, England. *Circa* 1680. Verge escapement, three-armed steel balance with spiral steel spring and regulator, fusee and chain, silver champlevé dial. Roman numerals for the hours 1-6 with Arabic numerals 7-12 superimposed, outer circle divided into sixty minutes for each hour, single blued steel serpentine hand, plain silver pair cases.

109a-c   THOMAS TOMPION, London, England. *Circa* 1680. Verge escapement. Fusee and chain. Later enamel dial with blued steel beetle and poker hands. Pair case, the inner case gilt, the outer case covered in shagreen. This watch probably dates from the period about 1680 just before Tompion started numbering his products.

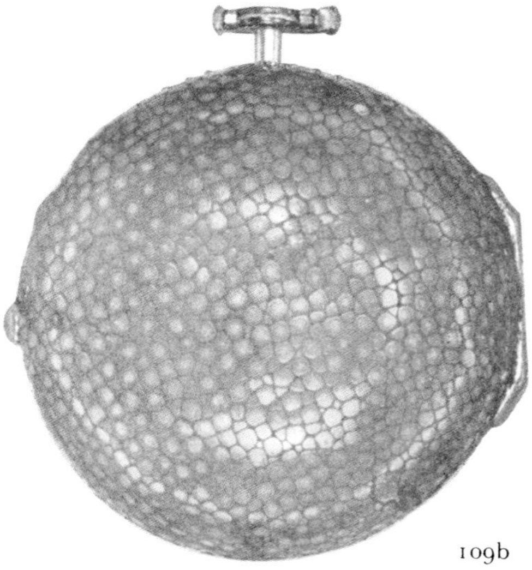

110a-b BENJAMIN BELL, London, England. *Circa* 1690. Verge escapement Fusee and chain. Champlevé dial with blued steel tulip and poker hands with subsidiary seconds dial. Plain silver pair case with square hinge. This watch, which is of very high quality, is of exceptionally large size and the presence of a seconds hand in the seventeenth century is unusual.

110b

110a

111a-b PETER GARON, London, England. *Circa* 1695. Spiral steel balance spring and regulator. Fusee and chain. Champlevé silver dial with blued steel tulip-type hour hand and poker minute hand. Heavily engraved silver pair case with square-ended hinge. This is an exceptionally large watch, 77 mm in diameter, and the movement is unusually decorative as to the cock foot, which is of unusually large size. The dial has a day-of-the-month aperture.

111a

111b

112a

112b

112c

112a-c GEORGE LYON, England. Probably last decade of the seventeenth century. Verge escapement. Fusee and chain. Silver 'sun and moon' dial with steel minutes hand and subsidiary seconds dial. Silver pair case, the outer case covered in tortoiseshell with silver inlay. For a discussion of the sun and moon dial see page 90. It was not unusual for the bottom part of these dials to be used for a seconds dial. The movement of this watch is highly decorative, even the rim of the contrate wheel being engraved.

113a   113b

113a-b FRANCISCO PAPILLION. Florence, Italy. Late seventeenth century. Verge escapement. Resting mainspring barrel without fusee. Alarum. Silver champlevé dial with rotating alarum set-dial in centre, single steel hand. Pierced silver case. Italian watches of this period are extremely rare. This watch is of fairly typical French lay-out, and it is rare but not unique to find such watches without a fusee. Note the irregular shape of the cock foot, typical of late seventeenth-century design, and stopwork above the plate to both going and alarum spring barrels.

114a-b HENRY POISSON, London, England. *Circa* 1700. Verge escapement with spiral steel balance spring and regulator. Fusee and chain. Engraved silver dial with eccentrically-mounted chapter ring. Subsidiary dial for regulator and mock pendulum swinging in aperture at the bottom of the dial. In this form of mock-pendulum watch with the pendulum showing through the dial, the back plate of the watch is usually engraved all over as in this example.

114a   114b

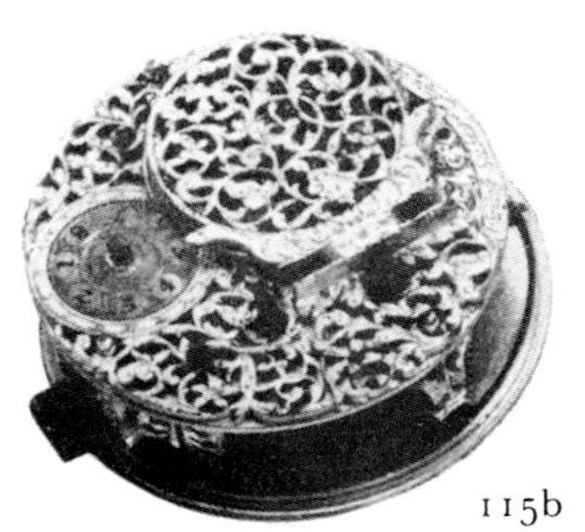

115a   115b

115a-b CHRISTOFF SCHÖENER, AUXBURG [*sic*], Germany. *Circa* 1700. Verge escapement with spiral steel balance spring and regulator. Fusee and chain. Metal-gilt sun and moon dial with subsidiary dial for seconds in the lower half of the dial, the whole surrounded by a calendar ring with a rotating pointer. Continental watches with unusual dials such as the sun and moon, fashionable at the end of the seventeenth century, are not very frequently found. It is however not unusual for sun and moon dial watches to have a seconds dial in the lower part of the dial. Note the all-over pierced decoration of the watch plate, sometimes found in Continental watches.

116a-b LANGLEY BRADLEY, CODE NO. HYY. London, England. *Circa* 1710. Verge escapement with spiral steel balance spring and regulator. Fusee and chain. Champlevé dial with beetle and poker blued steel hands. Silver pair case. This watch is unusual in that the dial is in mint condition, showing that the matt background of the silver champlevé dials were almost white in colour. This shows clearly in the illustration. Note also the unusual type of decorative pillars in which this maker seems to have specialised. See also col. pl. XIB.

116a

116b

117a

117b

117c

117a-c CHRISTOPHER EGLETON NO. 375. London, England. *Circa* 1700. Verge escapement. Fusee and chain. Champlevé silver dial with tulip hour hand and minute poker hand. Silver pair case, the outer case decorated with tortoiseshell and silver pinwork.

118a

118b

118c

118a-c LANGLEY BRADLEY, CODE NO. DBB. London, England. *Circa* 1700. Verge escapement. Fusee and chain. Silver dial with concentric calendar ring. Gilt pointer for hours and blued steel poker hand for minutes. Silver pair case. 118c shows the view under the dial. The watch, which is in Sir John Soane's Museum, Lincoln's Inn Fields, is said to have belonged to Sir Christopher Wren and to have been presented to him by William and Mary. The Royal monogram W & M is worked into the plate pillars (118b). Note the curved ends to the hinges of the outer case, which came in at about 1700.

119a

119b

119a-b GAUDRON, Paris, France. *Circa* 1700. Verge escapement. Spiral steel balance spring and regulator. Fusee and chain. Gilt-metal dial with separate enamel plaques for each hour and a white enamel ring for the hour divisions. Blued steel hand. Plain gilt case. This is a typical example of a single-handed oignon watch with bridge cock. The watch is wound through the centre of the dial. The case is unusual as these are usually engraved all over. It is interesting that this winding through the centre of the dial was later employed by Breguet for his souscription watches.

120a-b BROUNKER WATTS, London, England. *Circa* 1700. Verge escapement. Fusee and chain. Hour-striking clock-watch on a bell. Silver repoussé dial with blued steel hands. Silver pair case, both cases pierced and engraved. Note squared ends to hinges of outer case. Note also brass-gilt dust cap with revolving steel catch of which this is a very early example.

120a

120b

121a

121b

121a-b DANIEL QUARE NO. 4465. London, England. 1713. Verge escapement. Fusee and chain. Gold champlevé dial, blued steel tulip-type hour hand and poker minute hand. Gold pair case. The plain loose-fitting loop-type pendent typical of the third and early fourth quarters of the seventeenth century was retained by Quare long after it had been abandoned by most other makers. See also col. pl. I1.

122a-b   JOHN KNIBB, England. *Circa* 1700. Verge escapement. Fusee and chain. Champ-levé silver dial with blued steel beetle and poker hands. Plain silver pair case. The square-shaped hinges of the outer case are typical of the late seventeenth century. The signature on the dial is 'Knibb, London', but on the watch plate is engraved 'John Knibb at Oxford'; the word 'Oxford', however, has been erased as far as possible by being over-engraved with a foliate design. The inference may be that the watch was made by John Knibb for his brother Joseph in London.

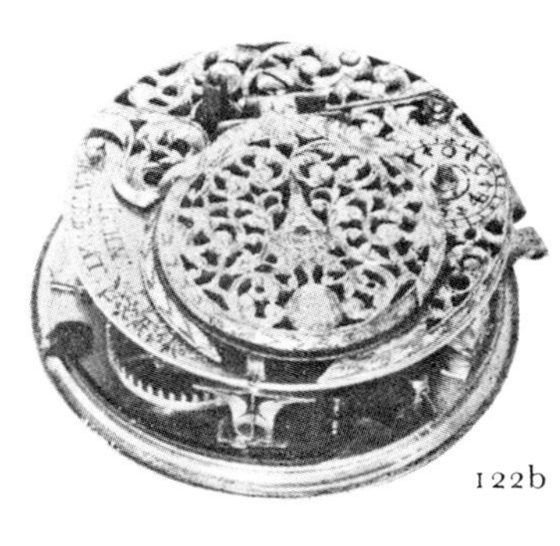

122b

122a

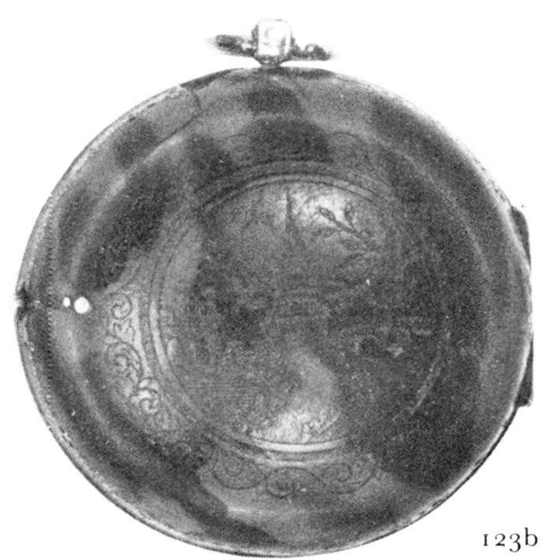

123a

123b

123c

123a-c   JOHN FINCH NO. 135. London, England. *Circa* 1700. Verge escapement. Fusee and chain. Silver differential dial (see pages 89 and 90). Pair case, outer case decorated with engraved tortoiseshell. Note Royal Arms engraved on balance cock, with supporters as part of pierced decoration of the cock-table.

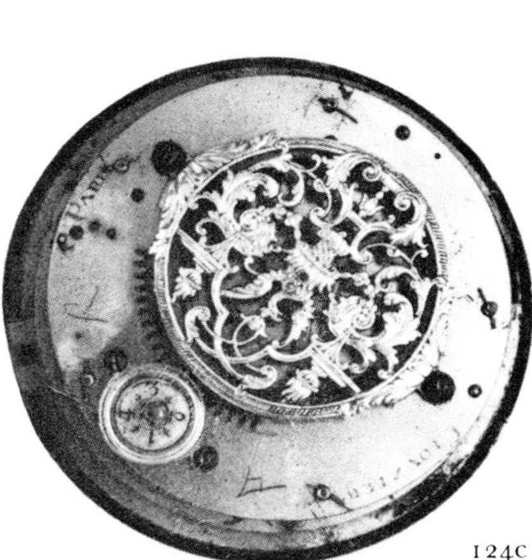

124a

124b

124c

124a-c   GLOVZIER, Paris, France. *Circa* 1700. Verge escapement. Plain steel balance. Spiral steel balance spring. Fusee and chain. Alarum. White enamel dial with gilt-metal central alarum disk. Later steel hand cast, pierced and engraved silver case.

125a

125b

125a-b GOTTFRIDT TORBOCH, Munich, Germany. *Circa* 1700. Verge escapement. Plain balance. Spiral steel balance spring. Fusee and chain. Silver wandering-hour dial. Silver case.

126a-b RICHARD STREET NO. 408. London, England. *Circa* 1700. Verge escapement. Plain steel balance. Spiral steel spring with worm and rack regulator. Fusee and chain. Silver champlevé dial. Blued steel hands, the minute hand turning once in fifteen minutes. Silver pair case.

126a

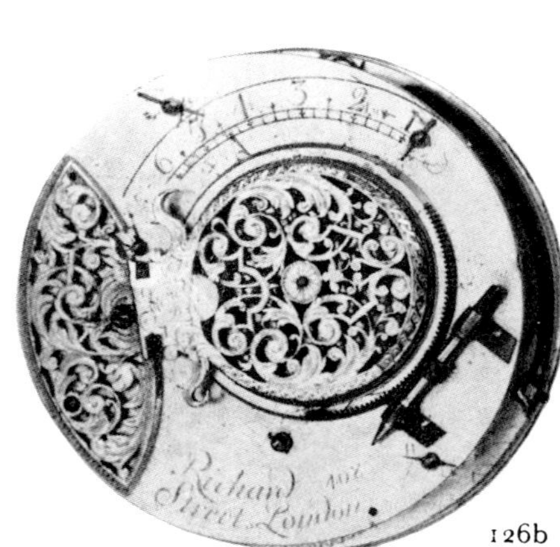

126b

127a

127b

127c

127a-c BARROW, London, England. *Circa* 1685. Verge escapement with spiral steel balance spring and regulator. Fusee and chain. Hour-striking clock-watch. Champlevé silver dial, blued steel hands. Pierced and engraved gold pair case. A typical clock-watch of the last quarter of the seventeenth century. Note design of balance cock, typical of early balance spring period, without mask and with irregular edge to cock foot. Note also locking plate on the back plate of the watch.

128a-c   CHARLES GRETTON, London, England. 1702. Verge escapement. Fusee and chain. Champlevé gold dial with blued steel beetle and poker hands. Gold pair case. The pair case is decorated in a form which appears to anticipate engine-turning but was probably done by punching with a dividing engine.

129a-b   DANIEL QUARE NO. 3720. London, England. *Circa* 1705. Verge escapement. Fusee and chain. Gold dial with blued steel tulip hour hand and minute hand cranked to clear winding square at 3 o'clock. Regulator, dial and operating square at 12 o'clock. Gold pair case. Quare made several watches of this kind in which evidently the emphasis was upon being able to wind and regulate the watch without hinging the movement out of the case. Some of his watches of this type have original enamel dials. In the picture of the movement, note also the early use of a diamond end stone to the balance staff.

130a-d   TOMPION & BANGER NO. 233. London, England. *Circa* 1710. Verge escapement. Steel balance wheel. Spiral steel balance spring with regulator. Fusee and chain. Quarter repeater on a bell. Gold champlevé dial. Steel hands. Gold pair case, the outer case pierced, engraved and repoussé.

131a

131b

131a-b JOHN LANAIS, signed 'London, England'. *Circa* 1720. Verge escapement, spiral steel balance spring with regulator. Fusee and chain. Silver champlevé dial. Pierced gilt hands. Plain silver pair case. This watch is typical of a large number of this period, ostensibly London-made, but in fact produced in Holland. Most of them, as in this case, have a bridge cock and an almost sure indication of Dutch manufacture in the arcaded minute ring round the edge of the dial. The pierced Continental-type gilt hands are also typical. These watches may nevertheless be of quite high quality, as is this example.

132a-b JOSEPH BANKS, Nottingham, England, First quarter of the eighteenth century. Spiral steel balance spring and regulator. Fusee and chain. Ornate silver differential-type dial with military and musical trophies, repoussé, chiselled and engraved. Plain silver pair case. For an explanation of this rare type of dial, see page 90. Provincial makers of this period were quite rare and work by Joseph Banks is usually characterised by ornate dials and the unusual position of the maker's signature on the balance cock. The Banks family seems to have been active in Nottingham from the end of the seventeenth throughout the eighteenth century.

132a

132b

133a

133b

133a-b BELLIARD, Paris, France. First quarter of the eighteenth century. Verge escapement. Fusee and chain. Metal-gilt dial with enamel cartouches for each hour and enamel hour ring. Blued steel hands wound through the dial with a male key. Metal-gilt cast and engraved case. A typical French oignon of the early eighteenth century with typical French bridge-type cock. The hands have no motion work and thus have to be set independently.

134a

134b

134a-b JOHN BUSHMAN, London, England. Early eighteenth century. Verge escapement. Fusee and chain. Silver wandering-hour dial. Silver pair case. The wandering-hour type of dial is found in the late seventeenth and early eighteenth century. For reasons not known nearly all these watches have one or more Royal emblems. In this case the Royal Arms are found engraved on the cock and a portrait of Queen Anne is found in the lower part of the dial surrounded by a pierced gilt fret. For an explanation of the workings of the wandering hour dial, see pages 89-90. See also col. pl. IJ.

135a

135b

135c

135a-c F. BERTRAND, Paris, France. First quarter of the eighteenth century. Verge escapement. Bridge-type balance cock. Alarum. Enamel dial with gilt hands and rotating alarum set-dial in centre. Silver pair case, both inner and outer cases pierced and engraved.

136a-b SIMPTON, London, England (but see comment below). *Circa* 1720. Verge escapement with spiral steel balance spring and regulator. Fusee and chain. Silver repoussé dial ornamented with semi-precious stones. Hands also decorated with semi-precious stones. Gold pair case, the outer case repoussé with free-standing decoration applied in two layers, the whole contained in a third case of shagreen. This watch is similar to that shown in 131, being made in Holland despite the London attribution. The arcaded minute chapter ring is typical of this type of watch. This form of free-standing repoussé decoration was fashionable for only a short period. See also col. pl. XVIJ.

136a

136b

137a

137b

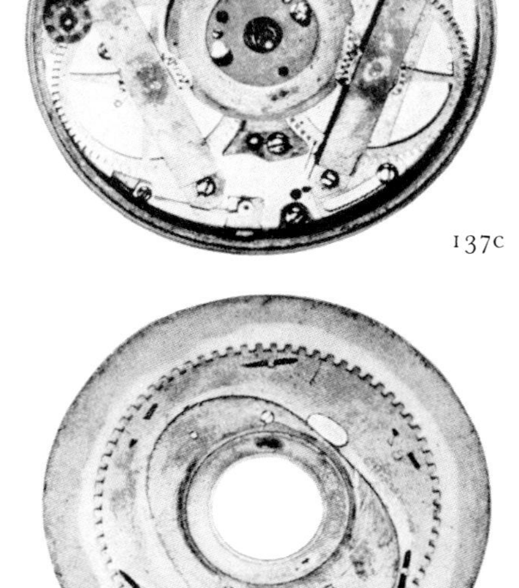

137c

138a

138b

139a

139b

139c

137a-d   JOHN ELLICOTT NO. 2700. London, England. 1742. Cylinder escapement with spiral steel balance spring and regulator. Fusee and chain with bolt and shutter maintaining power. Complicated calendar and astronomical trains. Equation work. Seconds dial on the back of the watch. Gold case. For a picture of the complicated dial, see col. pl. XVIG.

138a-b   ARCHAMBO NO. 1623. London, England. *Circa* 1740. Verge escapement. Steel balance wheel, spiral steel balance spring and regulator. Fusee and chain. Quarter-repeating clock-watch playing music (choice of two airs) on five bells, at each hour. Enamel dial. Steel hands. Movement only.

139a-c (*below*)   GEORGE GRAHAM NO. 883. London, England. 1745. Cylinder escapement. Fusee and chain. Quarter repeater on a bell, operated by depressing the pendent. Enamel dial, blued steel beetle and poker hands. Gold pair case, the outer case covered in leather decorated with gold pinwork. Note diamond end-stoned balance staff invariably used by Graham, and solid engraved foot to the balance cock, introduced by him in about 1725 and subsequently used in all his watches. The dust cap is gilt.

140a

140b

140c

140a-c   G. GRANTHAM NO. 2980. London, England. *Circa* 1750. Verge escapement. Fusee and chain. Enamel dial, blued steel beetle and poker hands. Pair case, the outer case painted enamel. The decoration is painted on a white background and a fly is painted on the inside of the back case.

141a

141b

141a-b   JULIEN LE ROY, Paris, France. *Circa* 1750. Verge escapement. Fusee and chain. Gold dial with ring of white enamel plaques for hour figures. Small plaque for minute figures. Two enamel plaques for name and Paris. Diamond-shaped inserts of green translucent enamel between minute plaques. Note typical mid-eighteenth century French type of bridge cock with steel coqueret as end bearing for balance staff.

142a

142b

142c

142a-c   CONYERS DUNLOP NO. 3383. London, England. 1750. Cylinder escapement with ruby end stones to balance staff. Fusee and chain. Dumb quarter repeater. Enamel dial with blued steel beetle and poker hands. Gold case, the back enamelled with a classical subject signed 'G. M. Moser'.

143a

143b

143c

143a-c   JULIEN LE ROY, Paris, France. *Circa* 1750. Verge escapement. Steel balance wheel, spiral steel balance spring with regulator. Fusee and chain. Enamel dial, gold hands. Inner case gold, outer case enamel with pierced and repoussé gold appliqué decoration. This is a very rare form of case decoration. The movement, almost certainly by Julien Le Roy, is also unusual in having steel pillars and a silver cock.

144b

144a-b   Mid-eighteenth century. Pair case chaise-watch. Outer case of leather with silver piqué decoration. Inner case silver repoussé with a landscape scene with classical columns in the foreground and signed 'CH'. The movement, quarter repeating, is by Joseph Martineau Senior, London, but the watch is illustrated solely for the exceptional quality of the case.

144a

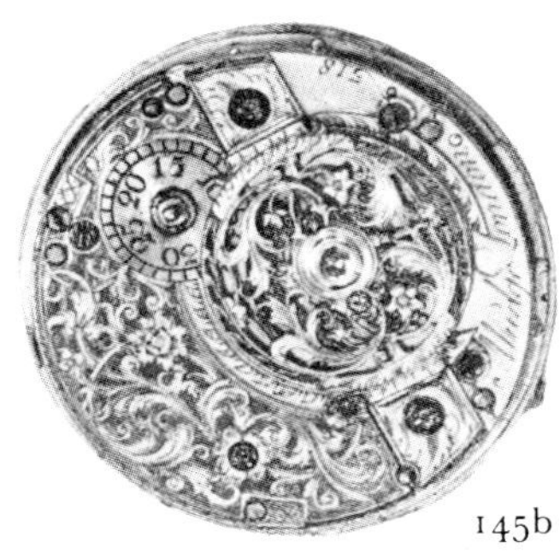

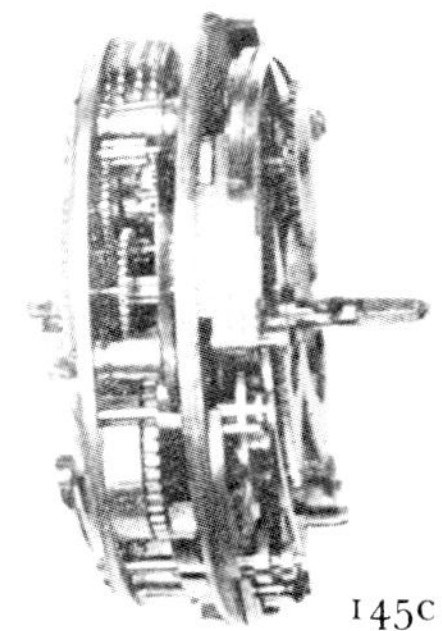

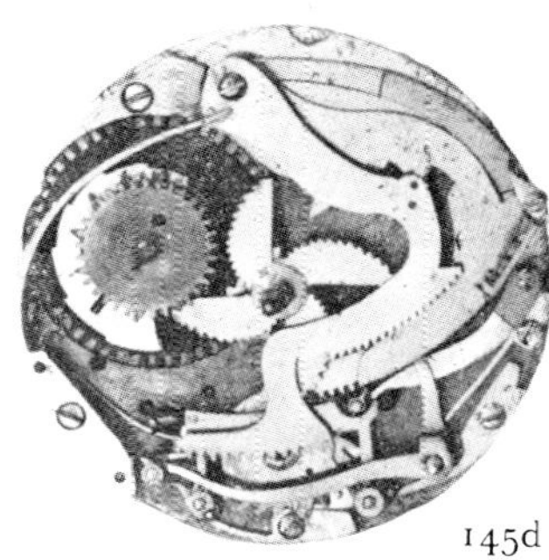

145b   145c   145d

145a

145a-d   THOMAS MUDGE NO. 318. London, England. *Circa* 1750. Cylinder escapement. Three-armed plain balance. Spiral spring with regulator. Fusee and chain. Minute repeating on a bell. White enamel dial. Blued steel hands. Repoussé gold case.

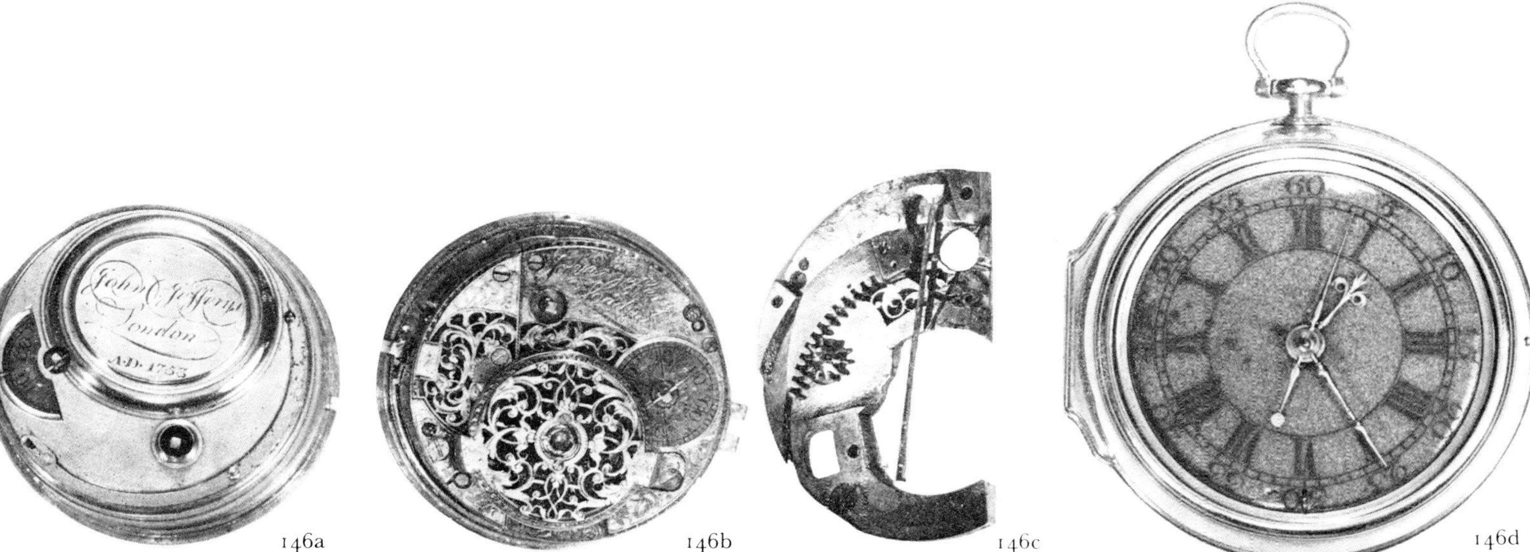

146a   146b   146c   146d

146a-d   JOHN JEFFERYS, London, England. 1753. John Harrison's isochronal verge escapement with diamond pallets. Plain steel balance. Steel spiral balance spring with straight bi-metallic compensation curb (see underside of regulator bridge, 146c). Fusee and chain with maintaining power. Enamel dial. Steel hands. Silver pair case. This watch was made by John Jefferys for John Harrison to test the effectiveness of his ideas to be incorporated in the successful marine timekeeper No. 4. In the 1939-45 war the watch was severely affected by heat in a fire, which turned the white enamel dial black. It is the first watch to contain maintaining power and temperature compensation.

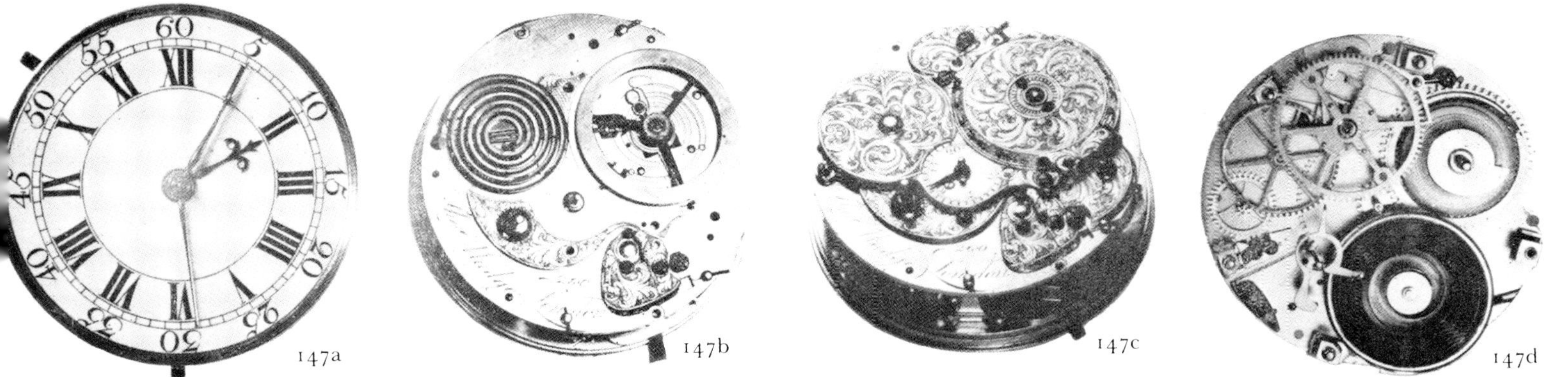

147a   147b   147c   147d

147a-d   THOMAS MUDGE NO. 260. London, England. 1755. Verge escapement with one-minute remontoir. Plain steel balance. Two spiral balance springs: one compensated by spiral bi-metallic curb; the other by a normal mean time regulator. Fusee and chain. White enamel dial. Blued steel hands and centre seconds hand. Movement only. The watch is said to have been made for the King of Spain.

148 JULIEN LE ROY NO. 4757. Paris, France. *Circa* 1775. Lever escapement. Steel and gold balance. Spiral steel balance spring with gridiron-type compensation curb. Fusee and chain. Movement only. For a discussion of this movement, see page 53.

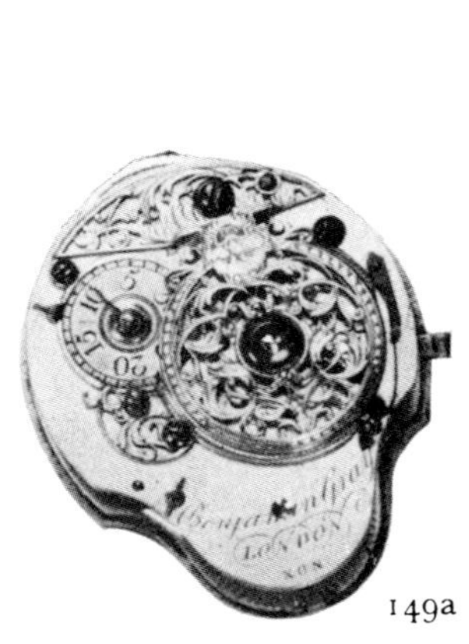

149a-b BENJAMIN GRAY CODE NO. XON. London, England. *Circa* 1760. Cylinder escapement. Spiral steel balance spring and regulator. Fusee and chain. Enamel dial, steel hands. Single gold case. Benjamin Gray may have specialised in watches of this shape as at least one other example is known to exist.

150a-b LAURENCE, Bath, England. 1763. Cylinder escapement. Fusee and chain. Enamel dial with blued steel beetle and poker hands. Silver-gilt case. A typical, if fairly rare, eighteenth-century skeleton movement.

151a-c MUDGE & DUTTON NO. 580. London, England. 1764. Cylinder escapement with spiral steel balance spring and regulator. Enamel dial. Blued steel hour and minute hands. Polished steel sweep centre seconds hand. Gilt dust cap. Plain gold pair case. This watch is virtually identical to the cylinder watches made by George Graham from before 1730.

152a

152b

152a-b FERDINAND BERTHOUD NO. 417. Paris, France. 1764. Verge escapement with spiral steel balance spring. Gridiron-type compensation curb with regulator. Fusee and chain. Enamel dial. Gold hour and minute hands. Steel sweep centre seconds hand. Plain gold case. With the exception of the Jeffreys-Harrison watch of 1753 this is the oldest surviving pocket watch with temperature compensation.

153a

153b

153c

153a-c RICHARD CARRINGTON NO. 326. London, England. *Circa* 1770. Verge escapement. Fusee and chain. Enamel dial with blued steel beetle and poker hands. Gold pair case, the outer case decorated with translucent enamel and an heraldic shield.

154a

154b

154a-b JAMES GREEN NO. 5969. London, England. 1776. Verge escapement. Plain steel balance. Spiral steel balance spring. Fusee and chain with epicyclic maintaining power. White enamel with subsidiary dial for sidereal and solar time and subsidiary dials for sidereal and solar seconds. Gold hour and minute and blued steel seconds hands. Silver case.

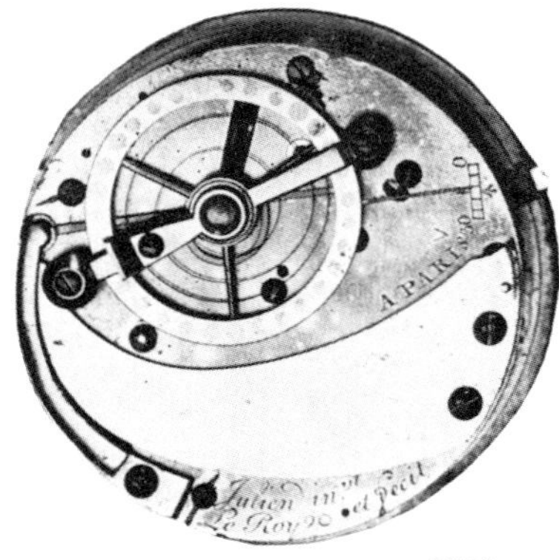

155a

155b

155c

155d

155a-d JULIEN LE ROY, Paris, France. *Circa* 1771. Sully-type frictional rest escapement. Steel balance wheel with gold inset weights. Spiral steel balance spring with bi-metallic compensation plate moving a pivoted regulator. Fusee and chain. White enamel dial with gold hour and minute hands and blued steel centre seconds hand. Gold case. The watch can be used in the pocket or fitted in the gimbals pivoting in the turned wood box. Although signed by Julien Le Roy the watch was made by Pierre who continued to sign his watches 'Julien' after his father's death. It is the type proposed by him for navigational use at sea and called by him 'La Petite Ronde', and is the sole known surviving example.

156a-b   JOHN ARNOLD NO. $\frac{2}{43}$. London, England. HM 1777-78. Spring detent escapement. 'Double-T' bi-metallic compensation balance. Helical balance spring with terminal curves. Fusee and chain. White enamel dial, blued steel hands. Silver consular case. This is Arnold's second type of compensation balance, with straight bi-metallic laminae pivoted in the rim and carrying gold weights. Originally with Arnold's pivoted detent escapement later converted in Arnold's workshop.

157a-b   JOSIAH EMERY NO. 781. London, England. HM. 1778. Pivoted detent escapement. Plain brass balance. Spiral steel balance spring with spiral bi-metallic compensation curb. Fusee and chain. White enamel dial. Gold hands. Gold case.

158a-c   JOSIAH EMERY NO. 939. London, England. *Circa* 1782-83. Two-plane lever escapement. 'Double-S' bi-metallic compensation balance. Helical steel balance spring. Fusee and chain. Enamel dial, with subsidiary thermometer dial. Blued steel hands. Silver-gilt case. This is the lowest-numbered surviving Emery lever watch and is unique among them in having a thermometer with bi-metallic strip (see 158b).

159a

159b

159a-b JOHN ARNOLD NO. 21. London, England. *Circa* 1775, but completely rebuilt by Arnold in 1797 with new silver case, dial and hands. Spring detent. Helical steel balance spring with terminal curves. Two-arm compensation balance.

160

160 ANONYMOUS, English. *Circa* 1780. No movement; illustrated solely on account of the unusual case which is of Battersea enamel.

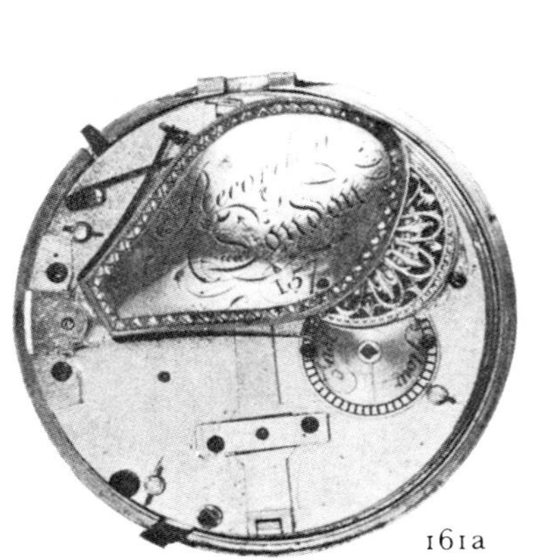
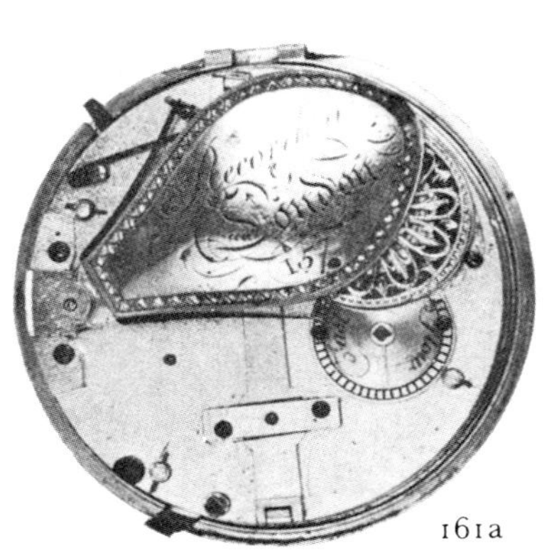

161a

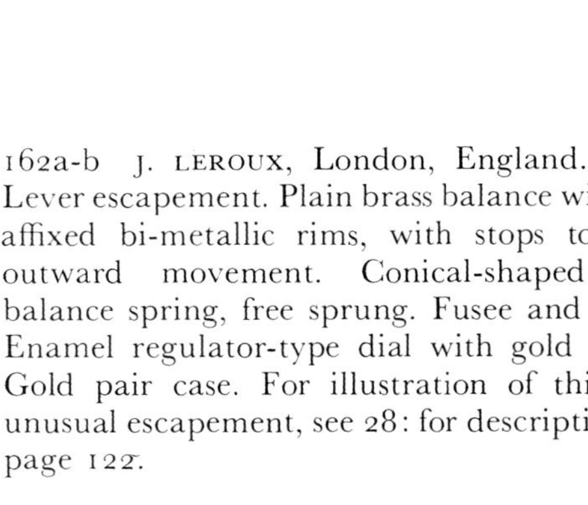

161b

161a-b RECORDON NO. 157. London, England. Probably 1780-90. Duplex escapement. Going barrel. Self-winding. See weight in illustration of movement. Enamel dial with gold hands and steel seconds hand. Plain silver case. In 1780 Recordon took out a patent in this country for the self-winding watch, of which this is probably the earliest surviving English example.

162a

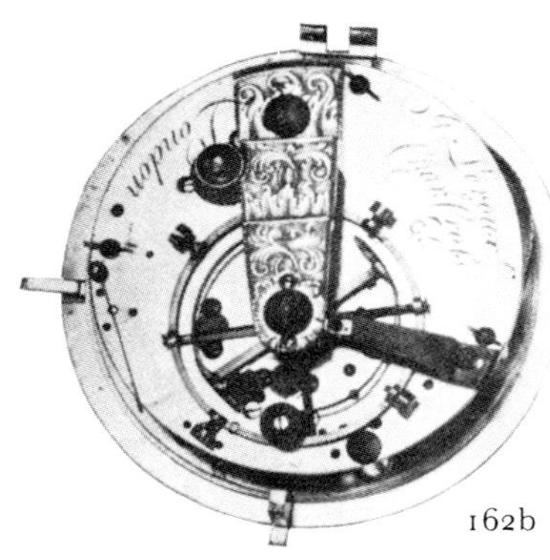

162b

162a-b J. LEROUX, London, England. 1785. Lever escapement. Plain brass balance with two affixed bi-metallic rims, with stops to limit outward movement. Conical-shaped steel balance spring, free sprung. Fusee and chain. Enamel regulator-type dial with gold hands. Gold pair case. For illustration of this very unusual escapement, see 28: for description, see page 122.

163a

163b

163a-b JOSIAH EMERY NO. 838. London, England. 1784. Pivoted detent escapement. Bi-metallic 'double-S' compensation balance. Helical steel balance spring with terminal curves. Fusee and chain. White enamel regulator-type dial, blued steel hands. Gold case. This watch has been modified to Emery's form of compensation balance. The silver regulator scale indicates that the watch had originally a bi-metallic curb similar to that shown in col. pl. IXB.

164a-b THOMAS WRIGHT NO. 2228. London, England. 1784. Spring detent escapement. Bi-metallic compensation balance with integral weights. Spiral steel balance spring and regulator. Fusee and chain. White enamel dial. Gold pierced hands. Gold pair case. Thomas Wright took out a patent on behalf of Earnshaw for his spring detent escapement and this is the earliest recorded example of a watch with the escapement.

164a

164b

165a

165b

165c

165a-c BREGUET À PARIS NO. 128$\frac{5}{85}$. Paris, France. 1785. Cylinder escapement with spiral steel balance spring with regulator. Fusee and chain. Minute repeater. Enamel dial, gold hands. Plain gold case. This is one of the earliest surviving Breguet watches. The fractional number $\frac{5}{85}$ indicates the date, May 1785. At this date Breguet always added 'à Paris' to his signature. The cylinder is of steel and not of Breguet's later pattern.

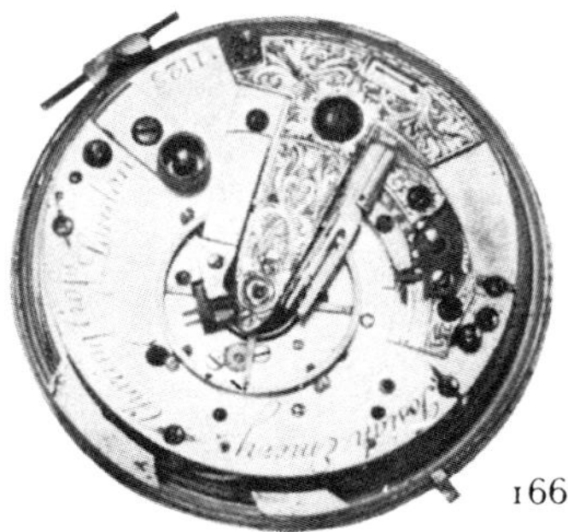

166

167a

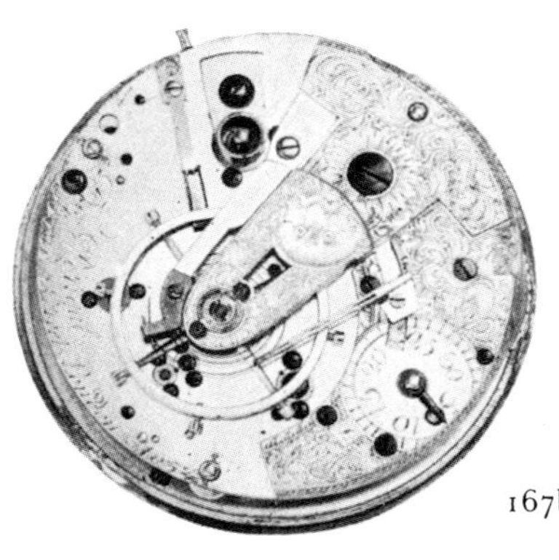

167b

166 JOSIAH EMERY NO. 1123. London, England. *Circa* 1786. Ruby cylinder escapement. Four-arm brass balance. Steel helical balance spring with regulator incorporating a compensation curb consisting of two bi-metallic strips. Movement only. While this movement bears no great resemblance to the lever watches of Emery, it bears a very close resemblance to the lever watch by Francis Perigal (see 167). Emery's watches, on the other hand, bear a close resemblance to those of Pendleton and Mudge & Dutton. The inference may be that all these makers used a common source for their ébauches.

167a-b FRANCIS PERIGAL NO. 1053. London, England. Probably late 1780s. Lever escapement. Plain four-arm brass balance with spiral steel balance spring and regulator carrying sugar-tong-type compensation curb with two bi-metallic arms. Fusee and chain. Enamel dial and gold hands with subsidiary dials for seconds and temperature. Plain silver case (recased). For illustration of escapement, see 24; for description, see page 121.

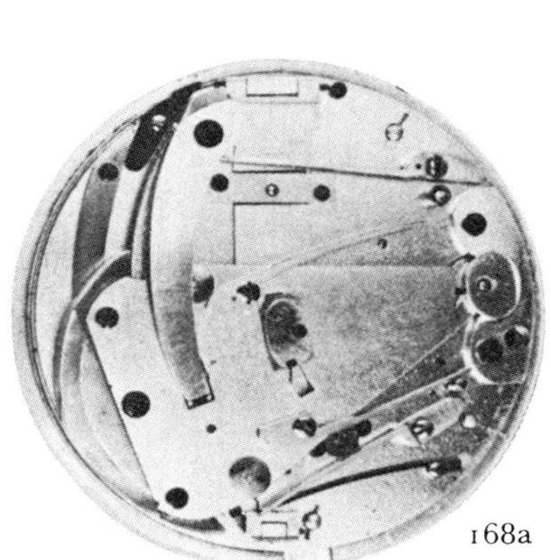

168a

168b

168a-b RECORDON NO. 180. London, England. Probably 1784. Duplex escapement. Regulator operated through the dial. Going barrel self-winding. See illustration of movement. See remarks in caption to 161 regarding Recordon and self-winding watches.

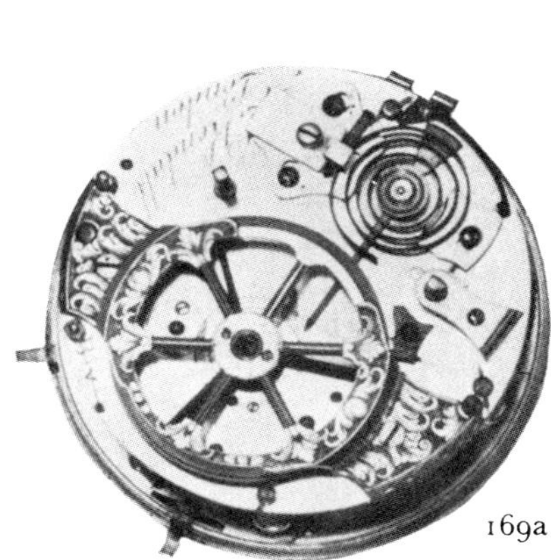

169a

169b

169a-b LARCUM KENDALL CODE NO. B+Y. London, England. Pivoted detent escapement. Three-arm brass balance; spiral steel balance spring with reverse curve at end for regulator, mounted on spiral bi-metallic compensation curb which can be adjusted for mean time by a square seen in the illustration between the cock foot and the cock of the compensation curb. Fusee and chain. Enamel regulator-type dial with blued steel hands. Silver case. For illustration of escapement, see 16; for description, see page 116.

170a

170b

170c

170d

170a-d   BREGUET NO. 19. Paris, France. *Circa* 1787. Lever escapement. Three-arm bi-metallic compensation balance. Helical steel balance spring with terminal curves and regulator. Two going barrels. Quarter repeating. Engine-turned silver dial with blued steel hands. Subsidiary dials for up-and-down and seconds. Engine-turned gold case with the arms of the Prince Regent. This is one of the first thirty watches appearing in Breguet's books which started 1787. All are perpetuelles – this is the simplest type. For a further discussion of the escapement, see page 125.

171a-b   BREGUET NO. 12. Paris, France. 1789. Pivoted detent escapement. Two-arm bi-metallic compensation balance. Helical steel balance spring without end curves, free sprung, the balance staff supported at each end by three friction rollers. Fusee and chain. Engine-turned silver dial with sweep centre minutes hand, subsidiary dials for hours and seconds. Plain silver case. This is the only example known of a watch by Breguet with friction rollers and the lay-out of the movement is in every way more typical of Ferdinand Berthoud.

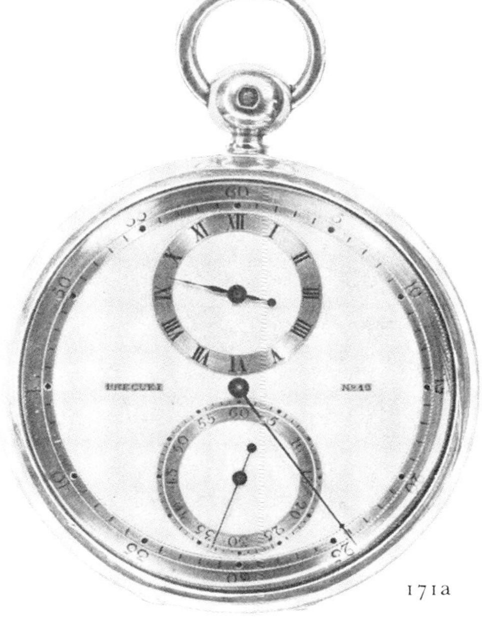

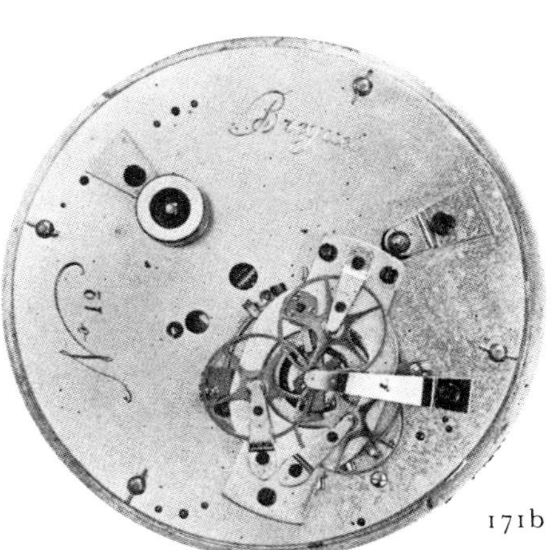

171a

171b

172a

172b

172a-b   JOHN ARNOLD NO. $\frac{16}{64}$. London, England. 1779. 'Double-S' bi-metallic compensation balance. Helical steel balance spring with terminal curves. Silver case. Enamel dial. Blued steel hands. For illustration of escapement see 11.

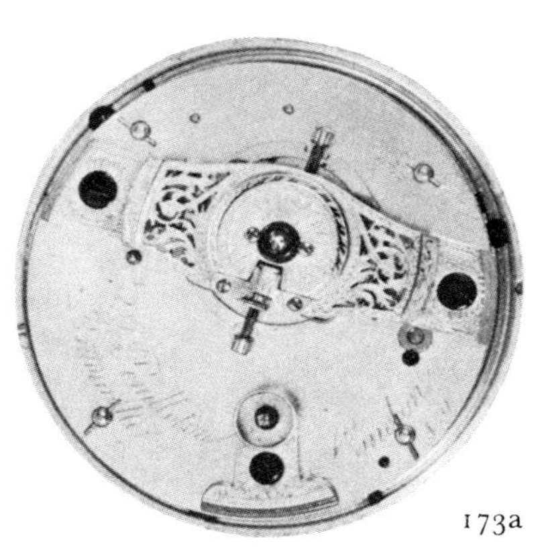

173a

173b

173a-b RICHARD PENDLETON NO. 180. Pentonville, London, England. Late eighteenth century. Lever escapement, plain brass balance with S-shaped bi-metallic compensation affixes. Helical steel balance spring with end curves free sprung. Fusee and chain. Enamel regulator-type dial with blued steel hands. Silver case. Pendleton is known to have made all or most of Emery's lever escapement watches which this watch greatly resembles, especially as to the bridge-type balance cock and balance wheel. The escapement, however, is quite different. For illustration of escapement, see 23; for description, see page 121.

174a

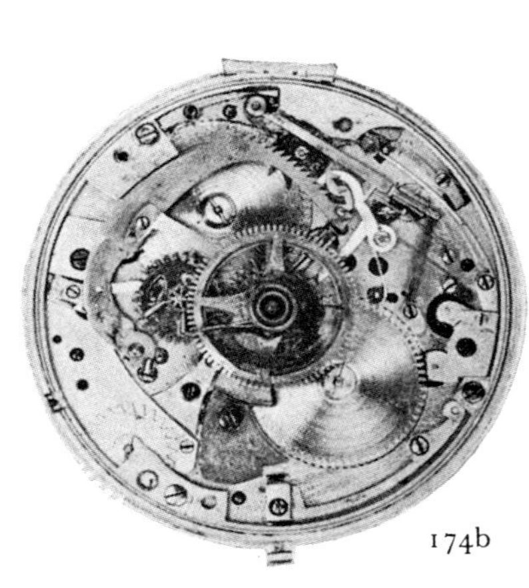

174b

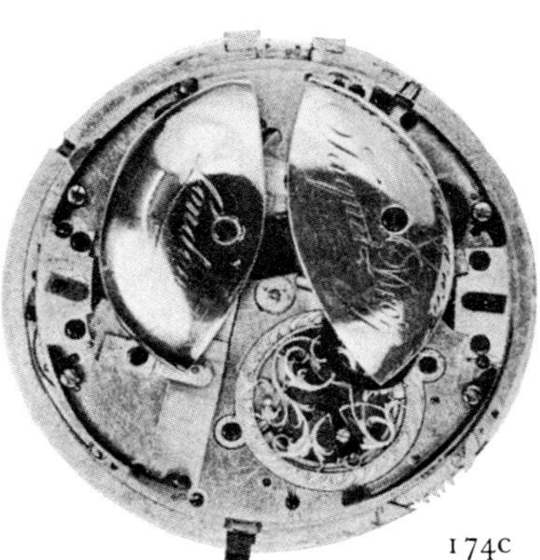

174c

174a-c JAQUET DROZ, marked London, but in fact Swiss; the firm had branches in several capital cities. *Circa* 1790. Cylinder escapement. Quarter striking clock-watch on one bell. Self-winding with separate weights for going and striking trains. Enamel dial with gold hands and sweep centre seconds hand. Case decorated round the dial with half pearls and green stones with painted enamel back. This watch is one of a pair and is also illustrated in col. pl. IE.

175a

175b

175a-b BREGUET NO. 84. Paris, France. 1798. Cylinder escapement with spiral steel balance spring and regulator. Going barrel. Enamel dial with blued steel hands. Subsidiary dial at 2 o'clock for seconds and also sweep centre jump seconds. Plain gold case. This is a very early example of Breguet's post-Revolution work. The script signature is typical of this earliest period. The seconds dial is of the kind usually used by Breguet on enamel dials and was formed by grinding away the glazed surface to form a recessed subsidiary dial.

176a

176b

176c

176a-c BREGUET NO. 62. Paris, France. Date, see below. Lever escapement. Three-arm compensation balance. Helical steel balance spring without end curves, with regulator. Two going barrels. Self-winding. Minute repeater. Enamel dial with blued steel hands. Subsidiary dials for thermometer, up-and-down, and seconds. Gold engine-turned case. The platinum weight of this watch bears the inscription 'Faite par Breguet pour M le Duc d'Orleans en 1780', but the date of the watch is certainly later. Possibly the weight may have been taken from an earlier watch.

177

177 FRENCH. *Circa* 1790. The movement is by Gide but is illustrated solely on account of the unusual case, which is of painted porcelain.

178a

178b

178a-b BROCKBANKS NO. 3499. London, England. 1791. Spring detent escapement of Earnshaw type. Three-arm bi-metallic compensation balance of typical Brockbanks type with heavy weights. Helical steel balance spring with end curves, free sprung. Fusee and chain. Enamel dial with gold hour and minute hands and steel seconds hand. Silver case.

179a

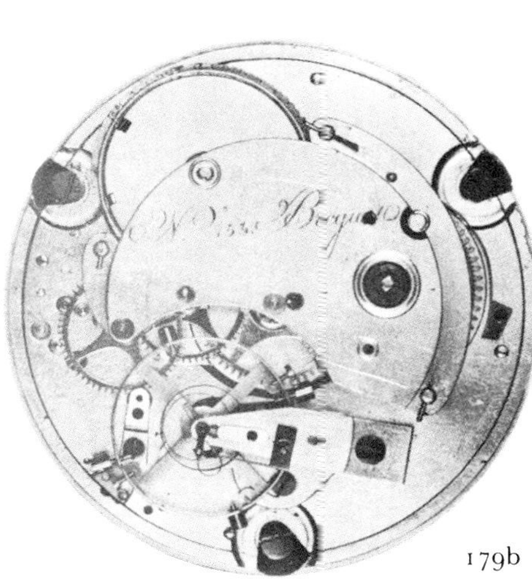

179b

179a-b BREGUET ET FILS NO. 153/4570. Paris, France. Pivoted detent escapement, four-arm bi-metallic compensation balance, helical steel balance spring with end curves, free sprung, fusee and chain. Engine-turned silver dial, blued steel hands, subsidiary dials for seconds and up-and-down indicator. Engine-turned gold case. This very early specimen of Breguet's work was taken back into the workshop and modernised with a new dial and probably a new case, at some time about 1825. It was then given the new number 4570 which appears on the cuvette. See also page 61 for information about Breguet's numbering.

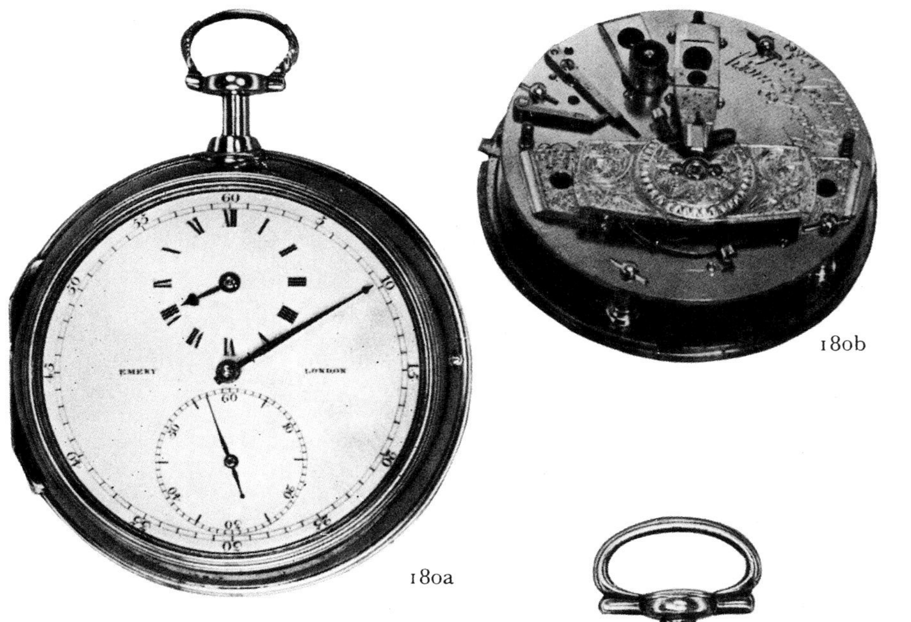

180a-b JOSIAH EMERY NO. 1289. London, England. 1792. Lever escapement. Brass balance with bi-metallic S-shaped compensation affixes. Steel helical balance spring, free sprung. Fusee and chain. Enamel regulator-type dial with blued steel hands. Plain gold case. This is an early example of Emery's second type of lever escapement with a pivoted impulse pin or cranked roller. The escapement is without draw.

180b

180a

181a-b BROCKBANKS NO. 3711. London, England. 1793. Peto cross-detent escapement. Three-arm bi-metallic compensation balance, helical steel balance spring with end curves free sprung. Fusee and chain. Enamel dial, gold hands, plain silver case. For illustration of escapement, see 20; for description, see page 117.

181a

181b

182a

182b

182a-b BREGUET NO. 147. Paris, France. Sold 1805. Spring detent escapement of Arnold pattern. Three-arm bi-metallic compensation balance, helical steel balance spring with end curves free sprung, fusee and chain. Enamel dial, with sweep centre minute hand and subsidiary dials for seconds and hours. Plain gold case. This watch is an important and unusual specimen of Breguet's pre-Revolution style; the script signature on the dial is rarely found after the Revolution. The watch is of the type described by Breguet as 'garde temps'.

184b

184a

184a-b   TAYLOR NO. 1143. London, England. *Circa* 1795. Lever. Pointed pallet without draw. Two spiral balance springs, one above balance and one below with regulator. Fusee and chain. White enamel dial, gold hands. Silver re-case. Nothing is known of Taylor. There is in the British Museum a watch movement made from the same ébauche as the movement shown in 184, but this is a cylinder with single balance spring. Taylor certainly had access to the work of Leroux, Pendleton and Grant, for this watch embodies in its escapement the principal features of their work. Thus the pallet stone arrangement is as used by Grant, the escape wheel is an exact copy of Leroux and the fork of the lever is carried round to the opposite side of the balance staff in the manner of Pendleton. For illustration of escapement, see 29.

183

183   Pedometer by Ralph Gout. *Circa* 1800. For use on horseback. Patent No. 2351 of November 4th, 1799. The gilt metal base is attached to the saddle and the rest of the 'watch' bounces up and down at each pace of the horse.

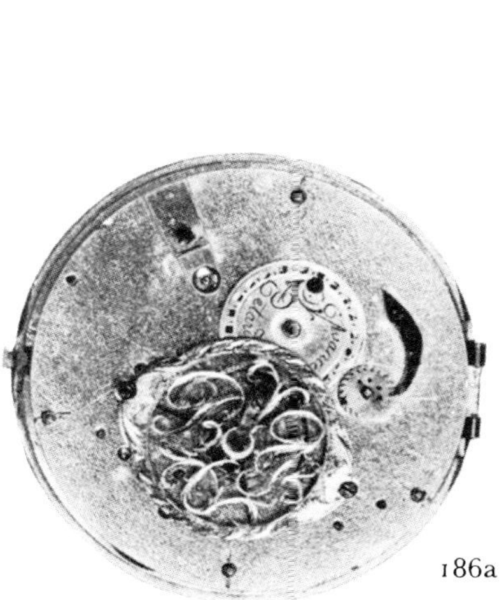

186a

186b

186a-b   ANONYMOUS, French. *Circa* 1795. Verge escapement with spiral steel balance spring and regulator. Fusee and chain. Enamel dial with decimal or Revolutionary time with subsidiary dials for days of the month and 12-hour time. Gold hands to Revolutionary dial, steel hands to calendar and 12-hour dial. Silver case. A typical late eighteenth-century French verge movement.

185

185   The Mudge lever replica watch made in 1795; see page 52.

187a

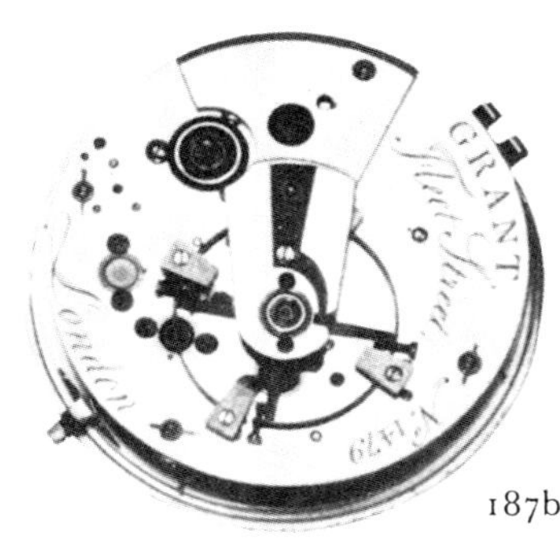

187b

187a-b JOHN GRANT NO. 1479. London, England. 1795. Lever escapement. Four-arm bimetallic compensation balance. Helical balance spring with end curves, free sprung. Fusee and chain. Enamel regulator-type dial, blued steel hands. Gold case. For illustration of escapement, see 25; for a description, see page 121.

188a

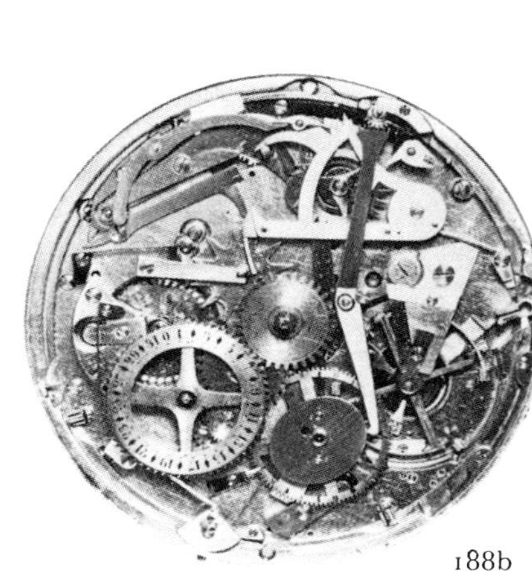

188b

188c

188a-c BREGUET NO. 217. Paris, France. Date probably just after the French Revolution. Lever escapement. Two-arm bimetallic compensation balance with two additional arms containing screws for regulation. Helical steel balance spring with end curves and regulator. Self-winding. Two going barrels. Equation work. Calendar. Quarter repeater. Engine-turned silver dial with gold hour and minute hands. Blued steel hands to subsidiary dials for equation, up-and-down, month, and seconds; the last two concentric, the date appearing through an aperture. Engine-turned gold case. This watch represents Breguet's highest level of workmanship, having considerable complications.

189a

189b

189c

189a-c ALEXANDER HARE NO. 440. London, England. Last quarter of the eighteenth century. Cylinder escapement. Fusee and chain. Enamel dial with gold beetle and poker hands. Gold pair case, the outer case enamelled with a scene with Cupid, surrounded by a band of transparent maroon-coloured enamel. The band of the case is of blue enamel with white spots.

190a-c   JAQUET DROZ, London, England. *Circa* 1793. Cylinder escapement. Diamond-set balance. Spiral steel balance spring. Going barrel. White enamel dial with subsidiaries for hours, minutes, seconds, quarter-seconds, age of moon and regulator. Aperture to reveal the balance. Gold hands. Gold case enamelled and set with pearls.

191a-b   GEORGE MARGETTS NO. 13. London, England. 1797. Earnshaw-type spring detent escapement. Two-arm bimetallic compensation balance. Helical steel balance spring with terminal curves. Fusee and chain. White enamel dial with overlapping circles for hours and seconds. Gold hands. Silver pair case.

191a

191b

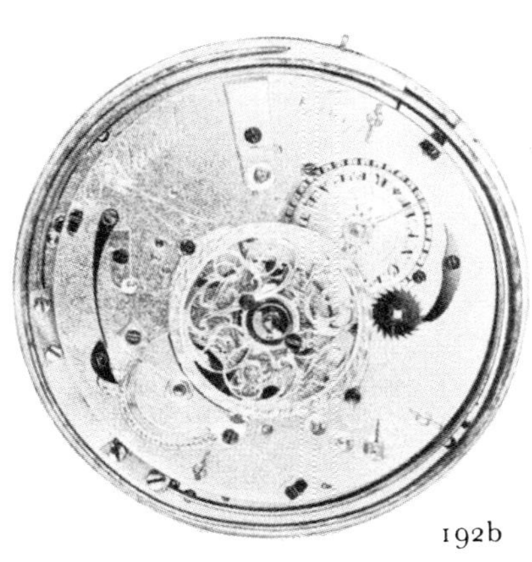

192a

192b

192a-b   VAUCHER NO. 10971. French. *Circa* 1800. Verge escapement. Fusee and chain. Quarter repeat on two gongs operated by pressing pendent. Enamel dial with blued steel hands, gold engine-turned case. This watch is typical of late eighteenth-century French verge watches. Typical bridge-type cock and winding through dial.

193a

193d

194a

193b

193c

193a-d   ANONYMOUS, Switzerland. *Circa* 1800. Cylinder escapement. Plain balance. Spiral steel balance spring. Going barrel. Pinned-drum musical mechanism and automata released at the hour or at will. Dancers move across a stage and flanking figures play their instruments. White enamel dial. Blued steel hands with gold tips. Gold case with raised gold decorated bezels. Pinned-drum mechanisms are usually earlier than pinned disks.

194b

194c

194a-c   J. HERBERT, Brighthelmstone (i.e. Brighton), England. Third quarter of the eighteenth century. Verge escapement. Mock-pendulum balance wheel. Enamel dial with Continental-type pierced gilt hands. Gilt pair case, the outer case covered in tortoiseshell. A fairly typical cheap watch of the third quarter of the eighteenth century and a late example of the mock-pendulum balance. The bridge-type balance cock has been engraved later with initials and a crest.

195a-b   DU BOIS ET FILS. Marked 'London' on the plate but certainly French or Swiss. *Circa* 1800. Debaufre-type, dead-beat escapement with two parallel escapement wheels acting on one pallet on the pallet arbor. The balance is geared up. Brass balance wheel, spiral steel balance spring and regulator. Fusee and chain. Enamel dial with sweep centre seconds beating seconds. Steel seconds hand, gold hour and minute hands. Aperture for mock-pendulum attached to pallet arbor. Silver-gilt case. For illustration of escapement, see 44; for description, see page 130.

195a

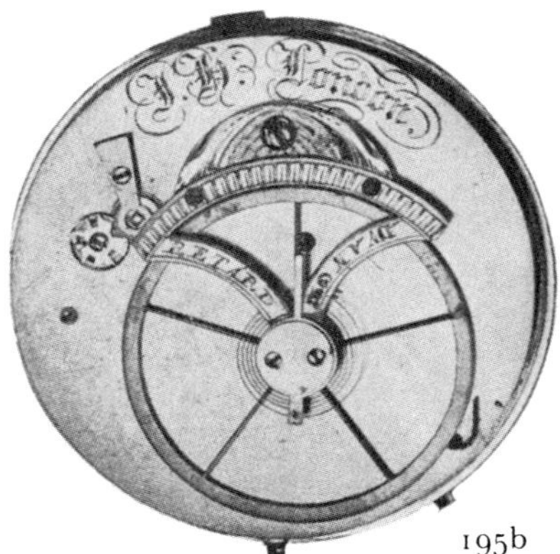

195b

196a-b Signed 'BREGUET À PARIS'.
France. *Circa* 1800. Pouzait-type detached lever. Very large brass balance.
Spiral steel balance spring with regulator. Fusee and chain. Enamel dial
with sweep centre seconds beating
seconds. Subsidiary dials for hours,
minutes and date. Steel hands for
date and seconds, gold hands for
hours and minutes. Silver case.
Pouzait's version of the lever escapement was invented in about 1785, but
the style of the case of this watch
suggests a date not earlier than 1800,
by which time Breguet had ceased
using the signature 'à Paris'. For an
illustration of this escapement, see 39;
for description, see page 124.

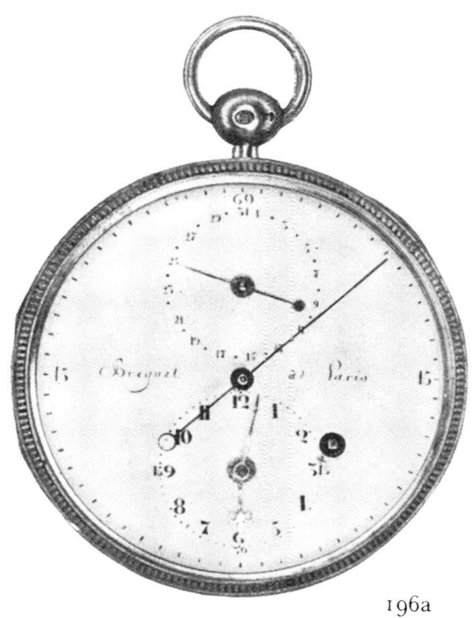

196a

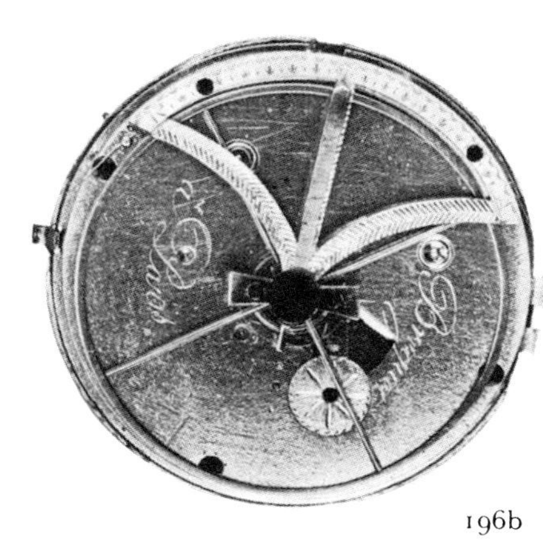

196b

197a

197b

197a-b BERTHOUD, Paris, France. *Circa* 1800. Cylinder escapement.
Going barrel. Self-winding. Gold hands with sweep centre seconds.
Silver case. The square visible above the 12 is for the regulator.

198

198 RICHARD COMBER, Lewes, England.
1776. Verge escapement with spiral
steel balance spring and regulator. Fusee
and chain. Enamel differential dial and
gold hand. Plain silver case. This is an
extremely late example of the differential dial which was very rarely employed
after 1710. It is also unusual in coming
from a provincial maker.

199a

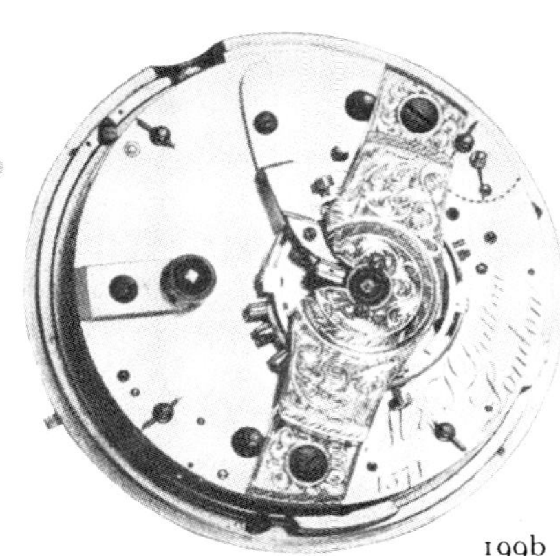

199b

199a-b MUDGE & DUTTON NO. 1571. London, England. 1800. Lever escapement,
similar to that of Pendleton No. 180 (see
23 and, for description, page 121). Two-arm bi-metallic compensation balance,
helical steel balance spring with terminal
curves, free sprung. Fusee and chain.
Enamel dial, gold beetle and poker hands
and steel seconds hand. Plain gold case.
Note bridge cock very similar to those used
by Emery.

200a

200b

200a-b   J. GRANT NO. 2315. London, England. *Circa* 1800. Lever escapement. Three-arm brass balance with two bi-metallic compensation affixes. Helical steel balance spring with end curves free sprung. Fusee and chain. Enamel regulator-type dial, steel seconds hand, gold hour and minute hands. Plain silver case (re-cased). For illustration of the escapement, see 26; for description, see page 121.

201a-b   BREGUET ET FILS NO. 898. Paris, France. *Circa* 1802. Verge escapement with jewelled pallets. Elastic suspension for verge. Reversed fusee and chain. Enamel dial, and blued steel hands. Turkish numerals. Secret signature 'Breguet et. mixte No. 898', together with a Turkish signature. Engine-turned silver case. For Breguet 'mixte' watches, see page 66.

201a

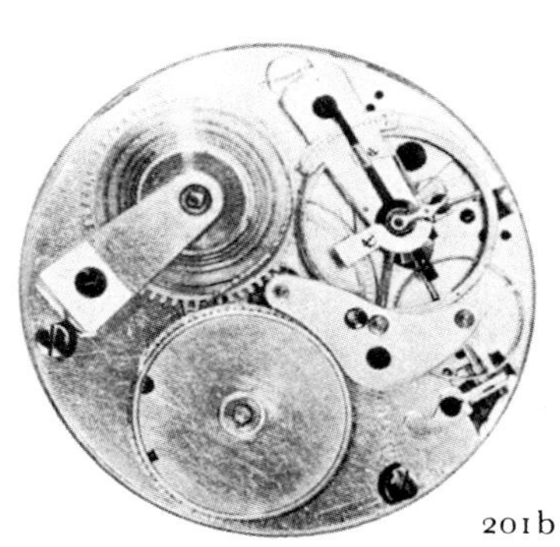

201b

202a

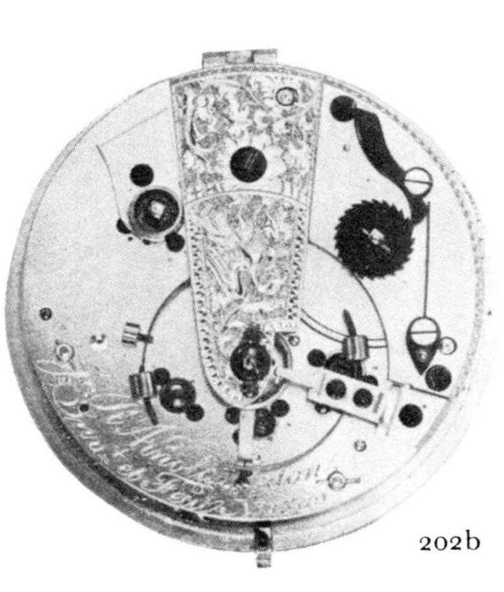

202b

202a-b   J. R. ARNOLD NO. 1869. London, England. 1802. Spring detent escapement Two-arm bi-metallic compensation balance, gold helical balance spring with end curves, free sprung. Fusee and chain. Enamel dial with blued steel hands. Silver case. Note blued steel mainspring ratchet set-up on the watch plate. The escapement is what is generally known as the Earnshaw type. J. R. Arnold made both Arnold and Earnshaw types of spring detent escapement.

203a

203b

203c

203a-d   BREGUET ET FILS NO. 1410. Paris, France. *Circa* 1804. Lever escapement. Three-arm bi-metallic compensation balance, the helical steel balance spring without end curves with regulator. Elastic suspension to the balance arbor. Two going barrels. Quarter repeating. Days of the month, phases of the moon. Self-winding. Engine-turned silver dial with blued steel hands, with auxiliary dials for moon work, up-and-down, seconds and day of the month. This represents Breguet's highest level of perpetuelle watch and incorporates all his improvements except for some incorporated in a few perpetuelles of somewhat different type made after 1820; see 265.

203d

204a

204b

204c

204a-c   BREGUET NO. 1225. Paris, France. *Circa* 1804. Cylinder escapement with spiral steel balance spring and compensation curb. Going barrel. Minute repeating clock-watch. Enamel dial with Turkish numerals. Blued steel hands. Gold pair case, the inner engine-turned and the outer decorated with red translucent enamel and gold shell pattern over engine-turning.

205a

205b

205a-b   J. HOVENSCHÖLD NO. 586. Stockholm, Sweden. Early nineteenth century. Sully escapement. Fusee and chain. Enamel dial with Arabic numerals and cut steel brilliants between each pair of hours. Gilt hands. Single brass-gilt case. This watch is fairly typical of Swedish and Danish design at about 1800. The bridge-type balance cock has the initials of the maker (J. H.) woven into the piercing of the cock. This also is fairly typical of Swedish makers. Sully's escapement is very rarely found in watches, as opposed to the Ormskirk escapement, widely used by the British Preston watch manufacturers at this time, which has a double escape wheel and a single pallet on the balance staff. Sully's escapement has a single escape wheel with double pallets on the balance staff. For illustration of escapement, see 45; for description, see page 130.

206a

206b

206a-b RECORDON (late EMERY) NO. 7392. London, England. 1807. Lever escapement, converted probably from duplex. Two-arm bi-metallic compensation balance also of later date. Spiral steel balance spring with regulator. Fusee and chain. Enamel dial and Breguet-type blued steel hands. Gold case.

207a-b MARGETTS NO. 1273. London, England. *Circa* 1805. Arnold-type spring detent escapement. Bi-metallic compensation balance with sliding weights. Helical steel spring. Fusee and chain. Enamel dial. Gold hands. Gold case.

207a

207b

208a

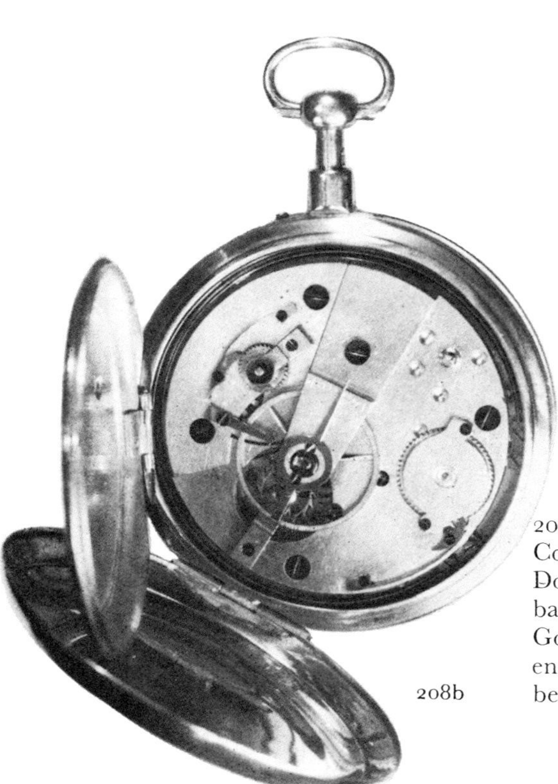

208b

208a-b URBAN JÜRGENSEN NO. 506. Copenhagen, Denmark. *Circa* 1805. Double-wheel duplex escapement. Plain balance. Spiral steel balance spring. Going barrel. Quarter repeating. White enamel dial. Blued steel hands. Pinchbeck case.

209a-b LITHERLAND & WHITESIDE
NO. 3682. Liverpool, England. 1806.
Rack-lever escapement with escape
wheel of thirty teeth. Plain steel
balance. Spiral steel balance spring.
Fusee and chain. Enamel dial. Gold
hands. Silver pair case.

209a

209b

210a

210b

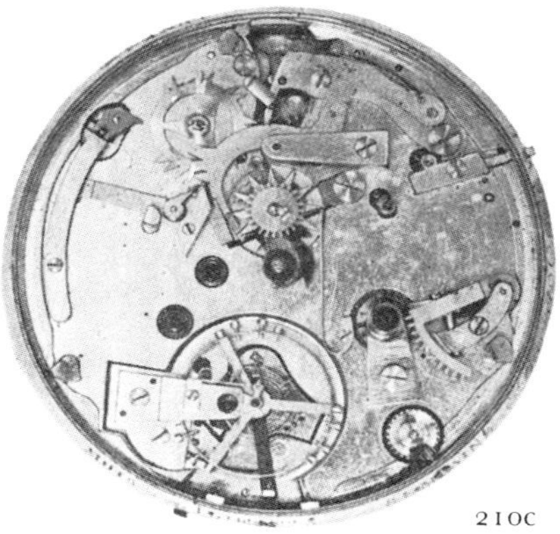

210c

210d

210e

210a-e BREGUET ET FILS NO. 1226. Paris, France. 1807. Double-wheel Robin escapement. Three-arm bi-metallic compensation balance. Spiral steel balance spring with overcoil. Regulator operated below the band of the dial. Going barrel. Quarter repeating. The watch has a dial on both sides. A gold engine-turned dial has normal concentric hours and minute hands, subsidiary dial for seconds at 8 o'clock, up-and-down indicator and a set-square for the calendar. This dial has secret signatures shown in enlargement in 210d. The other dial is of engine-turned silver, has two sweep centre hands indicating equation of time, and subsidiary dials for phases of the moon, the day of the month, and day of the week. The escapement is of a very unusual kind with only five teeth on the escape wheels.

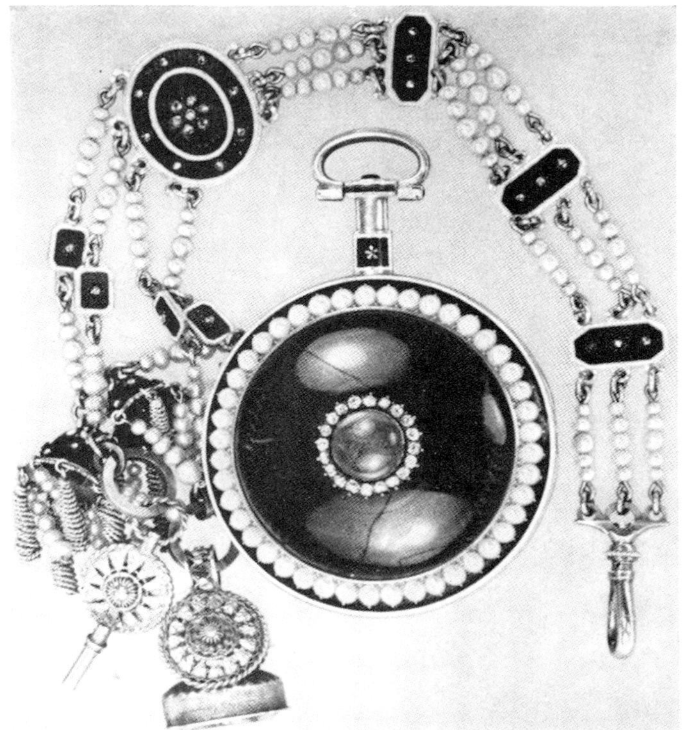

211a-b WIGHTWICK & MOSS, London, England. 1807. Verge escapement. Fusee and chain. Enamel dial with gold hands. Gold case, back of translucent blue enamel with ring of pearls; in centre plaited hair under glass with ring of diamonds. Bezel with blue enamel and ring of pearls. The châtelaine has blue enamel bars with white borders, studded diamonds joined by triple pearl strings, and fob hook.

212a-c BREGUET ET FILS NO. 1619. Paris, France. 1808. Double-wheel duplex escapement, three-arm bi-metallic compensation balance, spiral balance spring with overcoil and regulator, elastic suspension for balance staff, going barrel; quarter repeater operated by plunger in the band of the case at 11 o'clock. Engine-turned gold dial with blued steel hands and subsidiary dials for calendar and seconds. Engine-turned gold case. For illustration of escapement, see 9.

213a-b THOMAS EARNSHAW NO. 782-3391. London, England. 1808. Spring detent escapement. Plain steel balance. Spiral steel balance spring with 'sugar-tongs' bi-metallic compensation curb. Fusee and chain. White enamel dial. Gold hands. Silver hunting case.

214a-b BREGUET NO. 1890. Paris, France. 1809. Échappement naturel. Two-arm bi-metallic compensation balance. Spiral steel balance spring with overcoil, free sprung. Four-minute tourbillon. Reversed fusee. Engine-turned gold dial. Blued steel hands. Subsidiary dials for seconds, stop seconds, and up-and-down. Concentric central subsidiary dial for hours. This watch represents Breguet's highest achievement in the tourbillon watch. It has the added and unique peculiarity that in addition to the normal Breguet secret signature it has, signed in secret between the hour numerals 4, 5, 6, 7, 8, the name of the owner, the Comte Alexis de Razzoumofski. 214b shows the tourbillon carriage with the balance wheel removed. In general lay-out and appearance, the movement is almost identical to that of other Breguet tourbillons.

214a

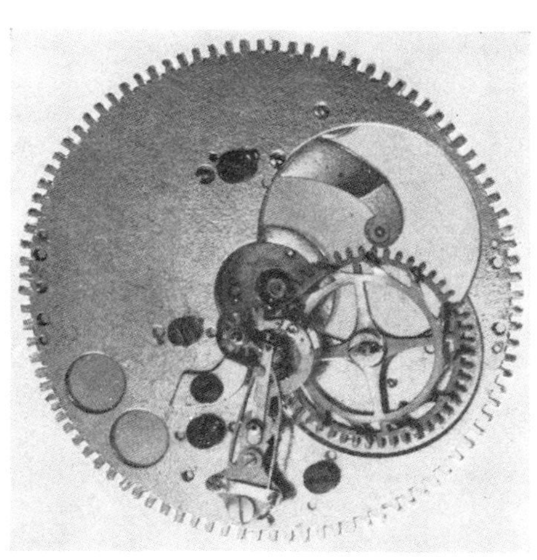

214b

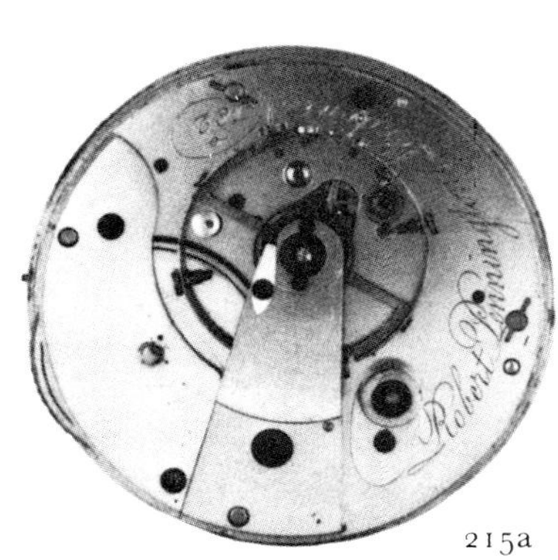

215a

215b

215a-b ROBERT PENNINGTON NO. 182. London, England. *Circa* 1810. Spring detent escapement. Bi-metallic compensation balance. Helical steel balance spring with terminal curves. Fusee and chain. White enamel dial. Gold hands. Gold case.

216a-b BREGUET ET FILS NO. 2555. Paris, France. *Circa* 1811. Spring detent escapement with Peto-type cross detent and six-minute tourbillon. Three-arm bi-metallic compensation balance, spiral steel balance spring with overcoil, free sprung. Fusee and chain. Engine-turned silver dial and blued steel hands. Engine-turned silver case.

216a

216b

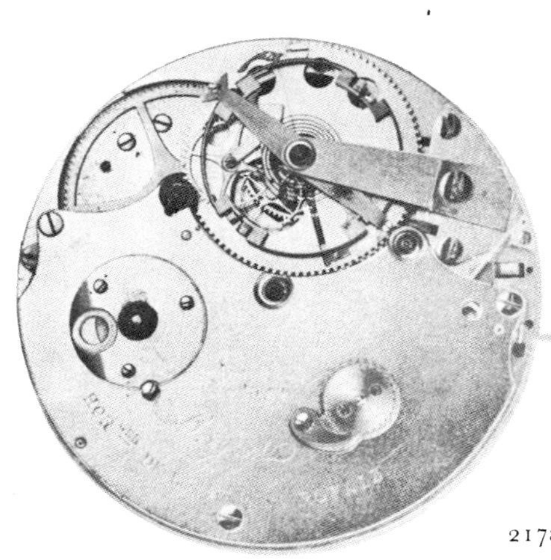

217a

217b

217a-b   BREGUET ET FILS NO. 2329. Paris, France. *Circa* 1811. Spring detent escapement with Peto-type cross detent and four-minute tourbillon. Two-arm compensation balance with four bi-metallic compensation arms. Spiral steel balance spring with overcoil free sprung. Fusee and chain. Engine-turned silver dial with blued steel hands. Stop, sweep centre seconds, subsidiary dials for hours, minutes and seconds. Engine-turned silver case.

218a-b   BREGUET ET FILS NO. 2571. Paris, France. *Circa* 1812. Double-roller lever escapement with minute tourbillon. Two-arm balance with four bi-metallic compensation arms. Spiral steel balance spring with overcoil, free sprung Fusee and chain. Enamel dial and blued steel hands. Silver case. The dial signature in script is fairly late for a Breguet. Signatures in script are usually restricted to watches made before the Revolution. The dial also carries a secret signature.

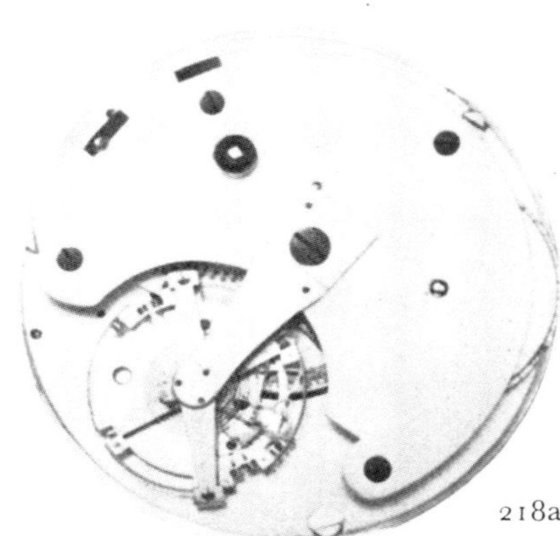

218a

218b

219a

219b

219c

219a-c   BROCKBANKS NO. 700. London, England. 1812. Earnshaw spring detent escapement. Three-arm bi-metallic compensation balance. Helical steel balance spring, free sprung. Fusee and chain. Quarter repeating on two gongs operated by depressing the pendent. Gold dial with engine-turned centre covered with translucent red enamel. Serpentine gold hands. Hunter case in three-colour gold.

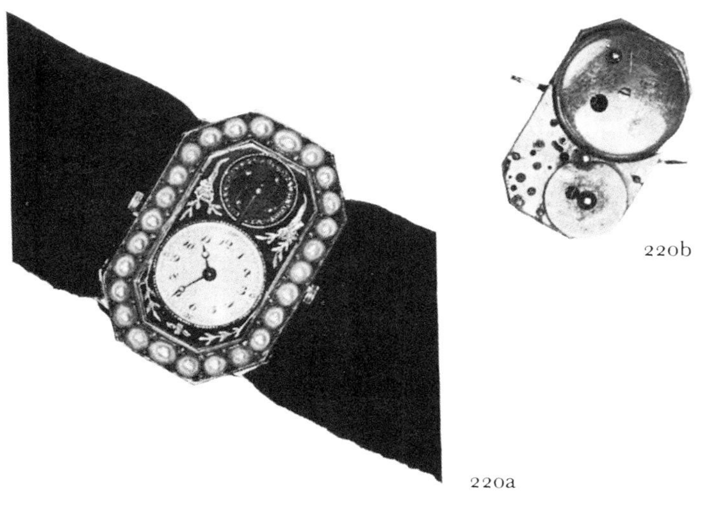

220a

220b

220a-b ANONYMOUS, Swiss or French. First quarter of the nineteenth century. Virgule escapement. Decorative four-armed steel balance wheel showing through the dial. Spiral steel balance spring and regulator. Going barrel. Grande sonnerie clock-watch. Enamel dial with blued steel Breguet-type hands surrounded by half pearls. Plain gold case. This must be one of the earliest watches designed to be worn as a wrist watch. It is also of exceptionally small size to include grande sonnerie striking, the dimensions of the movement being 25 mm by 15 mm.

221a

221b

221a-b RALPH GOUT NO. 267. London, England. *Circa* 1800. Verge escapement. Fusee and chain. Skeleton dial. Gilt-metal case. In addition to the watch train the watch contains a pedometer mechanism to indicate the distance walked in units of 10. The watch plate is engraved 'by the King's letters patent'. Gout patented his pedometer in 1799.

222

222 *Circa* 1800. Chinese duplex escapement, probably by Ilbury, made for the Chinese export market, with typically decorated case, and devils (for frightening away the evil eye) fixed to the balance wheel. See col. pl. XVI.

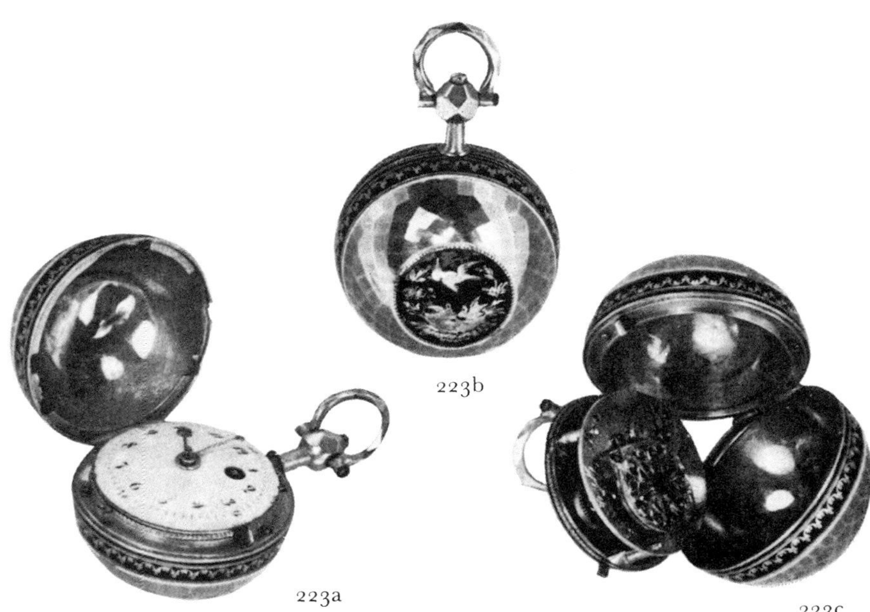

223b

223a

223c

223a-c ANONYMOUS NO. 4993. French or Swiss. Early nineteenth century. Verge escapement. Fusee and chain. Enamel dial with gold hands. Gold spherical case faceted all over and enamelled with a scene of a bird feeding young ones. (Slightly enlarged.)

224a

224b

224c

224a-c   BREGUET NO. 1310. Paris, France. First quarter of the nineteenth century. Steel cylinder escapement with steel escape wheel. Plain steel balance with spiral steel balance spring and regulator. Going barrel. Enamel dial with blued steel hands. Engine-turned silver case. This is a typical example of the Breguet 'mixte' watch. The cuvette is signed 'No. 1310. Etabl mixte de Breguet'. It is difficult to date Breguet 'mixte' watches as they do not appear in any surviving Breguet books.

225a

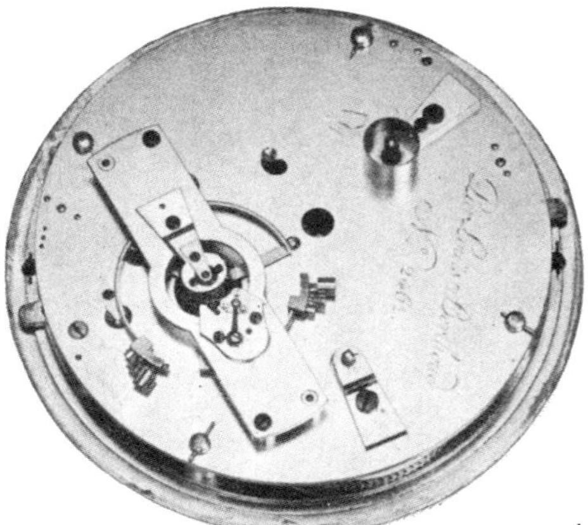

225b

225a-b   LOUIS BERTHOUD NO. 2462. France. Early nineteenth century. Pivoted detent escapement. Three-arm bi-metallic compensation balance, helical steel balance spring without end curves, free sprung. Fusee and chain. Enamel dial, blued steel hands, plain silver case. For illustration of escapement, see 13; for description, see page 115.

226a-b   LAMBERT. France. *Circa* 1815. Lever escapement, steel escape wheel, straight-line lever escapement without draw and with divided lift; double roller. Two-arm bi-metallic compensation balance with spiral balance spring free sprung. Going barrel. Engine-turned silver dial and blued steel hands, engine-turned silver case.

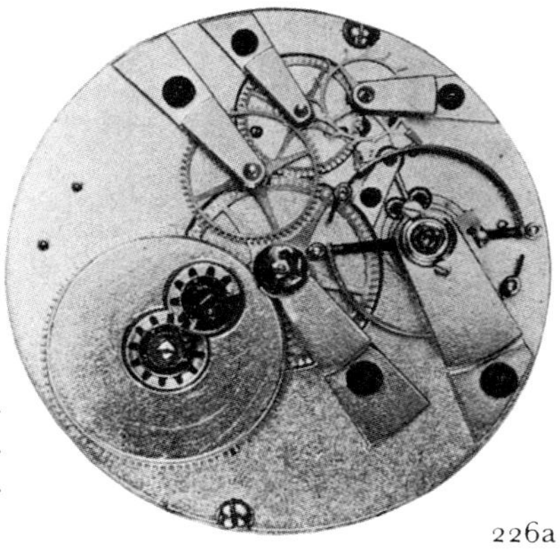

226a

226b

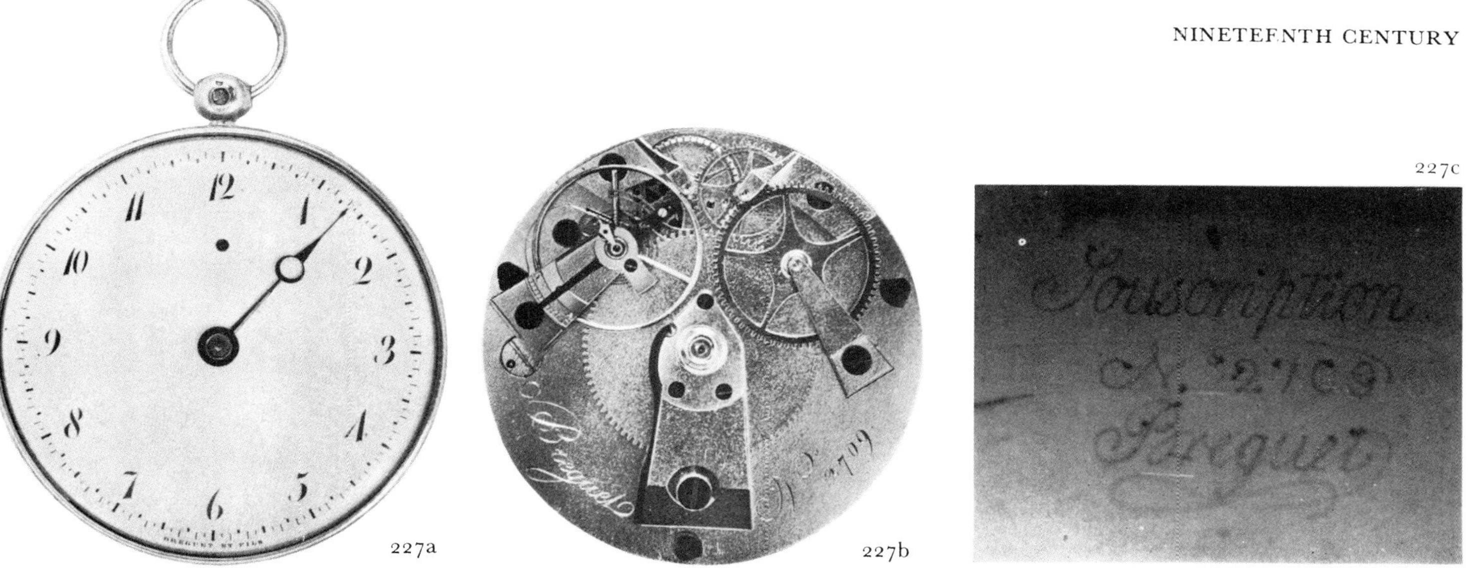

227a   227b   227c

227a-c   BREGUET ET FILS NO. 2709. Paris, France. 1815. Ruby cylinder escapement, plain gold three-arm balance. Spiral steel balance spring and regulator and elastic suspension to balance staff. Going barrel. Enamel dial with blued steel hand and secret signature. Engine-turned case, the band and back of silver and the rims and pendent of gold. This is Breguet's 'montre à souscription', see page 63. 227c shows the secret signature, greatly enlarged.

228a-b   BREGUET NO. 2732. Paris, France. 1815. Ruby cylinder escapement. Plain gold balance. Spiral steel balance spring with regulator. Bi-metallic compensation curb with elastic suspension. Going barrel. A small silver dial with blued steel hands. Montre à tact, the tact lever in gold on an engine-turned gold case covered with transparent blue enamel and pearl touch pieces. Short Breguet chain and key. (Slightly enlarged.)

228a   228b

229a

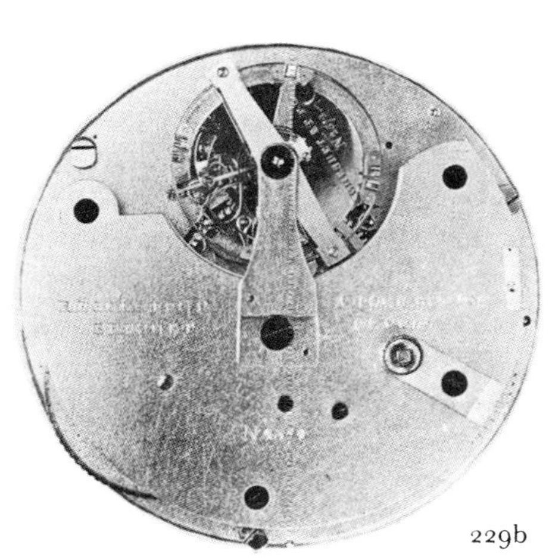

229b

229a-b   BREGUET ET FILS NO. 2572. Paris, France. 1811. Single-roller lever escapement with minute tourbillon. Two-arm compensation balance with quartered bi-metallic rim, spiral steel balance spring with overcoil, free sprung. Fusee and chain. Engine-turned silver dial and blued steel hands. Silver hunter case added in 1823, replacing an original gold case. For illustration of escapement, see 35; for description, see page 126.

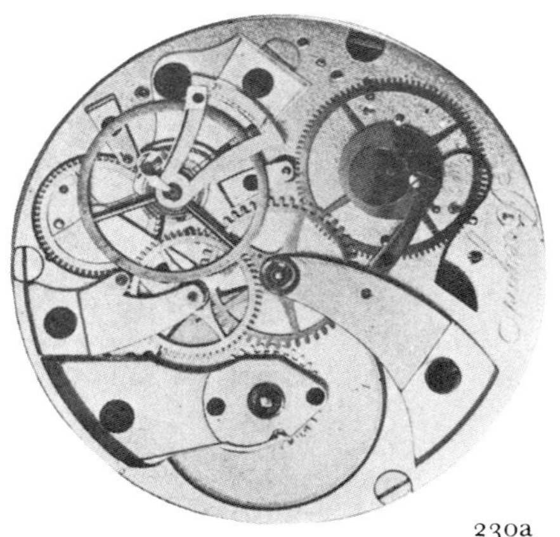

230a

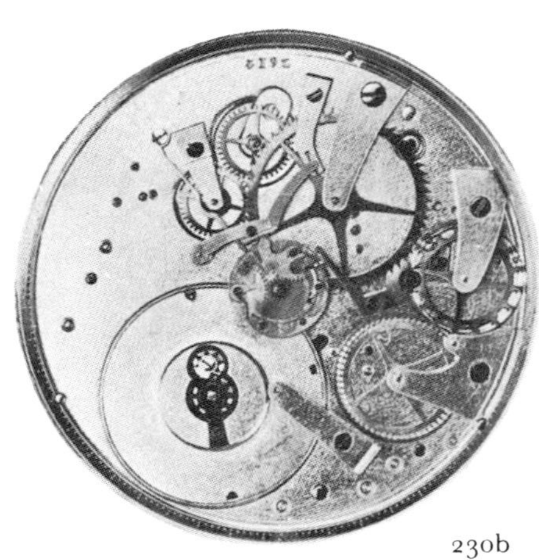

230b

230c

230a-c BREGUET NO. 2614. Paris, France. *Circa* 1815. Cylinder escapement. Plain brass balance with spiral steel balance spring and regulator with bi-metallic compensation curb and elastic suspension to the balance staff. Going barrel. Gold engine-turned dial engraved with Turkish numerals. Subsidiary dials for month and days of the month. Single gold hand. Second hand in blued steel for equation of time, mounted on a loose fitting collar over the cannon pinion: its differential movement is operated by a segmental rack mounted on the loose collar and whose position is determined by the equation lever. Engine-turned gold case.

231a

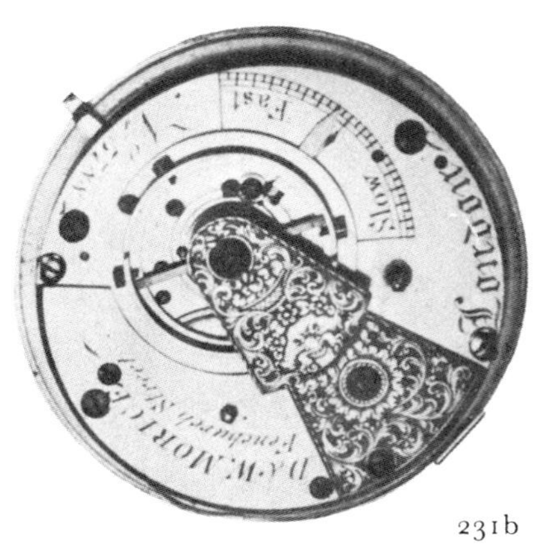

231b

231a-b D. & W. MORICE NO. 5788. London, England. 1815. Duplex escapement. Spiral balance spring and two-arm bi-metallic balance. Fusee and chain. Enamel dial with gold hands. Silver engine-turned case.

232a-b URBAN JÜRGENSEN NO. $\frac{24}{58}$. Copenhagen, Denmark. *Circa* 1815. Spring detent escapement of Arnold type, three-arm bi-metallic compensation balance, helical gold balance spring with end curves free sprung. Fusee and chain. Enamel dial with separate subsidiary dials for hours, minutes and seconds. Silver case.

232a

232b

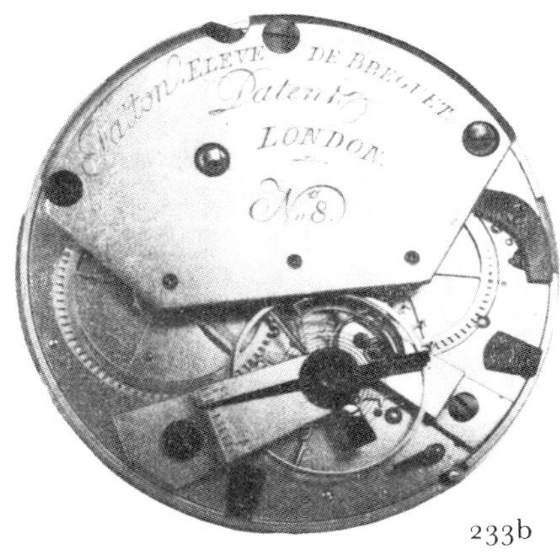

233a-b FATTON ÉLÈVE DE BREGUET NO. 8. London, England. *Circa* 1815. Cylinder escapement. Plain balance spiral steel balance spring with bi-metallic compensation curb. Going barrel. Enamel dial. Blued steel hands. Subsidiary dial for Fatton's patented ink-recording chronograph. Silver case.

233a

233b

234a-b LUTHER GODDARD & SON NO. 464. Shrewsbury, America. 1816. Verge escapement. Plain steel balance. Spiral steel balance spring. Fusee and chain. White enamel dial. Gold hands. Silver pair case.

234a

234b

235a

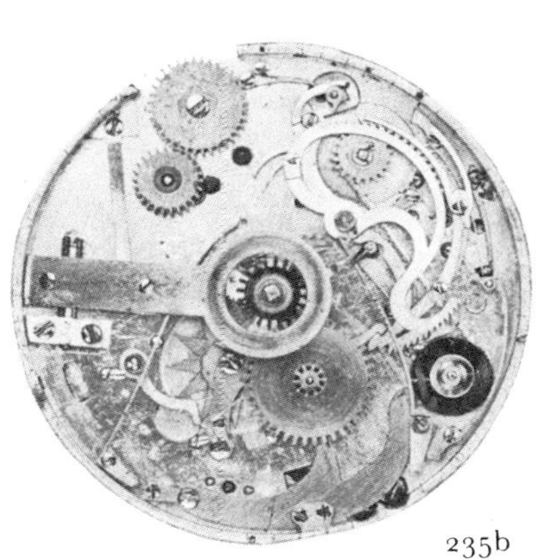

235b

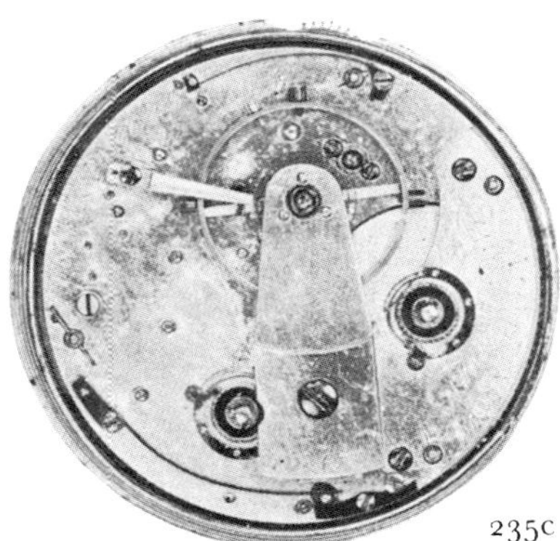

235c

235a-c VINER NO. 456. London, England. *Circa* 1816. Duplex escapement, two-arm bi-metallic balance, spiral steel balance spring free sprung. Fusee and chain; alarum and minute repeater; repeater operated by depressing pendent. Alarum set by rotating button at between 5 and 6 o'clock in the band of the case (Viner's patent). Matt gold dial with polished gold hour numerals and gold hands. Heavy cast gold case typical of English work towards the end of the first quarter of the nineteenth century. Note the very advanced design of the repeating work.

236a-b   RUNDELL, BRIDGE & RUNDELL, Ludgate Hill, London, England. 1817. No number. Cylinder escapement. Fusee and chain. Matt gold dial with polished raised Arabic numerals and gold hands. Embossed and engraved two-colour gold case.

236a

236b

237a

237b

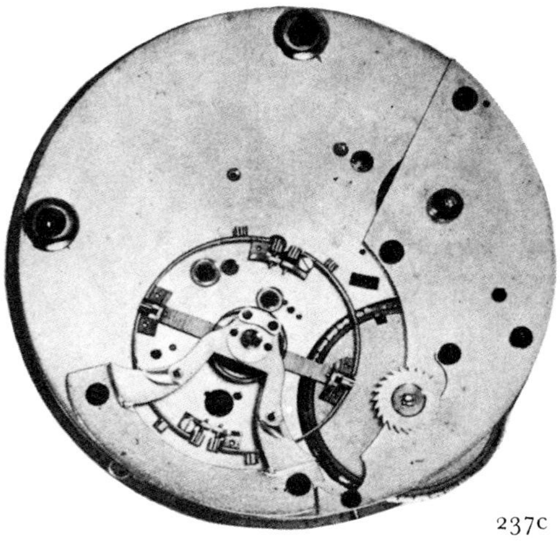

237c

237a-c   A. L. BREGUET NO. 2894. Paris, France. *Circa* 1817. Pivoted detent escapement with passing spring on the balance roller (for description see page 116). Two-arm balance with quartered bi-metallic compensation rim, helical steel spring with end curves free sprung, engine-turned silver dial with sweep centre seconds, subsidiary dials each with hour and minute hands for mean time and sidereal time and subsidiary dial for calendar. The hands of the mean time and seconds dials are blued steel. The hands of the sidereal time and calendar dials are gold. Plain silver case. 237b shows view under dial. For illustration of escapement, see 14.

238a

238b

238a-b   BREGUET ET FILS NO. 3261. Paris, France. *Circa* 1819. Ruby cylinder escapement. Three-arm brass balance. Bi-metallic compensation curb with elastic suspension to the balance arbor. Engine-turned silver dial with blued steel hands. Engine-turned gold hunter or savonette case. Note lay-out typical of most Breguet watches with cylinder escapement. Note also, plunger for operating repeater in the band of the case, and Breguet key with ratchet to prevent winding backwards. The hand-set is through a hole drilled in the glass, which is a common arrangement with Breguet savonette watches.

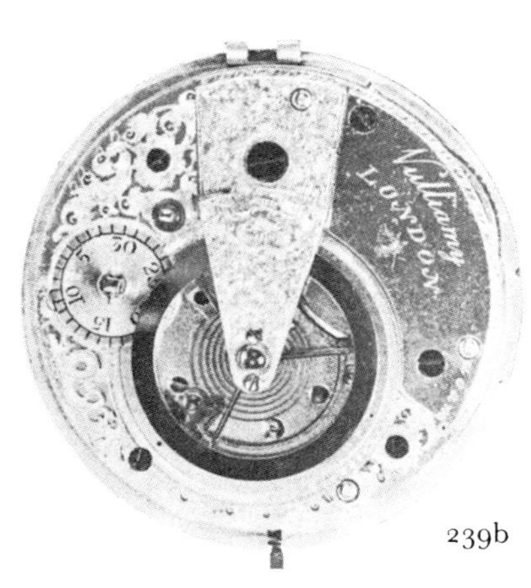

239b

239a

239a-b VULLIAMY CODE NO. UOAM. London, England. 1825. Duplex escapement, plain steel balance, spiral steel balance spring with regulator and compensation curb, fusee and chain. Silver dial with matt centre, gold hour and minute hands, blued steel seconds hand. Engine-turned gold case. Note the pierced gilt decoration to the watch plate typical of this type of Vulliamy duplex watch, which was made throughout the first quarter of the nineteenth century.

240a

240b

240a-b THOMAS EARNSHAW, London, England. *Circa* 1811. Spring detent escapement. Plain steel balance wheel. Spiral steel balance spring with regulator carrying 'sugar-tong' bi-metallic compensation curb. Fusee and chain. Enamel dial and gold hands. Silver case. For a discussion of Earnshaw's compensation curb, see pages 51 & 58. It is interesting that in this watch the balance spring stud has been moved so as partly to obscure the watch number which is 574. Earnshaw's watches also sometimes have a second number which in this case is 3017. The meaning of these numbers is not at present understood. This type of dial with radially-disposed Arabic numerals is one fairly commonly employed by Earnshaw. For illustration of escapement, see 18.

241a-b BAUTTE & MOYNIER, Geneva, Switzerland. *Circa* 1820. Cylinder escapement with spiral steel balance spring and regulator. Going barrel. Engine-turned silver dial with eccentric chapter ring for minutes. Subsidiary dial for seconds and an aperture through which the hour appears moving only at the hour. The case is of gold decorated in black enamel; this form of enamel and dial is characteristic of the decorative thin Swiss watch at this period.

241a

241b

242a-c WILLIAM ANTHONY NO. 1931. London, England. *Circa* 1820. Duplex escapement. Spiral steel balance spring and regulator. Going barrel. Enamel dial with separate dials for concentric hours and minutes and for seconds. Highly engraved and decorated movement. Oval gold case decorated with red and blue enamel and pearls.

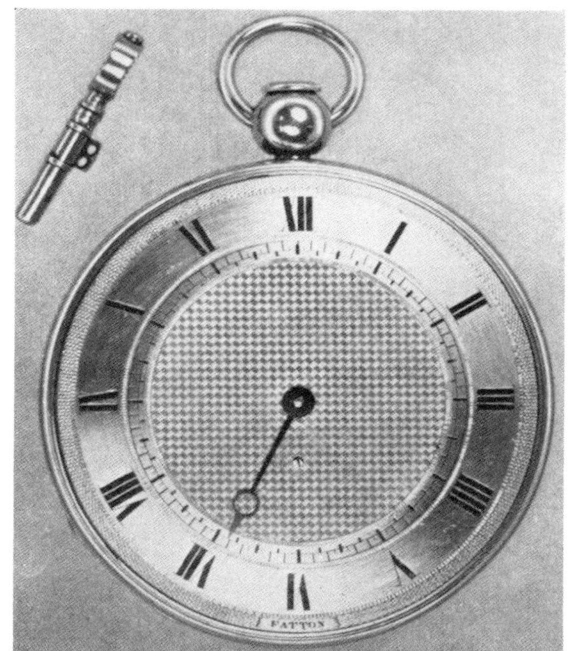

243a-b F. L. FATTON, Paris, France. *Circa* 1820. Cylinder escapement with spiral steel balance spring with regulator and Breguet-type bi-metallic compensation curb. Elastic suspension. Going barrel. Quarter repeating. Engine-turned silver dial, blued steel hand. Engine-turned silver case with gold bands and pendent. Fatton was a prominent pupil of Breguet's. This watch is almost identical to Breguet's souscription watches, with the addition of a repeating train, seen in one corner of the movement.

244a-c   LOPIN, Paris, France. *Circa* 1820. Cylinder escapement. Elastic suspension for balance staff. Spiral steel balance spring, with Breguet-type compensation curb. Going barrel. Quarter repeating on two gongs. Engine-turned silver dial with blued steel hands. Engine-turned gold case. A typical 'Elève de Breguet' product.

245a-d   T. D. PIGUET & P. MEYLAN, Geneva, Switzerland. 1820. Cylinder escapement. Brass balance. Spiral balance spring. Going barrel. White enamel dial, blued steel hands with gilt barking dog with articulated head chasing a swan on translucent blue enamel ground. Gold case with pearl-set bezels and coloured enamel bouquet of flowers.

246a-b THOMAS EARNSHAW NO. 581. London, England. 1811. Spring detent escapement. Two-arm bi-metallic compensation balance with heavy weights. Helical steel balance spring, free sprung. Fusee and chain. Enamel dial and blued steel hands. Silver case. It is interesting that the number of this watch is only 7 later than the watch in 240, showing that Earnshaw used compensation balances and compensation curbs at the same period on his spring detent watches.

246b

246a

247a-d ANONYMOUS, Switzerland. *Circa* 1820. Cylinder escapement. Plain balance. Spiral balance spring. Going barrel. White enamel dial. Blued steel hands not original. Gold and enamel case with pearl-set bezels. Pushing the pendent winds up a spring to release the singing bird which emerges through the band.

247a

247c

247b

247d

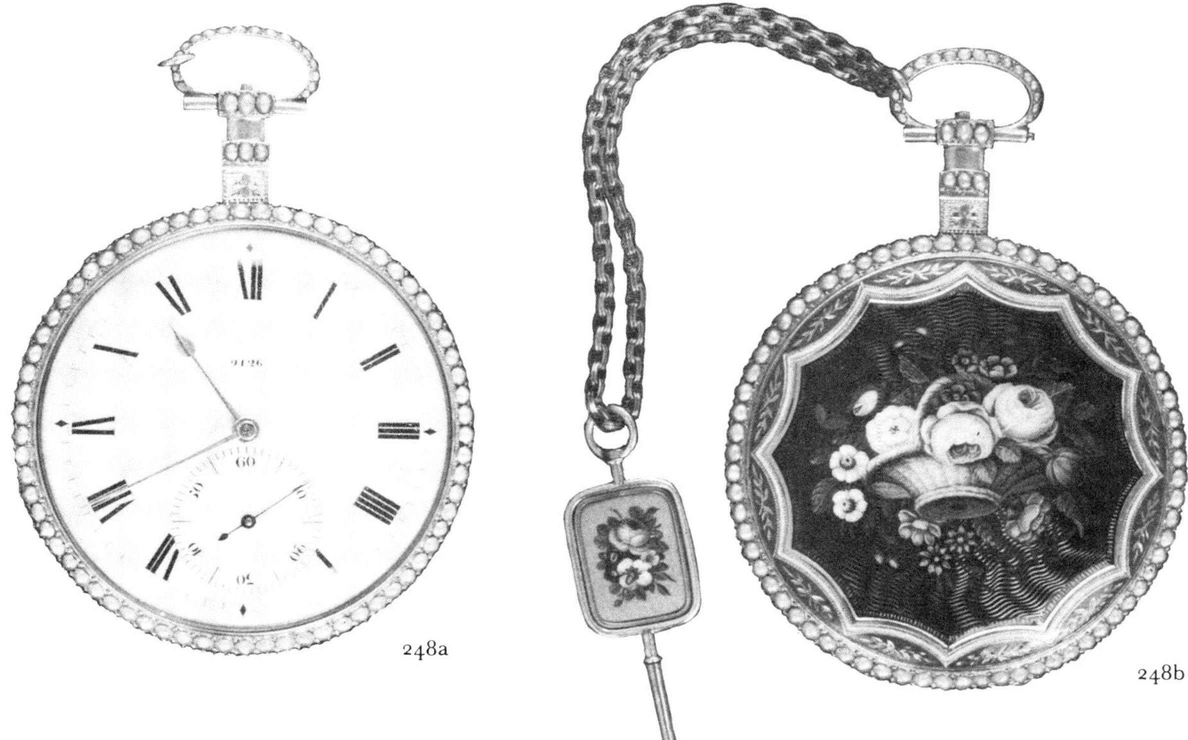

248a

248b

248a-d ANONYMOUS NO. 9126. Switzerland. *Circa* 1820. Ruby cylinder escapement. Plain balance. Spiral steel balance spring. Going barrel. Pinned-disk musical mechanism released at the hour or at will. White enamel dial. Gold hands. Gold case; pearl-set bezel with coloured enamel basket of flowers set in a field and translucent enamel over engine-turning. Short chain and matching enamelled key.

248c

248d

249a-b ILBERY NO. 5891. London, England. *Circa* 1820. Virgule escapement. Plain balance. Spiral steel balance spring. Going barrel. Automata released at the hour or at will. The scene is a blacksmith's workshop. Gold enamelled case. Swiss mechanism.

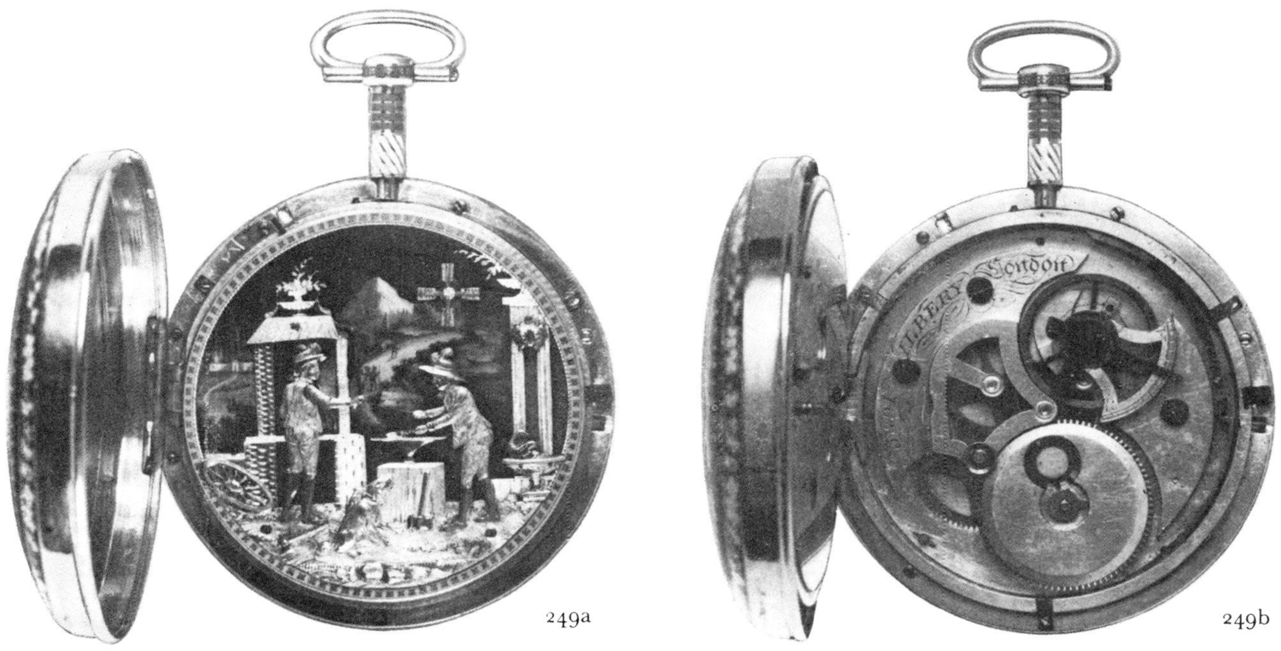

249a

249b

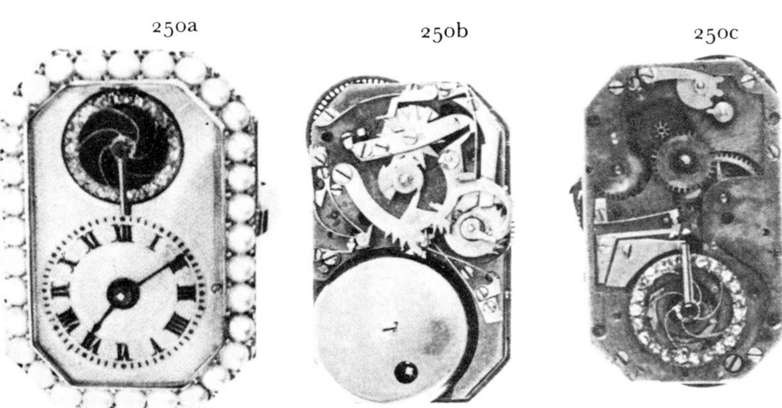

250a-c   ANONYMOUS, Switzerland. *Circa* 1820. Finger-ring, quarter-repeating watch (enlarged).

251   HUMBERT & MAIRET, Geneva, Switzerland. *Circa* 1820. Verge escapement. Plain balance. Spiral balance spring. Fusee and chain. The dial with three sectors for hours, minutes and date respectively. Depressing the pendent causes the old man, the boy and the monkey to raise their sticks, each pointing to the appropriate sector.

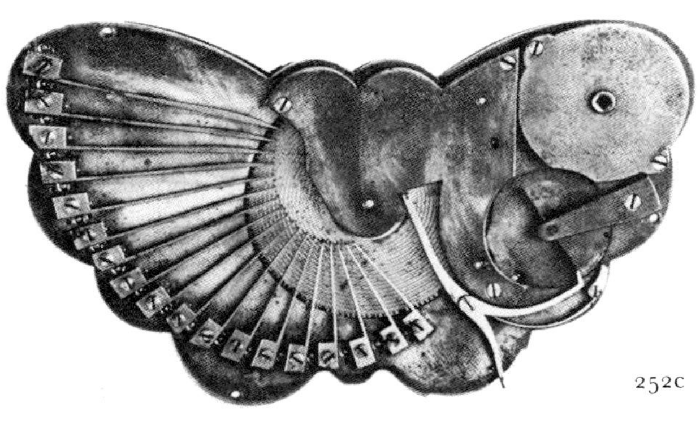

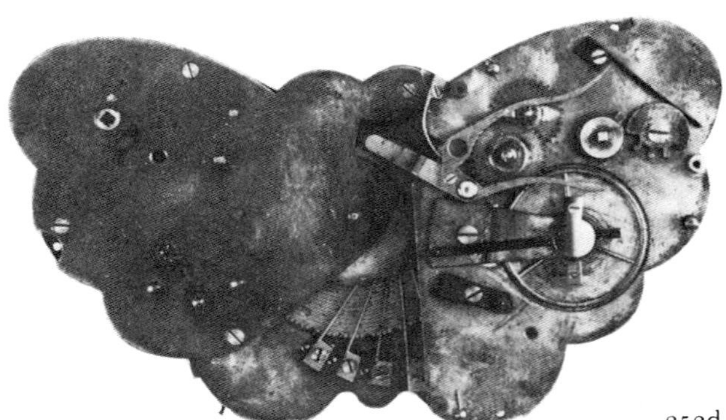

252a-d   PIGUET ET MEYLAN, Geneva, Switzerland. *Circa* 1820. Butterfly form watch with musical automata. White enamel dials for mean time and seconds. Automaton figures play in unison to music by pinned-disk mechanism.

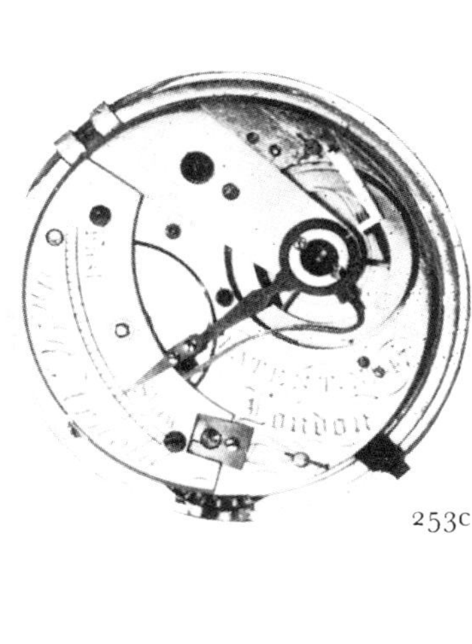

253a

253b

253c

253a-c   J. R. ARNOLD NO. 47. London, England. 1821. Cylinder escapement. Bi-metallic compensation curb consisting of a strip fixed to the regulator and carrying one curb pin. Going barrel with Prest keyless winding. Silver dial with blued steel hands with centre square for hand setting. Gold case. Prest's patent was the earliest effective system of keyless winding, but it made no provision for setting the hands, which was done in the old way by a square on the centre arbor.

254a-b   BREGUET ET FILS NO. 3679. Paris, France. *Circa* 1822. Spring detent escapement of Earnshaw type. Two-arm bi-metallic compensation balance. Helical steel balance spring without end curves. Two going barrels wound separately. Matt silvered dial with eccentric sweep minutes overlapping with the second dial, and subsidiary dial for hours. Plain silver case. This is a favourite lay-out for Breguet for deck watches, although this watch could be worn in the pocket. Breguet regarded two going barrels as a satisfactory substitute for the fusee and several such watches are wound from a single square mounted between the two barrels.

254a

254b

255a

255b

255a-b   ANONYMOUS but in the style of, and probably by, J. F. Cole, London, England. 1822. Pointed-pallet, right-angle, double-roller escapement. Bi-metallic compensation balance. Spiral steel balance spring. Going barrel. Silver-gilt dial. Later steel hands. Gold case.

256a

256b

256a-b RICHARD WEBSTER NO. 5479. London, England. 1823. Lever escapement. Plain gold balance with spiral steel balance spring and regulator. Fusee and chain. Engine-turned silver dial with gold hour and minute hands and blued steel sweep centre seconds hand. Engine-turned silver case. This watch is a very rare example of the transition from the full plate English movement to the three-quarter plate. This has a full plate movement, but the plate below the balance is recessed so that the balance wheel itself is on the level of the plate, thus producing the effect of the three-quarter plate movement.

257a-b R. HORNBY NO. 5537. Liverpool, England. 1823. Single-roller lever escapement. Fusee and chain. Matt gold dial with polished engraved decoration in centre and raised Roman numerals. Gold hands. Heavily decorated gold case. This is one of the earliest recorded examples of the English use of the single-roller lever escapement.

257a

257b

258a

258b

258a-b BREGUET NO. 4196. Paris, France. *Circa* 1824. Spring detent escapement, Earnshaw type. Two-arm bi-metallic compensation balance. Helical steel balance spring with end curves free sprung. Elastic suspension to the balance arbor. Going barrel. Silvered dial with overlapping chapter rings and steel hands. Silver case. This watch dates from just after the death of Abraham Louis Breguet and the use of a single going barrel in a chronometer watch of Breguet's is quite unusual. The going barrel is of exceptionally large size.

259a-b   J. F. COLE NO. 9Q. London, England. 1824. Single-impulse lever escapement. Bi-metallic compensation balance. Spiral steel balance spring. Fusee and chain. Enamel dial. Blued steel hands. Quarter repeating. Gold case.

259a

259b

260a-b   ANONYMOUS, England. 1825. Single-roller lever escapement. Brass balance, spiral balance spring. Going barrel. Keyless-wound by worm and wheel. Engine-turned gold dial. Blued steel hands. Gold case. An early example of keyless winding and of the single-roller escapement, with an unusual layout of the movement.

260a

260b

261a-b   D. WHITELAW NO. 596. Edinburgh, Scotland. 1825. Single-roller lever escapement. Plain balance. Spiral steel balance spring. Fusee and chain. White enamel dial. Blued steel hands-(minute hand replaced). Gold case. The Chester hallmark and 'Liverpool window' jewelling indicates that despite the signature this watch was made in Liverpool.

261a

261b

262a-b HENRI MOTEL NO. 48.
Paris, France. *Circa* 1825. Pivoted
detent escapement. Bi-metallic
compensation balance. Conical
steel balance spring. Fusee and
chain. White enamel dial. Blued
steel hands. Gold case.

262a

262b

263a

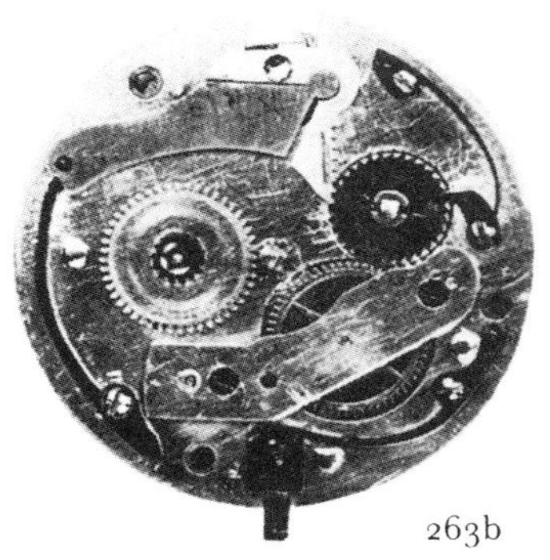

263b

263a-b WRIGHT NO. 440. England. 1816.
Massey-type lever escapement. Steel balance
wheel, spiral steel balance spring with regulator. Fusee and chain with pump wind. Gold
dial and hands. Plain gold case. The under-
dial illustration shows the operation of the
pump wind, which is Massey's patent No.
3854 of November 17th, 1814.

264a

264b

264c

264a-c BREGUET NO. 4764. Paris, France. *Circa* 1826. Lever escapement. Two-arm bi-metallic compensation balance, spiral
steel balance spring with overcoil and regulator. Going barrel. Half-quarter repeater. Silver engine-turned dial. Gold hands.
Plain gold case. The watch is keyless wound by a flat, un-knurled button. A smaller button concentric with the winding button
is pulled out and then turned in order to set the hands. The repeater is operated by a slide in the band of the case.

265a-c BREGUET NO. 4548. Paris, France. *Circa* 1826. Lever with draw. 20-tooth pierced escape wheel. Two going barrels. Perpetuelle à tact. Silver engine-turned dial with eccentric chapter ring for hours and minutes. Apertures for date and sectors for up-and-down indicator and regulator. Gold hands. This watch represents the final development of Breguet's perpetuelle. Extra thinness is achieved by setting the weight below the top plate. There is provision for winding the watch by hand.

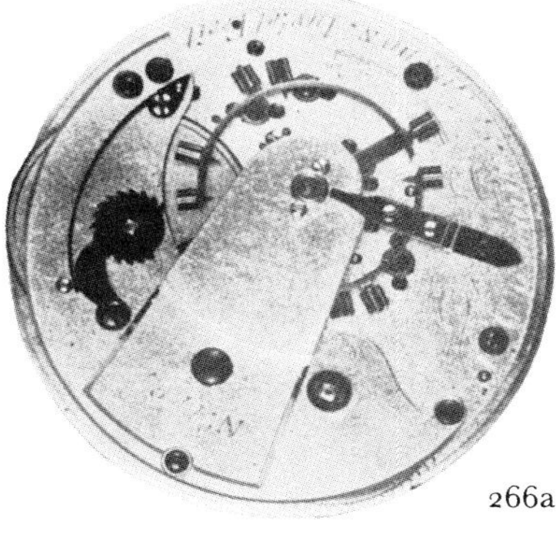

266a-b T. CUMMINS NO. 17-27. London, England. 1826. Lever escapement. Two-arm bi-metallic compensation balance. Helical steel balance spring, free sprung. Fusee and chain. Gold engine-turned, regulator-type dial. Steel hands. Gold engine-turned case. The escapement is a refined version of Massey's, and has resilient banking. There is no motion work and no square for setting the hands. For a discussion of Cummins' work, see page 70.

267a-c BREGUET NO. 4938. Paris, France. *Circa* 1829. Straight-line lever escapement with divided life and draw, two-arm bi-metallic compensation balance, spiral steel balance spring with overcoil and regulator, elastic suspension to balance staff. Going barrel, half-quarter repeater on one gong. Enamel dial with secret signature and gold hands. Engine-turned gold case.

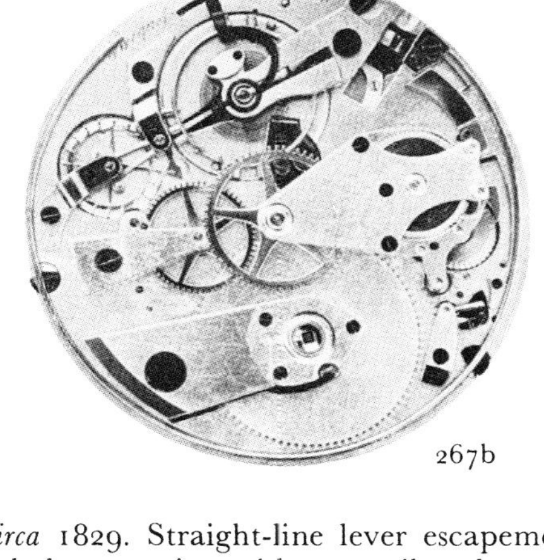

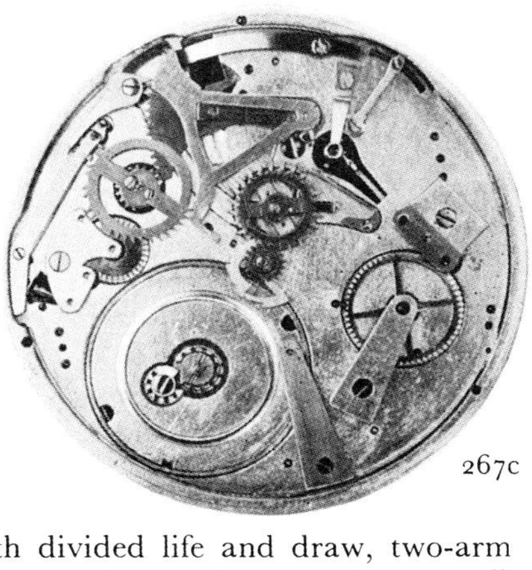

268a

268b

268a-b PARKINSON & FRODSHAM NO. 1804.
London, England. 1827. Spring detent
escapement. Bi-metallic compensation
balance. Spiral steel balance spring with
resilient stud. Fusee and chain. White
enamel dial. Gold hands. Gold case.

269a-b ANONYMOUS, probably Liverpool, England. 1828. Massey-type lever
escapement. Plain steel balance. Spiral
steel balance spring. Fusee and chain.
White enamel dial with gold hands.
Silver case. The dial signature 'Improved patent' refers to the escapement.

269a

269b

270a

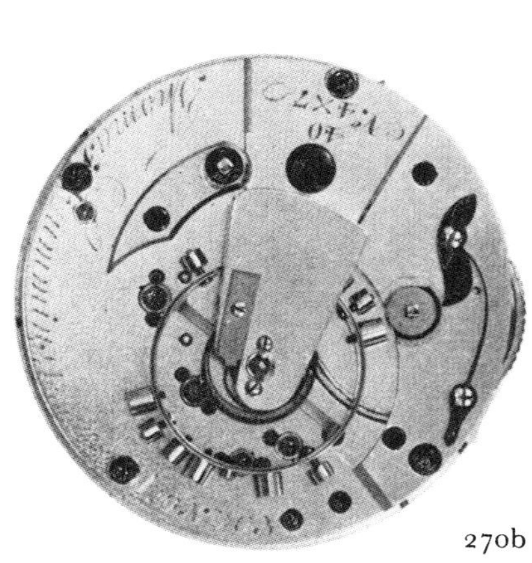

270b

271

271 CHARLES OUDIN & CIE, Paris, France.
Circa 1830. Breguet-type ruby cylinder
escapement. Crystal regulator. Going barrel. Movement of crystal with wheel-pivot
jewels in gold setting. Crystal dial with gold
hands. Case with crystal covers and gold
band and pendent.

270a-b THOMAS CUMMINS NO. $\frac{40}{4 \times 7}$. London, England. 1828. Massey-
type lever escapement. Bi-metallic compensation balance. Helical steel
spring with terminal curves. Fusee and chain. White enamel dial.
Blued steel hands without motion-work. Silver case.

272a
272b

272a-b SIGISMUND WRENTZSH. London, England. *Circa* 1830. Ruby cylinder escapement. Plain balance. Spiral balance spring, with bi-metallic compensation curb. Central going barrel. White enamel dial, blued steel hands: central hour and subsidiary dial for minutes and sector for regulator. Gold case. The mainspring is wound by turning the bezel.

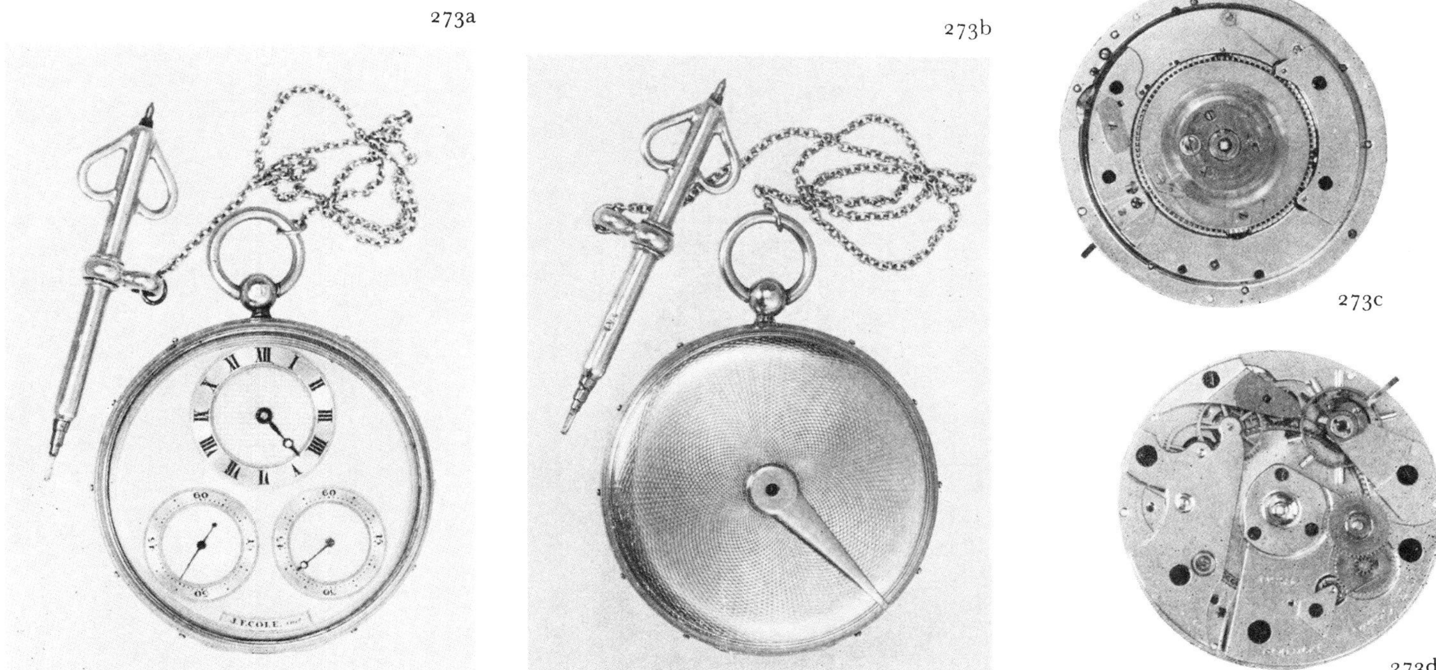

273a
273b
273c
273d

273a-d J. F. COLE, London, England. 1830. Lever escapement. Bi-metallic compensation balance with Cole's typical long-headed screws. Spiral spring with terminal curve and regulator. Central going barrel. Silver engine-turned regulator dial. Blued steel hands. Gold engine-turned à tact case with short chain and key.

274a
274b

274a-b DELAGE ET CIE, France. *Circa* 1830. Straight-line lever escapement with divided lift. Breguet-type bi-metallic compensation curb with Breguet-type elastic suspension to the balance staff. Fusee and chain. Enamel dial, blued steel hands. Engine-turned gold case. Nothing is known of the maker of this fairly high-grade, fairly early, Continental lever escapement watch.

275a

275b

275a-b ANONYMOUS. NO. 4212. Swiss. Second quarter of nineteenth century. Cylinder escapement, spiral steel balance spring and regulator. Going barrel. Engraved silver dial and blued steel Breguet-type hands. Heavily chiselled and engraved gold case. The unusual feature of this watch is its form of winding which is by a lever pivoted to the mainspring arbor and projecting through the side of the case. On being moved backwards and forwards it winds up the spring. The hands are set by turning a milled disc which projects through the cuvette.

276a-b L. GOLAY, Switzerland. Second quarter of the nineteenth century. Cylinder escapement. Bi-metallic compensation curb. Going barrel, quarter-striking grande sonnerie clock-watch. Silver engine-turned dial with blued steel hands, gold engine-turned case.

276a

276b

277a

277b

277a-b GEORGE COWLE NO. 342. London, England. *Circa* 1830. Spring detent escapement. Earnshaw type; two-arm bi-metallic compensation balance, helical steel balance spring with terminal curves, free sprung; fusee and chain. Matt gold dial with raised polished Turkish numerals, gold hands. Cast gold case with crescent incorporated in pendent.

278a

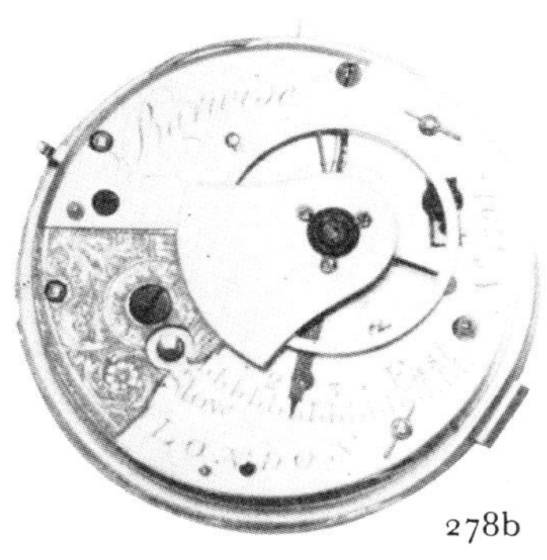

278b

278a-b BARWISE NO. 4791. London, England. 1829. Rack lever escapement. Fusee and chain. Enamel dial with gold hands. Plain gold case. A late and particularly finely finished example of the rack lever escapement. A watch of typical lay-out and appearance of the 1830 period.

279c

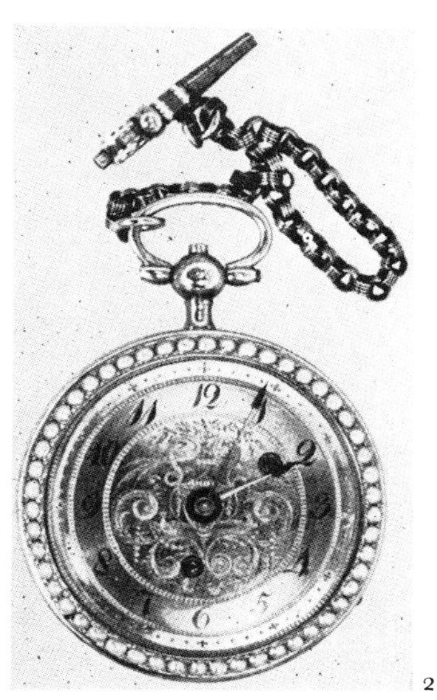

279a

279b

279a-c ANONYMOUS, probably Swiss. *Circa* 1830. Cylinder escapement. Going barrel. Gold dial and hands. Painted enamel case surrounded by a circle of half pearls. The figures in the painting are automata; a young man pushes a girl on a swing and another lady seated at the bottom left-hand corner of the picture plays a guitar. The Breguet-type key is enamelled to match the watch.

280a-b ANONYMOUS, French or Swiss. *Circa* 1830. Lever escapement. Breguet-type bi-metallic compensation curb. Going barrel. Engine-turned gold dial and blued steel hands. Engine-turned gold case. This is a remarkably thin watch for the period, especially in view of the lever escapement. The thickness of the movement is 3 mm and that of the entire watch is 6 mm.

280a

280b

281a

281b

281a-b ANONYMOUS, Swiss. First quarter of the nineteenth century. Verge escapement with spiral steel balance spring and regulator. Fusee and chain. Enamel dial, steel hands. The dial is surrounded by a ring of pearls and is set in a finger ring. (Slightly enlarged.)

282a-b TAVERNIER NO. 868. Paris, France. *Circa* 1830. Cylinder escapement. Two dials at front and back of the watch. The dial for hours, minutes and seconds is enamel and has steel hands. The other dial is silver. The outer ring is a table giving equation of time for the day; it has no equation mechanism. The subsidiary dials are for days of the week (centre); regulator; month; date; setter to determine at what time of day the date indicator shall change; phases of the moon.

282a

282b

283a

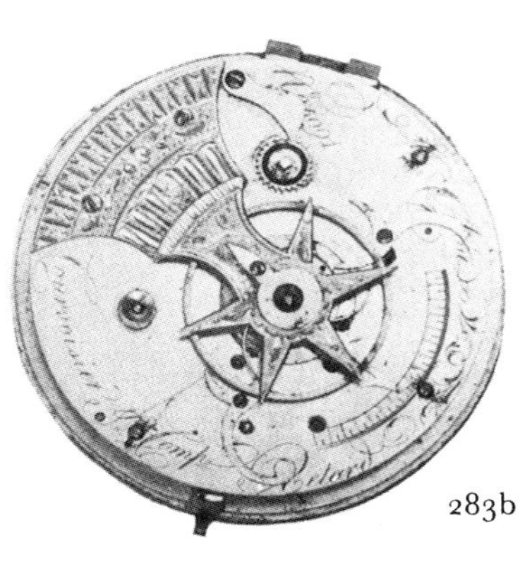

283b

283a-b COURVOISIER ET COMP. NO. 5091. Switzerland. *Circa* 1830. Verge escapement. Fusee and chain. Enamel dial with painted monochrome classical figures. Silver hunter case with no glass to dial. This watch is typical of middle-grade Swiss work at about this time.

284a-b HUNT & ROSKELL NO. 10514. London, England. 1836. Earnshaw-type spring detent escapement with one-minute tourbillon. Two-arm bi-metallic compensation balance. Spherical steel balance spring, free sprung. Fusee and chain. Enamel dial, steel hands. Sweep centre minute hand. Subsidiary dials for hours, minutes and thermometer. Gold engine-turned case. This watch was almost certainly made for Hunt & Roskell by Sylvan Mairet (compare the very similar dial in 298).

284a

284b

285b

285c

285a

285a-c ROBERT ROSKELL NO. 10030. Liverpool, England. 1832. Rack lever escapement. Spiral steel balance spring and regulator. Fusee and chain. Engine-turned gold dial and blued steel serpentine hands. Engine-turned gold case. This watch is unusual for its system of quarter repeating which is effected by turning the pendent to the left to obtain the hours and to the right to obtain the quarters. J. A. Berrollas's patent No. 3174 of October 31st, 1808.

286b

286d

286a

286c

286a-d BREGUET NO. 5009. Paris, France. 1836. Lever escapement. Two-arm bi-metallic compensation balance. Spiral steel balance spring with overcoil and regulator. Going barrel. Half-quarter repeater on one gong operated by a slide in the band of the case. Silver engraved dial with gold hands and subsidiary sectors for up-and-down and regulator. Gold case. This is the final development of Breguet's 'montre sympathique' and is here shown with the 'parent' clock (No. 128), which not only sets the watch to time but also winds it. As this is a precision watch the parent clock does not move the regulator, as was done with the earlier type of 'sympathique' watch, which had no temperature compensation.

287 WILLIAM WATSON, London, England. 1835. Duplex escapement. Plain steel balance. Spiral steel balance spring. Fusee and chain. Enamel dial. Blued steel hands with centre seconds. Gold case. A rare surviving example of a type of watch manufactured in pairs for the Chinese market, in their original fitted case.

287

288a-b CHARLES CUMMINS NO.589. London, England. 1837. Massey-type lever escapement. Bi-metallic compensation balance. Helical balance spring. Fusee and chain. Enamel regulator-type dial. Blued steel hands. Gold case. See also 270, Thomas Cummins. This watch differs only in that Charles added motion-work and a square for hand-setting. Also the movement is hinged to the case.

288a

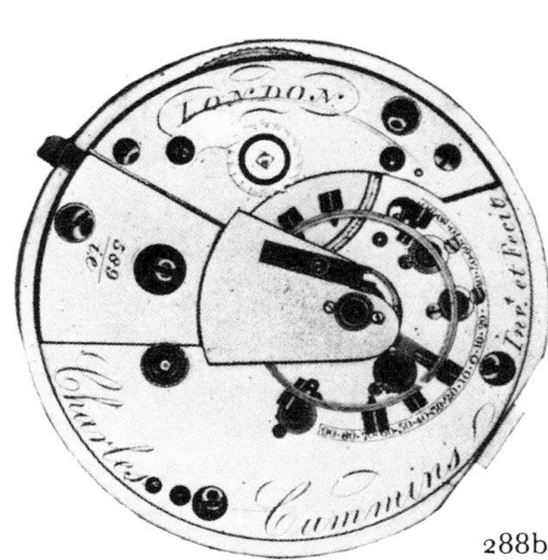

288b

289a-b J. F. COLE, London, England. *Circa* 1840. Pointed-pallet lever escapement. Bi-metallic compensation balance. Spiral steel balance spring with overcoil free sprung. Inverted reversed fusee and chain. Engraved gold dial with eccentric chapter blued steel hands. The segment let into the dial surface in the outer zone of the dial is of thinner metal to allow extra room for the balance pivot jewels. Gold case.

289a

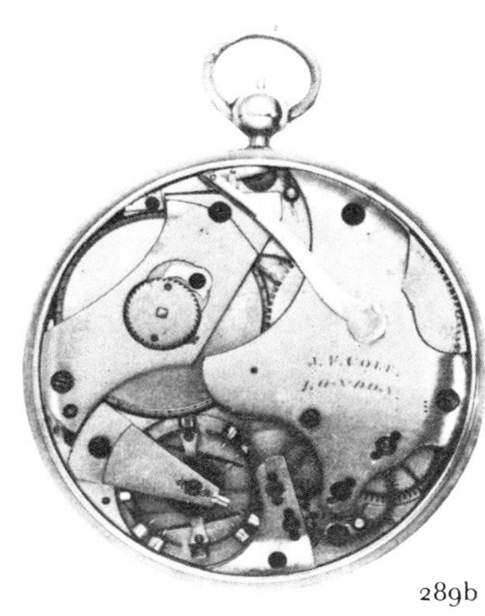

289b

290a

290b

290a-b E. J. DENT NO. 11425. London, England. *Circa* 1840. Spring detent escapement. Bi-metallic compensation balance. Helical steel balance spring with terminal curves. Fusee and chain. Enamel dial. Blued steel hands. Silver case.

291a-b LOUIS RABY, Paris, France. *Circa* 1840. Pivoted detent escapement. Bi-metallic compensation balance. Spiral steel balance spring with overcoil and regulator. Going barrel. White enamel dial. Blued steel hands. Gold case.

291a

291b

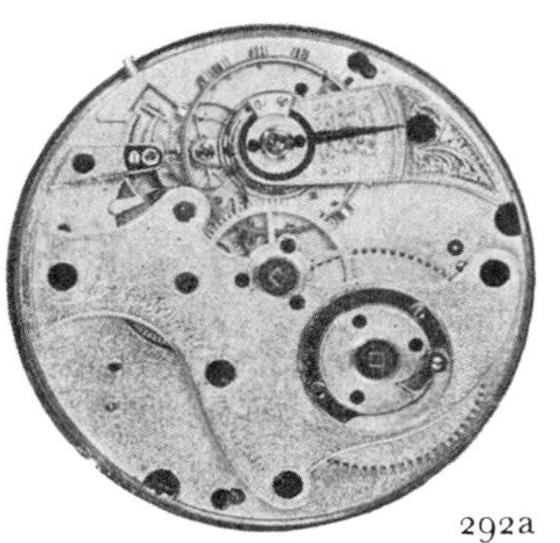

292a

292b

292a-b L. F. AUDEMARS, Brassus, Switzerland. *Circa* 1840. Robin escapement, two-arm compensation balance, spiral steel balance spring and regulator. Going barrel. Enamel dial and blued steel hands. Engine-turned gold case. The straight-line lay-out is unusual for the Robin escapement; the lever has the counterbalance arms popular in Swiss work at this time. See 40 and 41.

293a

293b

293a-b HUNT AND ROSKELL NO. 10413. London, England. *Circa* 1840. Lever escapement with divided lift, two-arm compensation balance, spiral steel balance spring with overcoil and regulator. Going barrel. Enamel dial with blued steel hands, sweep centre minutes hand, subsidiary dials for phases of the moon, hours concentric with days of the month, thermometer, seconds dial with two hands, one of which may be started and stopped at will. Plain gold case with à tact.

294a-b HUNT & ROSKELL. English, but an imported movement. Probably *circa* 1840. Lever escapement with divided lift, two-arm balance with bi-metallic rim but uncut, spiral steel balance spring with regulator and Breguet-type bi-metallic compensation curb, Breguet-type elastic suspension for the balance staff. Two going barrels, one for the independent jump seconds train. Engraved gold dial with blued steel sweep centre seconds hand. Two subsidiary dials, one for hours and one for minutes, one with steel and one with gold hands, driven off the same train but able to be set independently. Engraved gold case.

294a

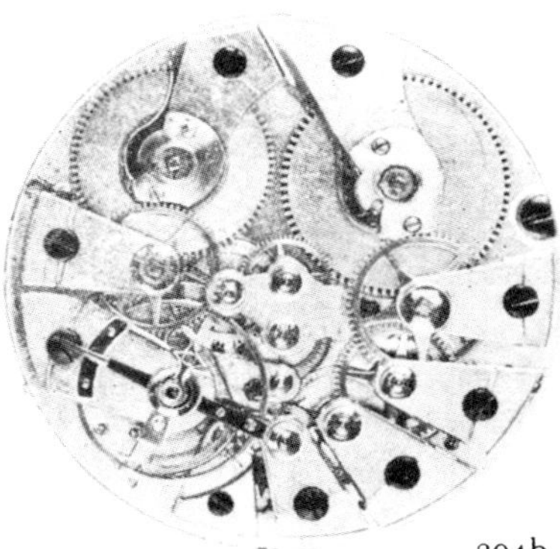

294b

295a   295b   295c

295a-c   VINER NO. 4913. Regent Street, London, England. 1840. Lever escapement, two-arm bi-metallic compensation balance, going barrel with pump wind. Enamel dial with blued steel hands. Engine-turned gold hunter case with à tact attached to back cover.

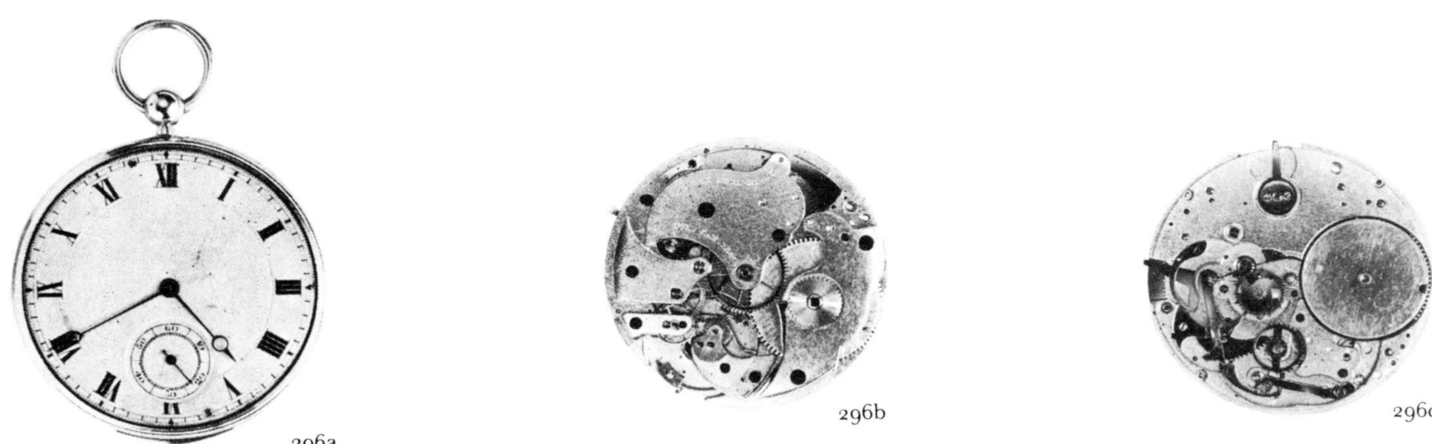

296a   296b   296c

296a-c   J. FERGUSON COLE, London, England. No number. January 1844. Escapement double-rotary detached chronometer. Two-arm bi-metallic compensation balance. Spiral steel balance spring with overcoil and regulator. Going barrel. Half-quarter repeater. Engine-turned silver dial and blued steel hands. Plain gold case. For illustration of escapement, see 21; for description, see page 118.

297a-b   ANONYMOUS. NO. 40757. English. 1845. Verge escapement. Fusee and chain. Painted enamel dial with gold hands. Silver pair case. This watch is typical of what is generally described as a 'farmer's watch' of the mid-nineteenth century. While the verge escapement continued to be made until the end of the century, this watch represents about its final appearance with classical lay-out and a pierced and decorated balance cock.

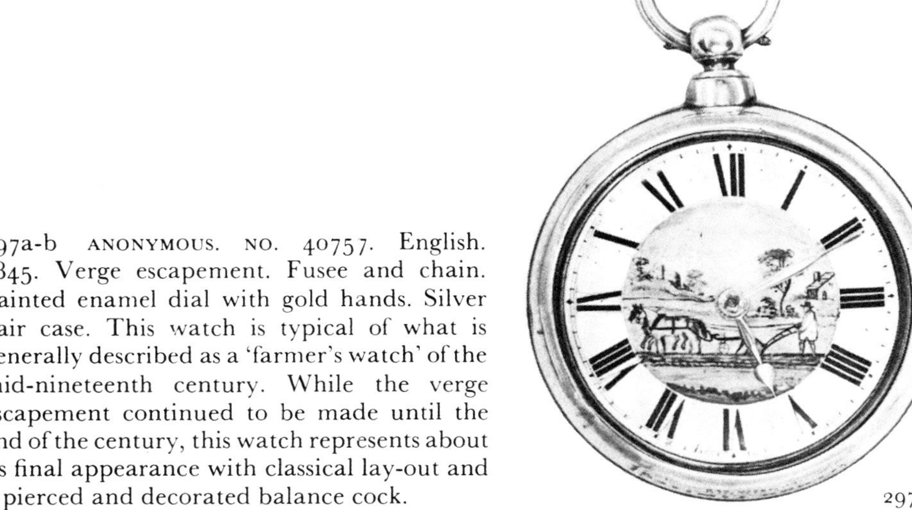

297a

297b

298a   298b

298a-b   HUNT & ROSKELL, London, England. 1846. Lever escapement. Two-arm bi-metallic compensation balance. Spiral steel balance spring with overcoil and regulator. Going barrel. Engine-turned gold dial with subsidiary dials for minutes and seconds. The hours appear through an aperture and pass a fixed pointer. Large segmental auxiliary dial for Fahrenheit temperature. Plain gold case. Although the watch is signed 'Hunt & Roskell', the movement is stamped 'S.M.' indicating that the movement was in fact made by Sylvan Mairet, the pupil of Breguet who worked for a time for Hunt & Roskell. It has his special form of lever escapement in which all the lift is on the escape wheel. This is clearly shown in the illustration. Note also the unusual form of fixed pendent. No photograph can do justice to the quite exceptional elegance of this watch.

299a-b   ANONYMOUS. NO. 224890. Swiss. Probably mid-nineteenth century. Lever escapement, two-arm uncut bi-metallic balance. Going barrel. Enamel dial, blued steel hands and sweep centre seconds hand. Chinese numerals with inner 24-hour chapter ring with European Roman numerals. Silver case.

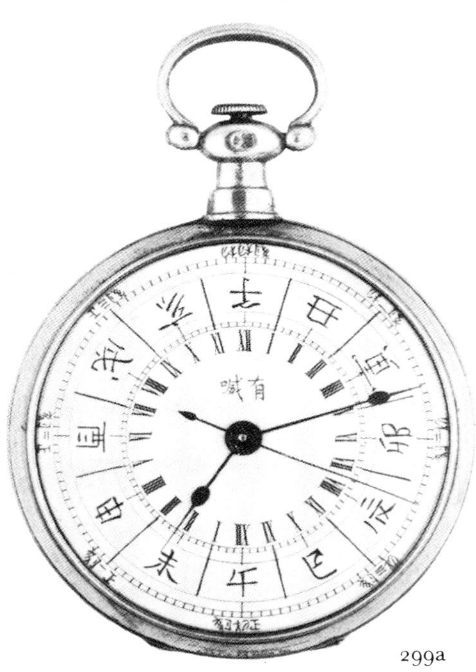

299a

299b

300a

300a-c   ANONYMOUS, Japanese. Mid-nineteenth century. Verge escapement. Brass balance, spiral steel balance spring. No regulator. Fusee and gut. Clock-watch with locking-plate and no warning. Gilt and painted dial with movable chapters for the varying length of day and night-time hours in the ancient Japanese time system. This is an inro watch with netsuke and ojime.

300b

300c

301a

301b

301a-b DENT NO. 15206. London, England. 1850. Movement signed 'E. J. Dent watchmaker to the Queen'. Single-roller lever escapement. Bi-metallic compensation balance. Spiral balance spring. Enamel dial. Blued steel hands. Gold case. An early example of both winding and hand-setting being effected by keyless-work; also, in an English watch, of a going barrel.

302a-b A. LANGE & SOHNE NO. 4384. Glashutte, Germany. *Circa* 1850. Lever escapement. Bi-metallic compensation balance. Spiral steel balance spring with terminal curve and regulator. Going barrel. White enamel dial. Blued steel hands. Gold hunting case.

302a

302b

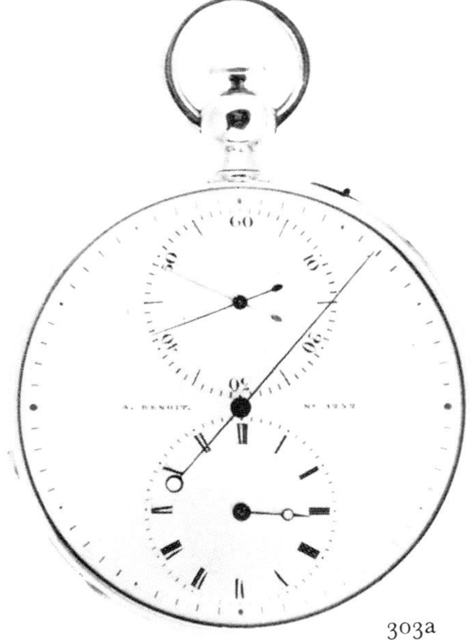

303a

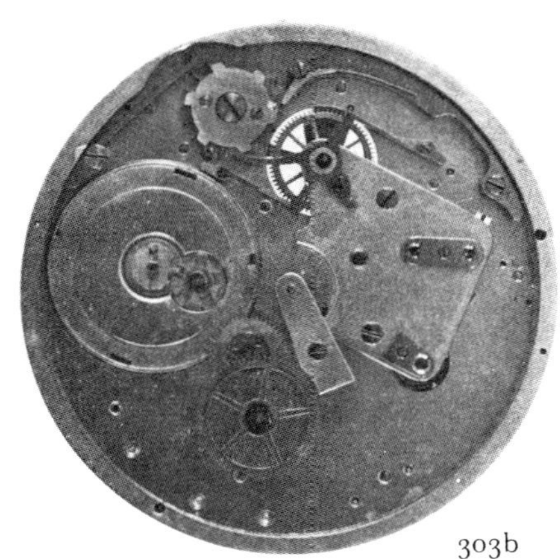

303b

303c

303a-c A. BENOIT NO. 1252. Versailles, France. *Circa* 1850. Lever escapement. Bi-metallic compensation balance. Spiral steel balance spring with overcoil and regulator. Going barrel. White enamel dial. Centre minute hand. Subsidiary dials for hours and for seconds. White enamel dial. Two concentric seconds hands with provision for stopping and starting one of them by depressing the pendent. Blued steel hands. Gold case.

304a-b J. NICOLAUS NO. 52. Vienna, Austria. *Circa* 1860. Spring detent escapement. Bi-metallic compensation balance. Helical steel balance spring with terminal curves. Going barrel. White enamel dial. Blued steel hands. Silver case.

304a

304b

305a

305b

305a-b ANONYMOUS, Swiss. *Circa* 1860. Cylinder escapement. Going barrel. Engraved gold dial, blued steel hands, painted enamel case.

306a-b RICHARD DOVER, STATTER & THOMAS STATTER NO. 1. Liverpool, England. 1862. Lever escapement. Two-arm bi-metallic compensation balance. Fusee and chain. Enamel dial with blued steel hands. Gold case. This is a very late example of the decimal dial. In the centre of the dial is marked 'Decimal timekeeper'. Ordinary 12-hour time is shown on a subsidiary dial. There is a sweep centre seconds, the minute being divided into 100 parts.

306a

306b

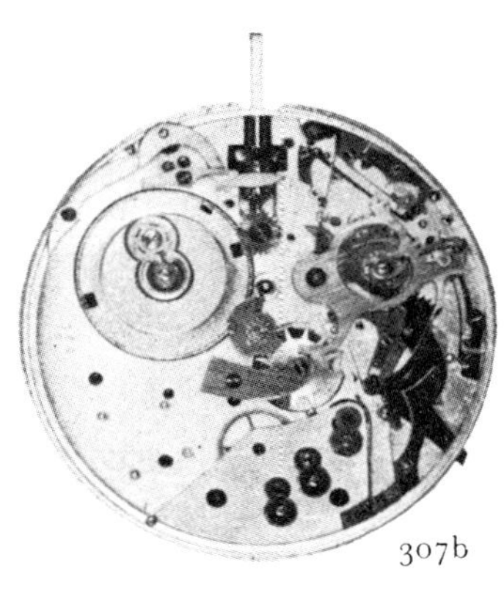

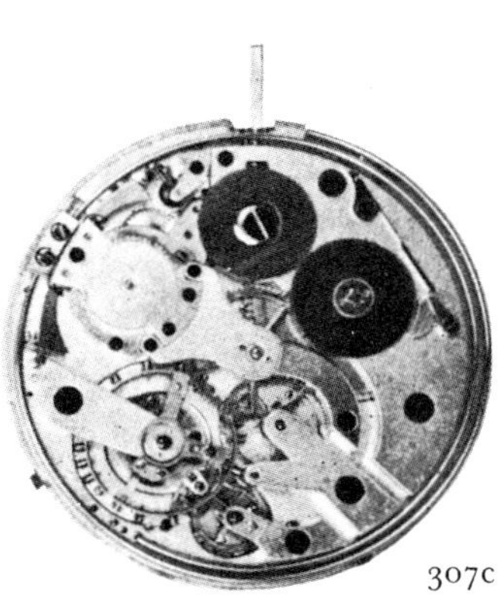

307a-c   BREGUET ET FILS NO. 931. Paris, France. *Circa* 1864. Lever escapement. Two-arm bi-metallic compensation balance. Spiral steel balance spring with overcoil and regulator. Going barrel. Gold engine-turned dial. Gold hands. Engine-turned gold case. This is an early example of keyless winding with pull-button for hand-setting. The watch number is of a later series than any used in A. L. Breguet's lifetime.

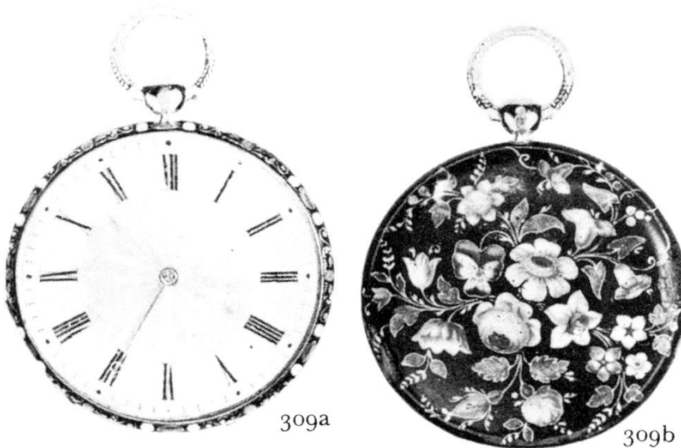

308a-b   J. FERGUSON COLE NO. 1848. London, England. 1867. Rotating detent escapement. Two-arm bi-metallic compensation balance. Spiral steel balance spring with terminal curve. Fusee and chain. Enamel dial and blued steel hands. Plain gold case. For illustration of escapement, see 47; for description, see page 131.

309a-b   ANONYMOUS, Swiss. 1850. Cylinder escapement, spiral steel balance spring and regulator, barred movement, going barrel. Engine-turned silver dial with gold hands. Gold case with painted enamel on black background; a typical decorated lady's Swiss watch of the mid-nineteenth century.

310a-b   G. K. ROSKOPF, Switzerland. *Circa* 1875. Pin-wheel lever escapement with spiral steel balance spring with regulator. Going barrel. Enamel dial. Gilt hands. Nickel-plated case. Roskopf was the first successful manufacturer of a really cheap watch, which he introduced in 1868.

311a

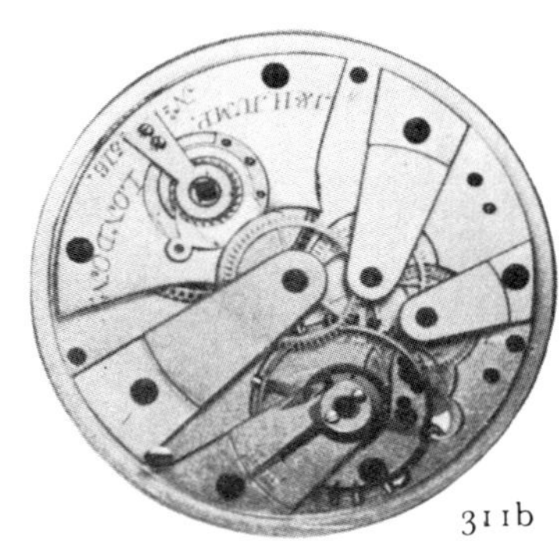

311b

311a-b J. &. H. JUMP NO. 516. London, England. *Circa* 1875. Single-roller lever escapement. Two-arm bi-metallic compensation balance, spiral steel balance spring with overcoil and regulator. Going barrel. Silver dial with matt centre, steel hands. Gold engine-turned case. Jump generally worked in the style of Breguet, and his watches are almost invariably of most elegant appearance.

312a

312b

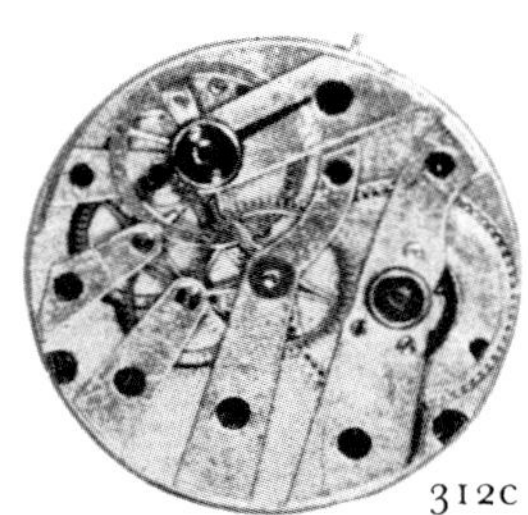

312c

312a-c ANONYMOUS, Swiss. Third quarter of the nineteenth century. Cylinder escapement. Barred movement, going barrel. Enamel dial with blued steel hands. Gold case, a good example of very small Continental work in the middle of the nineteenth century. The diameter of the case is 15 mm.

313a

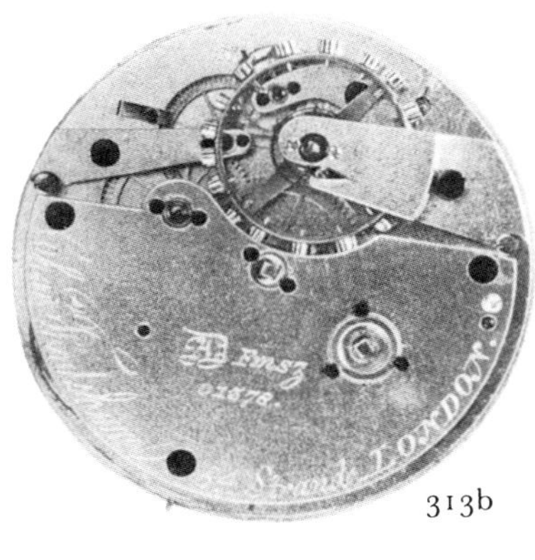

313b

313a-b CHARLES FRODSHAM NO. 01378. London, England. 1859. Spring detent escapement, two-arm bi-metallic compensation balance, helical steel balance spring with end curves, free sprung. Fusee and chain. Enamel dial with blued steel hands, subsidiary up-and-down dial. Plain gold case. This watch is typical of the highest grade of English work throughout the second half of the nineteenth century. For the interpretation of the inscription 'AD. FMSZ' on the dial and movement, see page 274.

314b

314a

314a-b CHARLES FASOLDT, Albany, New York, U.S.A. *Circa* 1880. Fasoldt escapement, compensation balance, spiral steel spring with Fasoldt regulator. White enamel dial, steel hands, silver hunting case.

315a-b A. JOHANNSEN NO. 6068. London, England. *Circa* 1880. Spring detent. Bi-metallic compensation balance. Helical steel balance spring with terminal curves. Fusee and chain. Enamel dial. Blued steel hands. Gold case.

315b

315a

316a

316b

316c

316a-c BARRAUD & LUND NO. $\frac{3}{3385}$. London, England. 1884. Single-roller lever escapement. Bi-metallic compensation balance. Spiral balance spring. Fusee and chain with keyless winding. Silver hunting case with à tact hand on back. The wheel and ratchet to the 'tact' is seen in the centre of the movement.

317a   317b

318a   318b   318c

317a-b DESOUTTER. No number. London, England. Late nineteenth century. Pivoted detent escapement. Bi-metallic compensation balance. Spiral steel balance spring with overcoil and regulator. Going barrel and keyless winding. Silver engine-turned dial. Blued steel hands. Gold case. Quarter repeating. Movement of Swiss manufacture.

318a-c JAMES MCCABE NO. 08400. London, England. Date: probably one of the last products of the firm, which closed in 1883. Lever escapement, two-arm compensation balance, helical steel spring with overcoil, free sprung, going barrels for going train and for minute-repeating grande sonnerie clock-watch train. Enamel dial with gold hands. Plain gold case.

319a   319b

319a-b HUNT & ROSKELL, London. Late nineteenth century. Swiss ébauche. Lever escapement. Spiral spring with terminal curve. Going barrel. Keyless wind. Enamel dial. Steel hands. Minute repeat. Fly-back chronograph. Perpetual calendar. This represents the final development of the Swiss precision complicated watch. Inner bezel opened to show calendar setting levers.

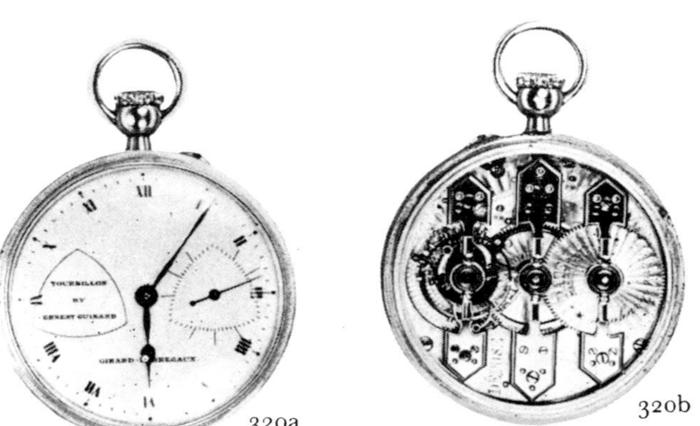

320a   320b

320a-b GIRARD PERREGAUX, Switzerland. Pivoted detent escapement. Two-arm bi-metallic compensation balance. Spiral steel balance spring with overcoil and regulator. Going barrel. Barred movement with tourbillon. Enamel dial also signed 'Tourbillon by Ernest Guinand'. Plain gold case with glass back and front. The movement is signed on the spring barrel 'Patent March 25th 1884', but the date of the watch is certainly somewhat later. It is also an exceptionally small watch for a tourbillon, the diameter of the case being 36 mm. Note also the elegant form of keyless-winding pendent.

321a

321b

321c

321a-c   WATERBURY WATCH COMPANY, Connecticut, U.S.A. *Circa* 1890. Duplex escapement with spiral steel balance spring and regulator. The mainspring is contained in the barrel and occupies the whole diameter of the movement. Enamel dial with steel hands. Nickel-plated case. The Waterbury watch was a pioneer in the production of a cheap watch. It was notorious for the extremely long time it took to wind it up. In some of them the entire movement revolved, thus producing in effect a tourbillon.

322a

322b

322a-b   JULES JÜRGENSEN NO. 8573. Copenhagen, Denmark (but in fact made by the firm then working in Switzerland). *Circa* 1890. Lever escapement with divided lift, two-arm bi-metallic compensation balance, spiral steel balance spring with overcoil and regulator, going barrel and separate barrel for independent jump seconds train. Enamel dial with blued steel hands. Plain gold case. This watch was probably made by the second Jules Jürgensen. The slender figuration of the dial and slender hands are typical of late Jürgensen work. These watches frequently had gold trains.

323a-b   CHARLES FRODSHAM NO. 08921. London, England. Lever escapement with one-minute tourbillon carriage. Bi-metallic compensation balance. Spiral steel spring with terminal curve. Split seconds chronograph operated by the button at XI o'clock. Minute repeating. Enamel dial with subsidiary dial for minute recording. Blued steel hands. Gold case.

323a

323b

324a-b ALBERT H. POTTER, Geneva, Switzerland. *Circa* 1890. Spring detent escapement with one-minute tourbillon. Bi-metallic compensation balance, spiral steel balance spring with overcoil, free sprung. Going barrel. White enamel dial. Blued steel hands. Gold case.

324b

324a

325a

325b

325a-b A. LANGE & SOHNE NO. 41000. Glashutte, Germany. *Circa* 1890. Spring detent escapement with one-minute tourbillon carriage. Bi-metallic compensation balances. Spiral steel balance spring with overcoil and regulator. Keyless-wound fusee. White enamel dial with up-and-down regulator. Pierced gold hands. Cast and engraved gold hunting case.

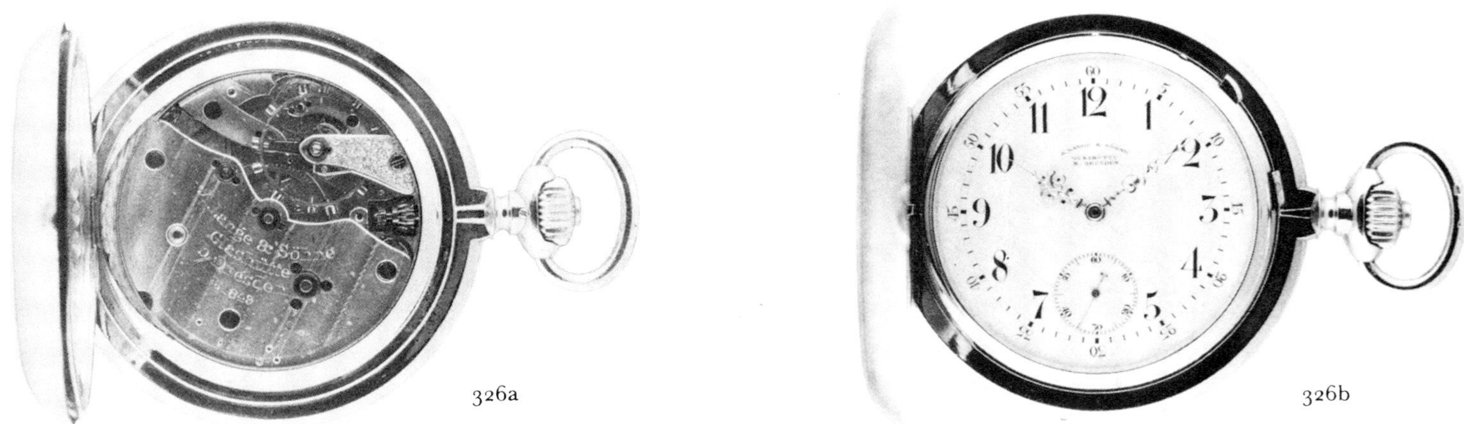

326a

326b

326a-b A. LANGE & SOHNE NO. 31849, Glashutte, Germany. 1890. Pivoted detent escapement. Bi-metallic compensation balance. Helical steel balance spring with terminal curves, free sprung. Keyless-wound going barrel. White enamel dial. Pierced gold hands. Gold hunting case.

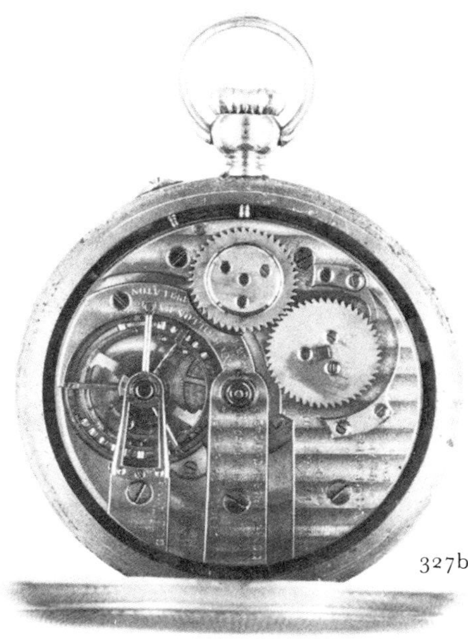

327a

327b

327a-b   GOLAY FILS & STAHL NO. 28766. Geneva, Switzerland. *Circa* 1890. Spring detent escapement with one-minute tourbillon. Bi-metallic compensation balance. Spiral steel balance spring with terminal curve and regulator. Going barrel. Silvered metal dial. Blued steel hands. Engine-turned gold case. The tourbillon carriage is by Albert Pellaton.

328a-b   A. LANGE & SOHNE NO. 45132. Glashutte, Germany. *Circa* 1895. Lever escapement. Bi-metallic compensation balance. Spiral steel balance spring with terminal spring and regulator. Going barrel. White enamel dial. Blued steel hands. Gold case.

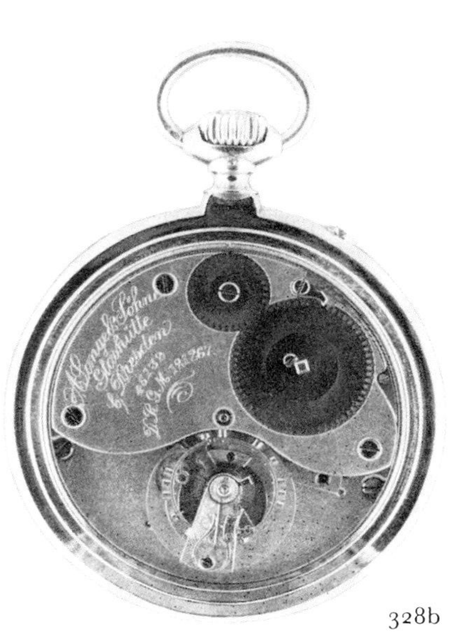

328a

328b

329a

329b

329a-b   CHARLES FRODSHAM by appointment to the Queen. NO. 08501 ADFMSZ. London, England. 1895. Lever escapement on 35-minute karussel carriage. Bi-metallic compensation balance. Spiral steel balance spring with overcoil and regulator. Going barrel. White enamel dial. Blued steel hands with centre seconds. Silver case.

330a-b A. LANGE & SOHNE NO. 45121. Glashutte, Germany. 1897. Spring detent escapement. Bi-metallic compensation balance. Helical steel spring with terminal curves. Keyless-wound fusee. Silvered dial with up-and-down indicator. Blued steel hands. Silver case with gold fittings.

330a

330b

331a

331b

331a-b A. LANGE & SOHNE NO. 8200. Glashutte, Germany. *Circa* 1900. Lever escapement. Bi-metallic compensation balance. Steel spiral balance spring with terminal curve and regulator. Going barrel. White enamel dial. Pierced gold hands. Clock-watch and minute repeating. Perpetual calendar with moon phases. Split-seconds chronograph. Gold hunting case.

332a

332b

332a-b A. LANGE & SOHNE. No number. Glashutte, Germany. *Circa* 1900. Lever escapement. Bi-metallic compensation balance. Spiral steel balance with overcoil and regulator. Going barrel. Clock-watch, grande and petite sonnerie, minute repeating. Perpetual calendar. Split-seconds chronograph. White enamel dial. Pierced gold hands. Gold hunting case.

333a

333b

333c

333a-d   J. W. BENSON, London, England. *Circa* 1900. English lever escapement. Bi-metallic compensation balance. Steel 'duo-in-uno' balance spring with regulator. Going barrel. Clock-watch with minute repeating and triple split-seconds chronograph. Enamel dial. Blued steel hands. Gold case.

333d

334a

334b

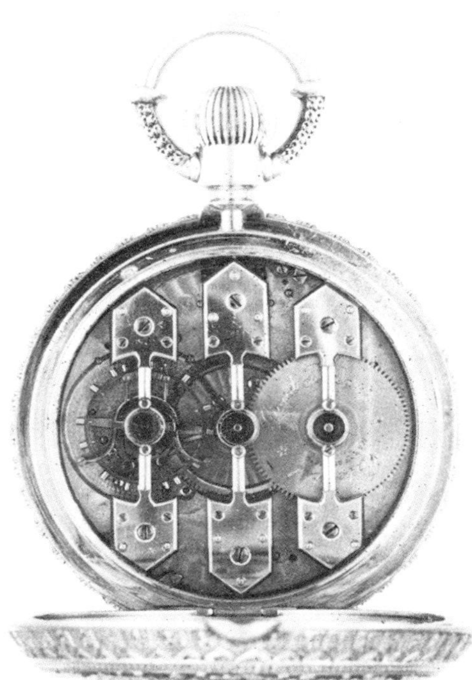

334a-b   GIRARD PERREGAUX NO. 7222. La Chaux de Fonds, Switzerland. *Circa* 1900. Pivoted detent escapement and one-minute tourbillon. Bi-metallic compensation balance. Spiral steel balance spring with overcoil and regulator. Going barrel. White enamel dial. Pierced gold hands. Multi-coloured gold case. The movement has engraved on it, in English, 'Patented March 24th 1884'.

335a-b BRIDGMAN & BRINDLE NO. aizz. London, England. 1900. Spring detent escapement. Bi-metallic compensation balance. Two steel spiral springs with resilient studs. Fusee and chain. White enamel dial with up-and-down indicator. Blued steel hands. Gold case.

335a

335b

336a-b LEONARD HALL NO. 37987. Louth, England. *Circa* 1900. Lever escapement with ten-minute tourbillon carriage. Bi-metallic compensation balance. Spiral steel balance spring with overcoil, free sprung. Going barrel. White enamel dial. Gold hands.

336a

336b

337a

337b

337a-b JULES JÜRGENSEN NO. 13946. Copenhagen, Denmark. *Circa* 1900. Geneva lever escapement. Bi-metallic compensation balance. Spiral steel balance spring with overcoil and regulator. Going barrel. Minute repeating. Chronograph. Perpetual calendar. White enamel dial. Blued steel hands. Gold case.

338a

338b

338a-b LANCASHIRE WATCH CO. NO. 695761. Prescot, England. *Circa* 1905. Lever escapement. Bi-metallic compensation balance. Spiral balance spring with terminal curve and regulator. Going barrel. White enamel dial with blued steel hands. Silver case. This watch is typical of the last of the Lancashire watches and the final attempt in England to make watches on a factory basis.

339a

339b

339a-b ALBERT H. POTTER & CO. NO. 4. Geneva, Switzerland. *Circa* 1890. Pivoted detent escapement, spiral steel balance spring with overcoil, free sprung, going barrel. Enamel dial with blued steel hands. Plain gold hunter case. Albert Potter was an American working in Geneva; he patented his pivoted detent escapement in 1875 and was known for his very high grade precision work.

340a

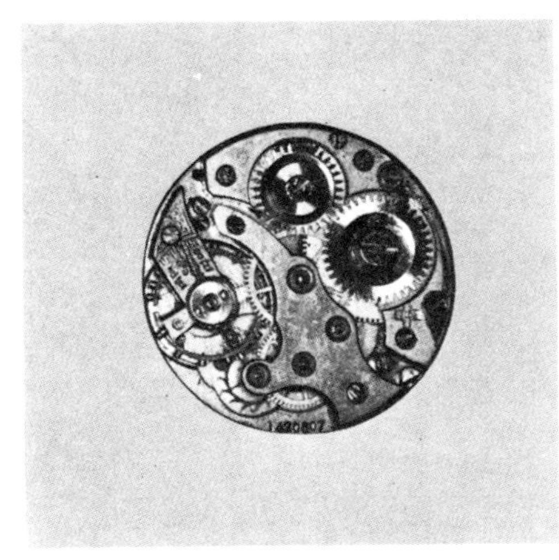

340b

340a-b ANONYMOUS NO. 1420607. Switzerland. 1907. Lever escapement. Bi-metallic compensation balance. Spiral balance spring. Going barrel. Enamel dial. Blued steel hands. English gold case and expanding bracelet set with turquoises. A typical Edwardian lady's wrist watch.

341a

341b

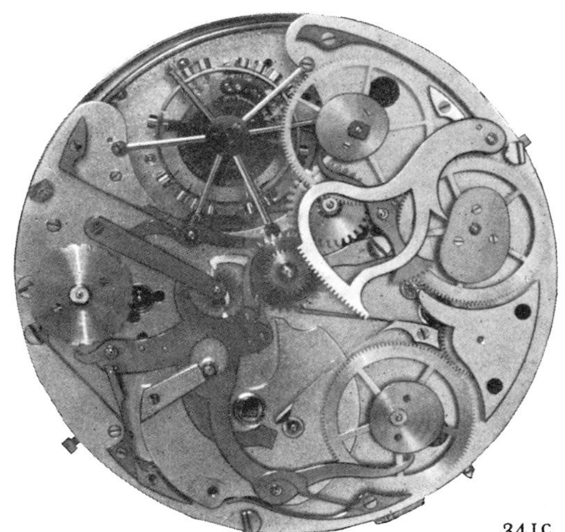

341c

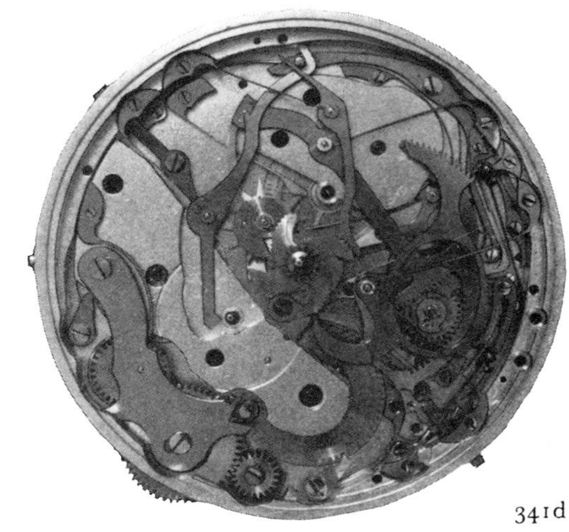

341d

341a-d J. W. PLAYER & SON, Coventry, England. *Circa* 1908. Lever escapement. Bi-metallic compensation balance with one-minute tourbillon carriage. Spiral steel balance spring with overcoil, free sprung. Going barrel. Quarter-striking clock-watch. Minute repeating. Double dials. One, enamel, for mean time and indicating date, sign and degrees of zodiac. Sun's declination. Age and phases of the moon. Time of high and low tide at any chosen place. Rising and setting of stars visible in northern hemisphere. The other dial is metal with engine-turned centre. It indicates equation of time, day of week. Subsidiary dial for sunrise and sunset. Aperture revealing tourbillon carriage. Outer circle for date of month. Gold and steel hands. Gold case.

342a-b DENT NO. 57374. London, England. 1910. Movement signed 'Watchmaker to the King'. Single-roller lever escapement with 35-minute karrusel. Bi-metallic compensation balance. Spiral balance spring with overcoil. Going barrel. Enamel dial. Blued steel hands with centre seconds. Silver case with gold bow and hinges. The watch has a National Physical Laboratory, Kew, Certificate 'Class A. Exceptionally good'.

342a

342b

343a-b S. SMITH & SON NO. 1899-1. London, England. 1910. Spring detent escapement with one-minute tourbillon. Bi-metallic compensation balance. Spiral steel balance spring with overcoil, free sprung. Going barrel. White enamel dial. Blued steel hands. Gold case. The movement is of Swiss manufacture with the escapement by Albert Pellaton. See 51.

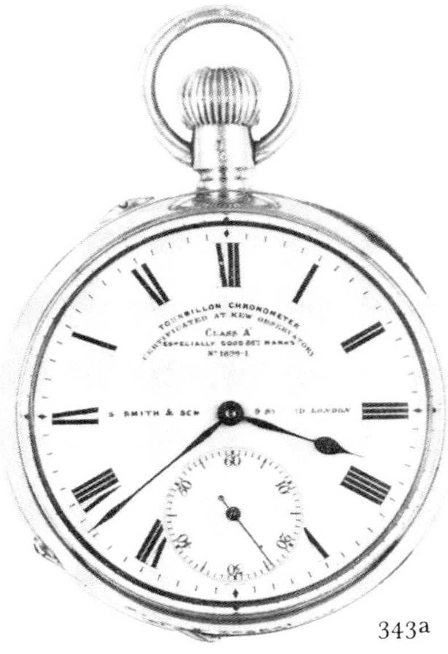

343a

343b

344a

344b

344a-b S. SMITH & SONS NO. 1901-20. London, England. *Circa* 1910. Lever escapement with one-minute tourbillon. Bi-metallic compensation balance. Spiral steel balance spring with overcoil free sprung. Going barrel. Split-seconds chronograph with minute recorder. White enamel dial. Blued steel hands. Gold case.

345a-b A. P. WALSH NO. 2101. London, England. *Circa* 1910. Spring detent escapement. Bi-metallic compensation balance. 'Duo-in-uno' steel balance spring, free sprung. Keyless wound fusee and chain. Blued steel hands. Gold case.

345a

345b

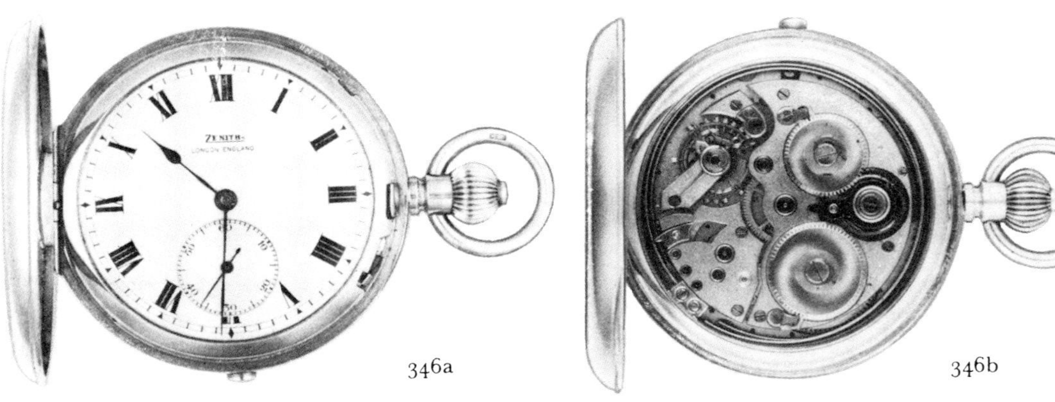

346a-b ZENITH WATCH COMPANY, Switzerland. *Circa* 1910. Lever escapement. Bimetallic compensation balance. Spiral steel balance spring with overcoil and regulator. Going barrel. Clock-watch. White enamel dial. Blued steel hands. Gold hunting case.

346a

346b

347a

347b

347c

347d

347a-d VICTOR KULLBERG NO. 7108. London, England. *Circa* 1910. Lever escapement. Bimetallic compensation balance. Helical steel balance spring with terminal curves. Keyless-wound fusee and chain. Minute repeating. Split-seconds chronograph. White enamel dial with subsidiary dial for minute recording. Up-and-down sector. Blued steel hands. Gold case. The movement is engraved 'Grand Prix, Paris, 1900'.

348a

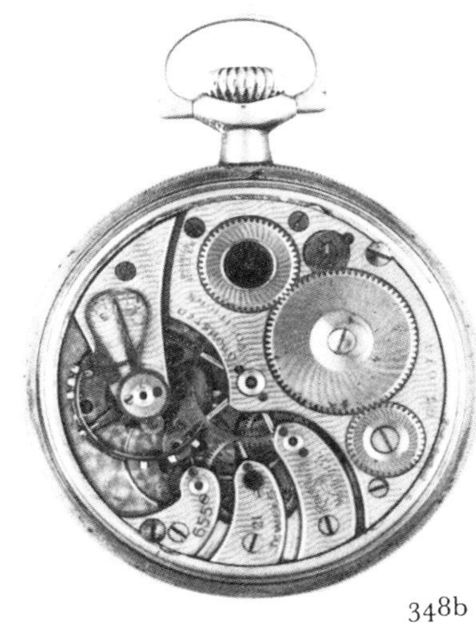

348b

348a-b ROCKFORD WATCH CO. NO. 858078. Rockford, U.S.A. *Circa* 1910. Lever escapement. Bi-metallic compensation balance. Spiral steel balance spring with overcoil and regulator. Going barrel. White enamel dial with up-and-down indicator. Blued steel hands. Gold case.

349a-b JULES JÜRGENSEN NO. 15172. Signed Copenhagen but see below. Lever escapement. Two-arm bi-metallic compensation balance. Spiral steel balance spring with overcoil and regulator. Going barrel. Split seconds chronograph. Enamel dial with steel hands. Gold hunter case. Jules was one of the two sons of Urban Jürgensen. He moved to Switzerland in 1834 but continued to sign his work 'Copenhagen'. The firm continued until 1912 and this is a late example of its work.

349a

349b

350a-b AMERICAN WALTHAM WATCH COMPANY NO. 3006260. Waltham, Mass., U.S.A. *Circa* 1910. Lever escapement, two-arm compensation balance, spiral steel balance spring with overcoil and regulator, going barrel. Enamel dial and blued steel hands. Gold case. The words 'Riverside Maximus' on the movement indicate that this is the highest quality work of the Company.

350a

350b

351a

351b

351a-b CHARLES FRODSHAM NO. 09551. London, England. 1913. Lever escapement. Two-arm bi-metallic compensation balance. Spiral steel balance spring with overcoil and regulator. Going barrel. Engine-turned silver dial with gilt chapter ring to the seconds dial and gilt plaque for the signature. Blued steel Breguet-type hands. Engine-turned gold case. This is a remarkably late date for a watch externally almost completely of Breguet style. Note also the elegant form of keyless winding, not unlike earlier forms of repeater pendents. Note also typical three-quarter plate English lay-out. The watch bears the stamp ADFMSZ, characterising Frodsham's highest grade work from 1850 onwards. The movements of Frodsham's watches at this time were made by Nicole Nielsen and while this example retains the English single-roller escapement the escape wheel has divided lift. See also col. pl. XVII.

352a-b FRODSHAM NO. 09655 and the additional code NO. A. D. FMSZ (see page 274). London, England 1913. Lever escapement and one-minute tourbillon. Two-arm, bi-metallic, Guillaume-type compensation balance. Spiral steel balance spring, free sprung. Key-less-wound fusee and chain. Silvered dial, with up-and-down regulator. Steel hands. Plain gold case. This watch has on four separate occasions between 1913 and 1946 taken a Kew certificate.

352a

352b

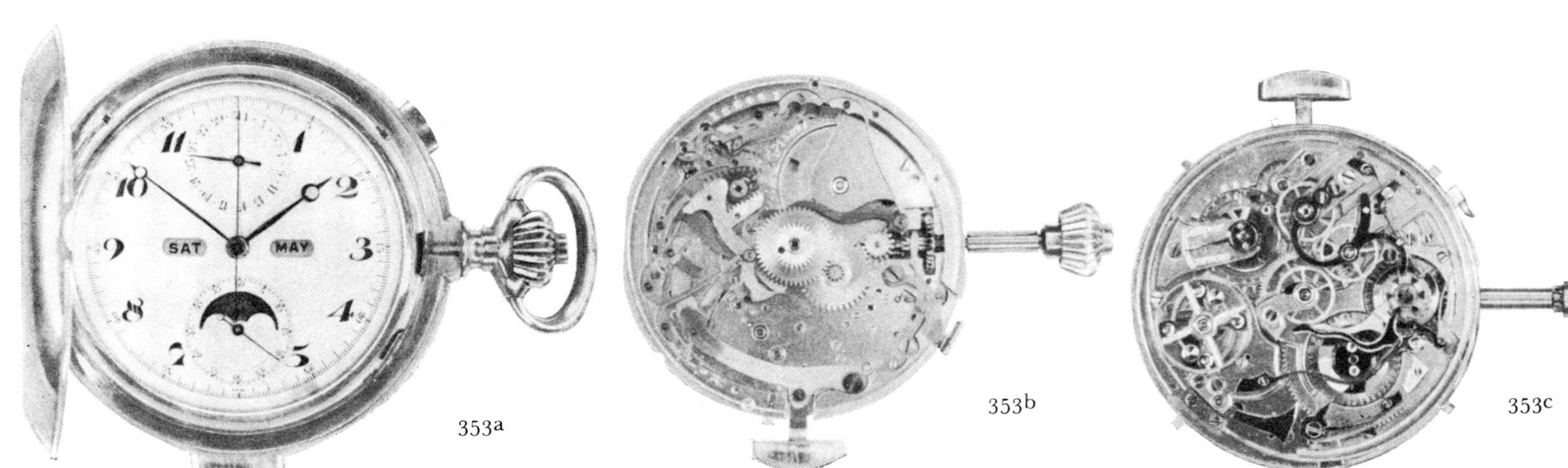

353a

353b

353c

353a-c ANONYMOUS. First quarter of the twentieth century. Lever escapement, two-arm bi-metallic compensation balance wheel, spiral steel balance spring and regulator, going barrel. Simple calendar with moon work; fly-back sweep centre seconds; minute repeater on gongs, operated by a plunger in the band of the case. Enamel dial with blued steel hands, plain gold hunter case. This watch is typical of complicated Continental work in the first quarter (and a little later) of the twentieth century.

354a-b   EDWARD HOWARD NO. 76. Boston, U.S.A. 1912. Lever escapement. Bi-metallic compensation balance with eccentric weights for mean time regulation. Spiral steel spring with overcoil free sprung. Going barrel with wolf-teeth winding. White enamel dial. Blued steel hands. Gold case. This watch is accompanied by a certificate stating that it is No. 76 of the 350-dollar watch series. Among the conditions is 'this watch shall not be sold to anyone designated by the manufacturer as objectionable'.

354a

354b

355a-b   ANONYMOUS NO. 71582. Switzerland. 1918. Lever escapement. Bi-metallic compensation balance. Spiral balance spring with overcoil and spring-tensioned regulator. Going barrel. Enamel dial with luminous figures and hands. English half-hunting gold case. A typical high-quality man's wrist watch of the period.

355a

355b

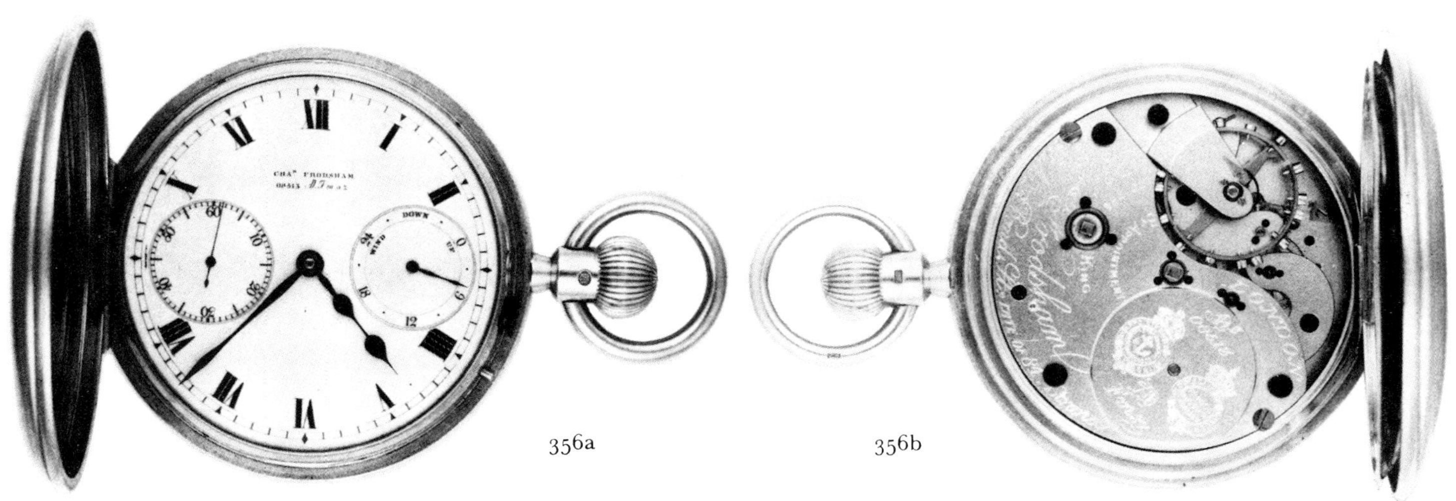

356a

356b

356a-b   CHARLES FRODSHAM NO. 08513. ADFMSZ. London, England. *Circa* 1920. Spring detent escapement. Bi-metallic compensation balance. Helical balance spring with terminal curves. Fusee and chain, keyless wound. White enamel dial with subsidiary dial for up-and-down indicator. Blued steel hands. Gold half-hunting case.

357a

357b

357a-d   S. SMITH & SON NO. 304-2. London, England. *Circa* 1920. Lever escapement. Bi-metallic compensation balance. Spiral steel balance spring with over-coil, free sprung. Going barrel. Clock-watch, minute repeating. Perpetual calendar and phases of moon. White enamel dial. Blued steel hands. Gold case.

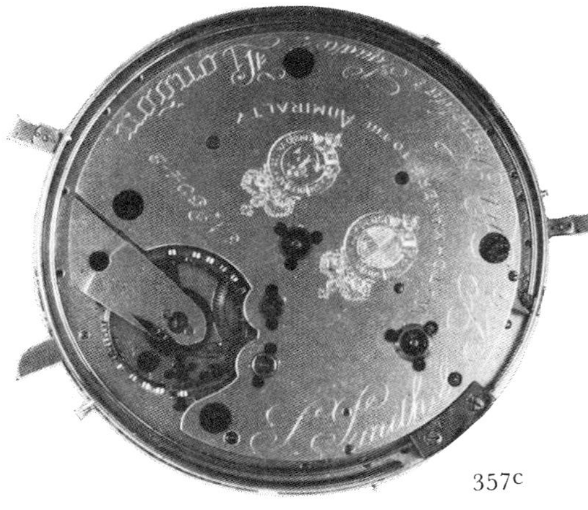

357c

357d

358a

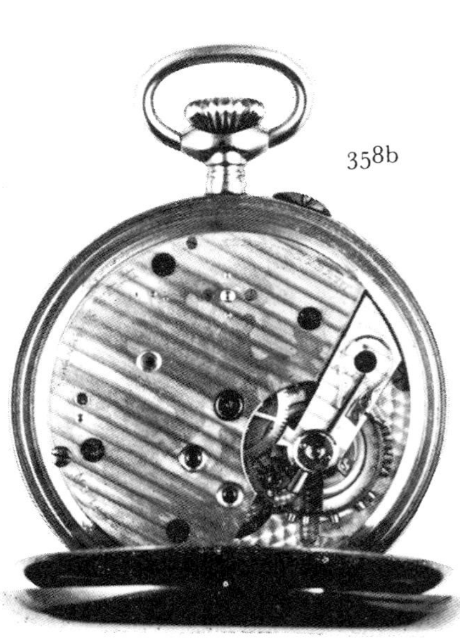

358b

358a-b   ULYSSE NARDIN NO. 14788. Le Locle and Geneva, Switzerland. Spring detent escapement. Bi-metallic compensation balance. Spiral steel spring with terminal curve and micrometer regulator. Enamel dial. Blued steel hands. Gold case.

359a-b ILLINOIS WATCH COMPANY NO. 2652203. U.S.A. *Circa* 1920. Lever escapement. Bi-metallic compensation balance. Spiral steel spring with terminal curve and micrometer regulator. Enamel dial with blued steel hands. Gold case.

359a

359b

360a

360b

360a-b WALTHAM PREMIER MAXIMUS NO. 17057029. Massachusetts, U.S.A. Lever escapement. Bi-metallic compensation balance. Spiral steel spring with terminal curve and micrometer regulator. Silvered metal dial with up-and-down indicator. Blued steel hands. Gold case. The Waltham Premier Maximus was the most expensive production watch manufactured by the Waltham Company and this may be the reason why it is generally accepted as their finest product. Since the Riverside Maximus (see 350) performs equally well and displays a higher grade finish to the movement, the authors have no hesitation in adhering to the final sentence in the caption to the illustration just cited.

361a-b WALTHAM PREMIER VANGUARD NO. 26745264. Massachusetts U.S.A. Lever escapement. Bi-metallic compensation balance. Spiral steel spring with Lossier's patent terminal curve. Enamel dial with up-and-down indicator. Blued steel hands. Gold case. Note the damascened nickel plates typical of most American high grade watches.

361a

361b

362a

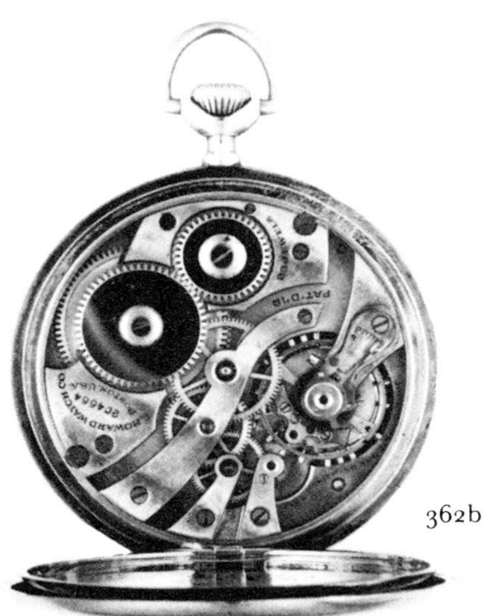

362b

362a-b E. HOWARD NO. 204664. Roxbury, U.S.A. Twentieth century. Lever escapement. Bi-metallic compensation balance, spiral steel spring with terminal curve, micrometer regulator. Enamel dial with blued steel hands. Gold case.

363a-b JAMES C. PELLATON NO. 123232. Made for Ulysse Nardin, Le Locle, Switzerland. Right-angle lever escapement with one minute tourbillon carriage. Bi-metallic compensation balance. Spiral steel spring with terminal curve. Enamel dial. Blued steel hands. Gold case.

363a

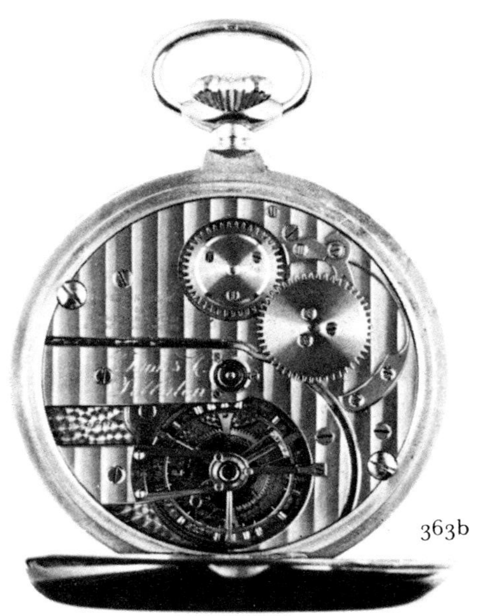

363b

364a

364b

364a-b ZENITH WATCH COMPANY, Switzerland. Spring detent escapement. Bi-metallic compensation balance. Spiral steel spring with terminal curve and micrometer regulator. Enamel dial with gold hands. Gold case.

365b

365a

365a-b   A. LANGE & SOHNE NO. 89533. Glashutte, Germany. 1920. Lever escapement. Bi-metallic compensation balance. Spiral steel balance spring with overcoil and regulator. Going barrel. Minute repeating. White enamel dial. Pierced gold hands. Gold case.

366a-b   VACHERON & CONSTANTIN NO. 345435. Geneva, Switzerland. *Circa* 1920. Spring detent escapement with one-minute tourbillon. Bi-metallic compensation balance. Spiral steel balance spring with overcoil and regulator. Going barrel with wolf-teeth winding. White enamel dial. Steel hands. Gold case.

366a

366b

367a

367b

367a-b   ILLINOIS WATCH COMPANY NO. 5563228. Springfield, U.S.A. *Circa* 1920. Lever escapement. Plain balance with Elinvar balance spring with overcoil and regulator. Going barrel. Enamel dial. Blued steel hands. Gold case. 'Bunn Special', denoting very high quality.

368a   368b   368c

368a-c   RECORD WATCH CO. NO. 27951. Tramelan, Switzerland. *Circa* 1920. Lever escapement. Bi-metallic compensation balance. Spiral steel balance spring with overcoil and regulator. Going barrel. White enamel dial with fly-back, blued steel hands. Cast silver case signed 'Holy Frères'.

369a-b   PAUL DITISHEIM NO. 51126. La Chaux-de-Fonds, Switzerland. *Circa* 1920. Lever escapement. Bi-metallic compensation balance. Spiral steel balance spring with terminal curve and regulator. Going barrel. Enamel dial. Blued steel hands. Gold case.

369a

369b

370a

370b

370a-b   CHARLES FRODSHAM NO. $\frac{010295}{3583}$. London, England. 1924. Movement signed 'Watchmaker to the King'. English double-roller lever escapement. Spiral balance spring with overcoil free sprung with one-minute tourbillon carriage. Fusee and chain, keyless wound. Silvered metal dial with up-and-down indicator. Blued steel hands. Gold case. This watch is of the 'FMSZ' series started in 1850.

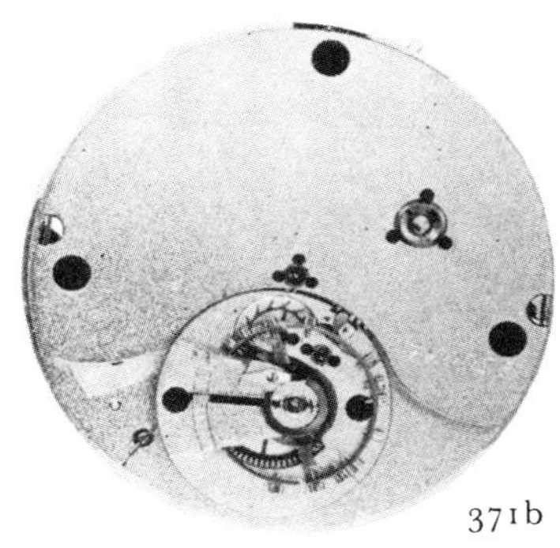

371a

371b

371a-b ROTHERHAM NO. 250024. London, England. 1923. Lever escapement and karrusel. Two-arm bi-metallic compensation balance. Spiral steel balance spring with over-coil and regulator. Going barrel. Enamel dial and blued steel hands. Gold case. For a discussion of the karrusel watch, see page 109; for a discussion of the invention of the karrusel by Bonniksen, see page 79. This example is the last karrusel watch made by Rotherham.

372a-b EPVJ (École Professionelle Vallée de Joux). Switzerland. *Circa* 1930. Two lever escapements and bi-metallic compensation balances with spiral steel balance springs with overcoil and regulators. Single going barrel. Silver engine-turned dial. Blued steel hands. Gold engine-turned case.

372a

372b

373a

373b

373a-b A. LANGE & SOHNE (unsigned and made for the firm by Alfred Helwig). Glashutte, Germany. *Circa* 1930. Spring detent escapement and one-minute 'flying' tourbillon carriage. Bi-metallic compensation balance. Spiral steel balance spring with overcoil and regulator. Keyless-wound fusee with epicyclic maintaining power. Silvered dial with gold batons. Blued steel hands. Gold case.

374a

374b

374a-b PATEK PHILIPPE NO. 198240. Geneva, Switzerland. *Circa* 1930. Lever escapement. Bi-metallic compensation balance. Spiral steel balance spring with overcoil and regulator. Going barrel. Clock-watch and minute repeating. Split-seconds chronograph. Perpetual calendar with fly-back sector date of month.

375a-b JOHN HARWOOD, England. *Circa* 1930. Lever escapement, compensation balance, spiral spring with regulator. Self-winding mechanism without provision for manual winding, hand-setting by rotating the bezel. Silvered dial with aperture to indicate disengagement of hand-set. Gold case.

376a

376b

376a-b SIDNEY BETTER NO. 2096. London, England. Made for the Northern Goldsmiths' Company. Lever escapement with one-minute tourbillon carriage. Bi-metallic compensation balance. Spiral steel spring with terminal curve, free sprung. Enamel dial. Blued steel hands. Gold case. Sidney Better worked alone in Clerkenwell until the mid-1930s. Using Swiss components for the escapements he finished and fitted them to his own carriages. See 52.

377a-e PATEK PHILIPPE NO. 198385. Geneva, Switzerland. 1932. Lever escapement. Bi-metallic compensation balance. Spiral steel balance spring with overcoil and regulator. Grande sonnerie clock-watch. Minute repeating on four gongs. Double enamel dials, one showing split-seconds chronograph with hour and minute recording; mean time. Perpetual calendar. Age and phases of moon. Up-and-down indicators for striking and going trains. The other dial showing 24-hour sidereal time with aperture for the rising and setting of stars in the northern hemisphere. Sidereal seconds. Sunrise and set. Equation of time. Blued steel hands. Gold case. This watch was commissioned by Mr Henry Graves, Jr, of New York. Its diameter is 75 mm and the thickness is 37 mm. The illustrations show successive layers of the movement. See also col. pl. XIIIc.

378a-b BREGUET NO. 3356. Paris, France. 1939. Lever escapement. One-minute tourbillon; two-arm bi-metallic compensation balance, spiral steel spring with overcoil, free sprung. Going barrel. Engine-turned silver dial with blued steel hands, subsidiary up-and-down dial. Plain gold case. The movement of the watch is of Swiss manufacture, finished and adjusted by Breguet.

378a

378b

379a

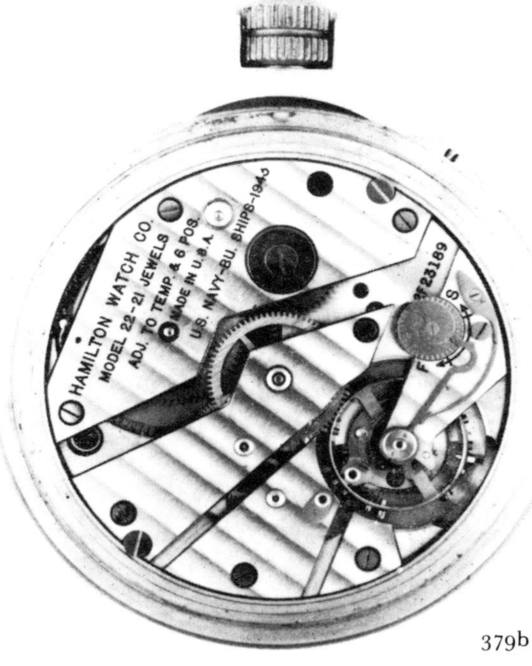

379b

379a-b HAMILTON WATCH CO. NO. 37615. Lancaster, U.S.A. Lever escapement. Hamilton patent stainless steel balance with Invar cross-bar for subsidiary temperature compensation. Hamilton patent alloy balance spring incorporating temperature and isochronal compensation with terminal curve and micrometer regulator. Silvered metal dial with up-and-down indicator. Blued steel hands. Nickel case. These watches were specially developed by the Hamilton Company for use by the American services during the Second World War.

380a-b STOPFER & CO. NO. 281374. Switzerland. *Circa* 1950. Double-roller lever escapement. Bi-metallic compensation balance. Spiral balance spring with overcoil. Going barrel. Enamel dial with centre seconds chronograph and subsidiary dial for minute recording. Blued steel hands. Silver case. The Stopfer stamp on the movement adjoins the balance cock.

380a

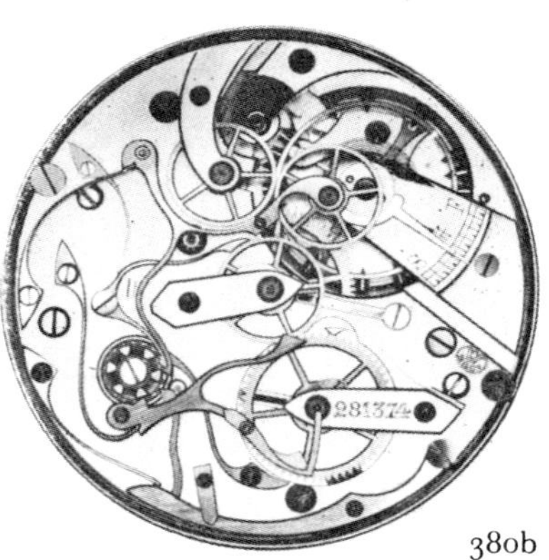

380b

381a-b PATEK PHILIPPE NO. 681367. Geneva, Switzerland. 1965. Lever escapement. Bi-metallic compensation balance. Spiral steel balance spring with overcoil and regulator. Clock-watch. Minute repeating. Perpetual calendar. Split-seconds chronograph. Silvered dial. Gold and steel hands. Gold case.

381a

381b

382a

382b

382a-b BREGUET NO. 5026. Paris, France. *Circa* 1960. Lever escapement, two-arm compensation balance, spiral steel balance spring with overcoil and regulator, going barrel. Engine-turned silver dial with blued steel hands with subsidiary dials for phases of the moon, months, days of the month, up-and-down indicator, equation of time. Plain gold case. The movement of this watch is of Swiss manufacture, finished, adjusted and cased by Breguet, and is an example of the latest work of the firm.

383a

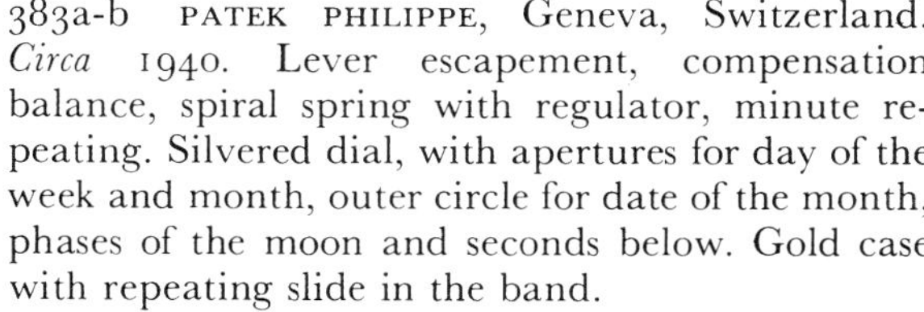

383b

384a

384b

383a-b PATEK PHILIPPE, Geneva, Switzerland. *Circa* 1940. Lever escapement, compensation balance, spiral spring with regulator, minute repeating. Silvered dial, with apertures for day of the week and month, outer circle for date of the month, phases of the moon and seconds below. Gold case with repeating slide in the band.

384a-b OMEGA, Bern, Switzerland. *Circa* 1977. Lever escapement, mono-metallic balance, spiral alloy balance spring. Chronograph, automatic winding, metal dial, centre seconds hand with subsidiary dials for seconds and minutes and hours recording. Night and day indicator. Stainless steel case with buttons for the chronograph.

385a

385b

385a-b GEORGE DANIELS, London, England. 1972. One-minute tourbillon with Earnshaw spring detent escapement. Stainless steel balance with eccentric gold adjusting weights, Ni-span-C balance spring with terminal curve free sprung, two going barrels. Silver engine-turned dial with eccentric minute circle, sector for retrograde hour hand, seconds below. Gold engine-turned case. See col. pl. XVA.

386a-b GEORGE DANIELS, London, England. 1974. One-minute tourbillon with fifteen-second remontoir, Daniels' spring detent escapement with detent in tension. Stainless steel balance with recessed adjusting screws, Ni-span-C balance spring with terminal curve free sprung, two going barrels, equation of time cam. Silver engine-turned dial with eccentric chapter circle, seconds below, sectors for the state of winding and equation of time. Gold hands. See col. pl. XVB.

386a

386b

387a

387b

387a-b GEORGE DANIELS, London, England. 1976. Daniels' pivoted detent escapement with two escape wheels driven by separate trains and barrels. Stainless steel balance with eccentric adjusting weights. Ni-span-C balance-spring with terminal curve free sprung. Seconds zeroing lever. Silver engine-turned dial with eccentric chapter circle, seconds above, sectors for the state of winding and thermometer. Gold engine-turned case. See col. pl. XVC.

Appendix

Biographical notes

The following selection of biographical notes is in no way exhaustive. It is confined to makers or men who in some way advanced or influenced the progress of watchmaking, and mostly to facts which are likely to be of assistance in identifying or dating their work. Their watchmaking exploits only are described; clocks are not mentioned.

For more numerically complete lists the reader is referred to Britten's *Old Clocks and Watches and their Makers* (the current, Eighth Edition, by G. H. Baillie, C. Clutton and C. A. Ilbert, London, 1973) and *Watchmakers and Clockmakers of the World* by G. H. Baillie (the Second Edition, London, 1947). Baillie's immense list of 37,000 names is practically confined to dates and towns of origin, with only occasional references to specimens in well-known collections. Britten's list, only about a third as long, gives a good deal more information, including addresses in a large number of cases. The book also includes about two dozen fairly full biographies of the most famous makers.

ARNOLD, JOHN and JOHN ROGER. John, the father, was born in 1736 and apprenticed to his father in Bodmin. In about 1760 he moved to London and established himself in Devereux Court. In 1764 he presented to George III a half-quarter repeating watch set in a ring. He started work on marine timekeepers in about 1770 and in 1775 patented (No. 1113) a bi-metallic compensation balance and helical balance spring. A patent for terminal curves followed in 1776. All early chronometers had the pivoted detent escapement, but between 1780 and 1782 Arnold developed his spring detent and patented it in 1782 (No. 1328).

At some time before 1776 he left Devereux Court for Adelphi Buildings and in about 1785 he moved to 112 Cornhill. He also had a chronometer factory at Chigwell. He took his son, John Roger, into partnership in 1787 and the work was then signed 'Arnold & Son' until John's death in 1799. Curiously, John was not admitted to the Clockmakers' Company until 1783.

John's son, John Roger, was admitted to the Clockmakers' Company in 1796 and was Master in 1817. In 1820 he moved from Cornhill to 27 Cecil Street, and in 1830 to 84 Strand, where he continued until his death in 1843, although latterly he took little active part in the business. In 1830 he entered into partnership with E. J. Dent with power to break up in 1840, which Dent then did (Arnold having been almost inactive) and set up on his own at 82 Strand. Products of the partnership are signed 'Arnold & Dent'.

The numbering of Arnold's early watches is not always consistent. The numbering of his pocket chronometers seems to have started in about 1773 and in 1778 he continued the series with an additional number set out as a fraction. These fractional numbers relate to watches with the 'double-T' and 'double-S' balances and continue until the first spring-detent watches in 1782. The last recorded number is $\frac{34}{89}$. From about this date another series of fractional numbers started, probably commencing with $\frac{1}{302}$ and continuing with a constant difference of 301 between the two numbers. The last recorded is $\frac{518}{819}$, made in 1796. John Roger Arnold used a series of plain numbers of which the earliest recorded is 1681 and was made in 1800. The series continued at any rate to the end of the Arnold & Dent partnership in 1843 with No. 6401.

For further details of the Arnolds' work see pp. 49–52. The subject is dealt with in great detail in *John Arnold and Son* by Vaudrey Mercer, published by The Antiquarian Horological Society. The above details of numbers are taken from Dr. Mercer's book.

BARRAUD. Of the several makers of this name the most famous was Paul Philip, Clockmakers' Company 1796–1813 and Master 1810 and 1811. He traded at 86 Cornhill.

His sons, John and James, traded as 'Barraud & Sons' during 1813–36 and at 41 Cornhill until 1838 after which the firm became known as Barraud & Lund.

BARROW, NATHANIEL, was the most famous of several makers of this name. He was among the pioneers of the balance spring and invented a form of regulator called after him. This was a worm and nut operating on the end of the balance spring for which purpose the end length of the spring had to be straight. He was apprenticed to Job Betts in 1653, admitted to the Clockmakers' Company in 1660 and was Master in 1689.

BERTHOUD, FERDINAND and LOUIS. Ferdinand was born in 1729. He moved to Paris in 1745 and in 1762 was appointed Horloger de la Marine. He did much experimental work on marine timekeepers and from crude beginnings arrived independently at the spring-detent escapement between 1780 and 1782. A fine collection of his marine timekeepers is in the Musée des Arts et Métiers in Paris. Although he made fine watches they were almost without exception conventional in design. He appears to have made very few, if any, pocket chronometers. He was a prolific writer, largely about his own experiments and methods. In 1802 he published his *Histoire de la Mesure du Temps par les Horloges* which is a general work, but also gives a summary of his own work, especially as to marine and pocket chronometers. This important work has been re-published in facsimile (Paris, 1900, two volumes) and it is intended that his other works shall be similarly reproduced. Ferdinand died in 1807 and was succeeded by his nephew Louis (1754–1813) who made both pocket and marine chronometers of superb quality and performance, all with pivoted detents. He made about 150 chronometers in all. Like his uncle, he was appointed 'Horloger de la Marine'. He died in 1813 and was succeeded in business by his sons Charles-Auguste and Louis-Simon-Henry. For further details of the Berthouds' work see pp. 46–49 and 71.

 BONNIKSEN, BAHNE (1859–1935), was a Dane who moved to England and was naturalised as a British subject in 1910. In 1890 he began to develop a type of watch intended to combine the timekeeping qualities of a tourbillon with the robustness of an ordinary lever watch. The resulting watch he called a karrusel which he patented in 1894 and of which he brought out an improved pattern in 1903. The mechanism of this revolving escapement is explained on p. 109.

For many years Bonniksen's karrusels carried all before them in the Kew Certificate trials, but with the growing demands of industrial mechanisation he turned his attention to other fields. He died in 1935.

Other makers were permitted to use Bonniksen's karrusel under licence, including Hector Golay, a Swiss who settled in 1870 at Spencer Street, Clerkenwell, where he made very high-quality watches. Golays had their ébauches made in Switzerland but did the finishing themselves; but they also supplied unfinished movements, largely to Rotherhams, who then finished them and sold many karrusel watches. The firm of Golay continues in business in Hatton Garden as sole agent for Cyma watches.

BOOTH, EDWARD (1636–1716), changed his name to BARLOW after his godfather. He invented rack-striking for clocks, followed by repeating mechanism, in 1676, which he applied to watches in 1686. Applying for a patent, he was defeated by Daniel Quare on the grounds that Barlow's repeater had two push pieces, for hours and minutes respectively, whereas Quare operated the whole mechanism with a single push piece. Barlow, in conjunction with William Houghton and Thomas Tompion, devised a precursor of the virgule escapement in 1695 (patent number 344). He died in 1716.

BREGUET, ABRAHAM-LOUIS, was born in 1747. In 1762 he moved to Paris, but the details of his early career are obscure. It is likely that after completing his articles he worked for a time for Berthoud. His earliest identifiable work dates from 1782, and at any rate up to 1787, and to some extent into the early nineties, he used a system of fractional numbering: thus $17\frac{10}{85}$ would be watch number 17 completed in October 1785. In 1787 the books of the firm start and have all survived. These have a straightforward system of numbering which continued up to Breguet's death in 1823 and subsequently and by which, with some exceptions, his watches may be fairly closely dated (see p. 61). Any number higher than 5021 is a fake.

In 1791 Breguet was forced to leave France to escape the Revolution, living first in Switzerland and then in England, and he returned in 1795. Up to 1791 his work is almost, if not quite universally, signed 'Breguet à Paris', but practically never after 1795. Nearly all watches manifestly later than 1795 and signed 'à Paris' are fakes. In 1807 Breguet took his son Louis Antoine into partnership after which the work is signed 'Breguet et fils'. Soon after 1813 he was appointed Horloger de la Marine in succession to Louis Berthoud, after which important work is so signed.

After Breguet's death in 1823 the style of the firm continued; for a short time in about 1830 the work was signed 'Breguet et Neveu'.

For further details of Breguet's work see pp. 55–56, 125–26, 136–38. Every known aspect of his life and work is described in the copiously illustrated *The Art of Breguet* by George Daniels, published in 1975.

BROCKBANK. Of the several clockmakers of this name, the most important were the brothers John and Myles who traded at 6 Cowper's Court, and were among the pioneers of chronometer making. Myles was admitted to the Clockmakers' Company in 1776 and John in 1779. Earnshaw claimed to have revealed his version of the spring detent escapement to them and that they passed on the information to Arnold. One of their principal workmen, Peto, invented the cross detent escapement which combined the theoretical advantages of Arnold's and Earnshaw's escapements, and several examples by the Brockbanks survive, coupled with an overbanking device which limits the expansion of the helical spring. Myles died in 1821 and the brothers were succeeded by two nephews, also John and Myles, who traded as 'John Brockbanks & Company'. From 1815–35 the style of the firm was 'Brockbank & Atkins', still at 6 Cowper's Court.

BULL, RAINULPH or RANDOLPH, is the earliest identifiable British watchmaker. A large oval watch by him, dated 1590, is in the British Museum. The dates of his birth and death are not known, but he was certainly alive as late as 1617.

CARON, PIERRE AUGUSTIN (1732–99), won early fame as a watchmaker and invented the double-virgule escapement. (He was a brother-in-law of Lepine (*q.v.*)). Later, under the name of Beaumarchais, he enjoyed a successful and varied career as a playwright.

COLE, JAMES FERGUSON (1798–1880), was a most prolific inventor of escapements, some of them practical, but others of the wildest improbability. His work is nearly always of the highest order and elegance. In addition to the experimental or one-off pieces he produced many ordinary pocket watches of high grade and mostly with the lever escapement. He also made fine and elegant travelling clocks.

CUMMING, ALEXANDER (*c.* 1732–1814), improved the cylinder escapement, including curved teeth on the escape wheel. He was also a famous chronometer maker and in 1766 published *The Element of Clock and Watch Work*. He was at the 'Dial and 3 Crowns' in Bond Street until 1777; then 12 Clifford Street until 1794; and finally had a shop in Fleet Street until his death, after which it was occupied by his nephew John Grant *(q.v.)*. Cumming was a Fellow of the Royal Society.

CUMMINS, T. This virtually unknown maker was almost certainly the first English maker to use the lever escapement in watches of the highest quality, when this escapement was revived after its almost complete neglect in England during 1800–20. He appears to have started using the lever escapement soon after 1820, with a highly sophisticated form of Massey's escapement. He used a form of double numbering, thus: 17–27. The 27 almost certainly refers to the year of manufacture, and 17 is probably the watch number in that year. The earliest recorded specimen was made in 1821. Another name and address, presumably that of the owner of the watch, sometimes appears on the balance cock. Almost identical watches were made by his successor, C. Cummins.

DANIELS, GEORGE (b. 1926). Over a period of several years he has built up a reputation for restoring complicated and valuable watches, especially those by Breguet. By 1967 he felt he had advanced technically to the point where he could start making his own watches. The complete collapse of the Clerkenwell industry necessitated his executing every part himself, excepting the balance spring and mainspring, and any engraving on the plates and dial. A pivoted detent tourbillon with two going barrels was completed in 1969 and was followed by a series of spring detent tourbillons and a further series of double chronometers.

DEBAUFRE, PETER, traded at Church Street, Soho. After working for a period in Paris, he became a member of the Clockmakers' Company 1689–1722. In 1704 he was associated with Nicholas Facio (*q.v.*) in obtaining a patent for watch jewelling. He also invented a dead-beat escapement with two parallel-mounted escape wheels, often called after him. It did not become popular until revived by Litherland and the Preston makers at the end of the century. Sir Isaac Newton spoke favourably of his escapement and Sully used a variant of it, with one wheel and two pallets, in his marine timekeeper of 1724.

DENNISON, AARON (1812–95), (not to be confused with E. B. Denison, later Lord Grimthorpe) is sometimes known as the 'Father of American Watchmaking'. He started business on his own as a repairer and retailer, in 1839, and in 1849 he interested E. Howard and D. P. Davis in setting up a watch factory at Roxbury, where he was the superintendent until 1861. After varying vicissitudes it became the Waltham Watch Company, but by this time Dennison had left it. After a period in Switzerland he settled in England and in 1874 set up a successful case-making industry at Handsworth, Birmingham, where he continued until his death. He formed a superb collection of watches which his son bequeathed in 1937 to the Waltham Watch Company. There it was brutally neglected and ill-treated until the mangled remains were returned to this country and auctioned at Christie's in 1961.

DENT, EDWARD JOHN (1790–1853). After starting life as a tallow chandler, Dent went over to watchmaking, and during 1815–29 he was employed by the Vulliamys and the Barrauds. In 1830 he went into partnership with John Roger Arnold and in 1840 set up on his own at 33 Cockspur Street. The year before his death he secured the contract for making 'Big Ben'. He made many fine chronometers and watches as well as others of a high commercial grade.

DUTTON, WILLIAM. Clockmakers' Company 1746–94. He was apprenticed to Graham and afterwards in partnership with Mudge, as 'Mudge & Dutton', 1759–90, at 148 Fleet Street. Mudge's name was dropped at his death in 1794 and after Dutton's death the business was carried on by his son Matthew at the same address. He made fine watches including at any rate one before 1800 with the lever escapement; was Master of the Clockmakers' Company and continued in business until shortly before his death in 1843.

EARNSHAW, THOMAS (1749–1829). On coming to London he first worked for the trade. In about 1781, simultaneously with Arnold and Berthoud, he devised a spring detent escapement. The first examples were unsuccessful, and all his detent escapements were subject to rapid cutting of the impulse pallet, so that very few of his marine chronometers, and relatively few of his pocket watches have survived. Nevertheless, it is Earnshaw's form of the spring detent escapement, subject only to minor development, which has survived to this day in universal

271

use. He also devised the modern method of fusing together the brass and steel laminae of the compensation rims, as opposed to the earlier methods of riveting or soldering them together. In pocket watches he employed either compensation balances and helical springs, or alternatively flat steel balances with a spiral spring and a pincer-type compensation curb, commonly known as 'sugar tong compensation'. His superficial finish is generally perfectly plain, and he only engraved the watch plates and jewelled throughout to special order.

As he could not afford a patent for his spring detent one was taken out for him by Thomas Wright, and early examples were stamped 'Wright's Patent'. However, Wright was very dilatory in taking out the patent after Earnshaw had approached him, and in the meantime Arnold had patented his own. Earnshaw always claimed that the Brockbanks had told Arnold of his design; but no real evidence of this was produced, and Arnold's design was completely different, as also was Berthoud's.

Earnshaw was perhaps the first leading maker to make quite cheap, single-roller, lever escapement watches (in the last year or two of his life).

In 1794 or 1795 Earnshaw took over the business of Wm. Hughes at 119 High Holborn. He died in 1829, when the business was carried on until 1842 by his son, also Thomas, first at High Holborn, and subsequently at 87 Fenchurch Street.

EAST, EDWARD (1602–97), had such a long life that it has been suggested that he was really two people of the same name, but the researches of the late Alan Lloyd have shown fairly conclusively that there was only one of him. In 1631 he was a founder member and junior assistant of the Clockmakers' Company. He was Master in 1645 and 1652. He was clockmaker successively to Charles I and II. His work was of an extremely high order, both mechanically and decoratively. Particularly beautiful is the tall, sloping script signature which he put on all his work until about 1670. After about 1670, his later work is coarser in design, partly following the fashion, but suggesting that he had virtually retired at the age of seventy, and was no longer directly responsible for the work sold under his name. Balance spring watches, even nominally by him, are exceedingly rare.

ELLICOTT, JOHN. The first John Ellicott was a creditable maker. He was admitted to the Clockmakers' Company in 1696 and died in 1733. He was an enterprising watchmaker, having been one of the first to use a centre seconds hand, and he made one very thin watch, dating from soon after 1700, which measured only $\frac{1}{5}$ inch between the plates. He often signed his name under the cock or balance.

His son John was much more famous. Born in 1706, he set up in business in Sweeting's Alley in about 1728. Also about this time he began to use the cylinder escapement, very little later than Graham. In some of his later watches he used a ruby cylinder. All his watches are finely made and some are in highly-decorated cases.

He seems to have used only one series of numbers and Baillie gives the following sequence of dates and numbers.

Year		*No.*	
	1728		123
	1730		400
	1740		1800

1750	3250
1760	4770
1770	6435
1780	7620
1790	8450
1800	8760
1810	9074

Special watches, however, were not numbered.

Until about 1750 watches were signed 'Jno Ellicott' but thereafter 'Ellicott' or 'John Ellicott & Son'.

Ellicott was a Fellow of the Royal Society and clockmaker to the King. He died in 1772, being succeeded by his son Edward who died in 1791. He in turn was succeeded by his son, also Edward, at about which time the style of the firm became 'Edward Ellicott & Sons'. The second Edward went into partnership with one Taylor, when their work was signed 'Ellicott & Taylor', during about 1811–30. Subsequently the signature was 'Ellicott & Smith', up to 1840, Edward Ellicott having died in 1835.

EMERY, JOSIAH, was a Swiss, born in about 1725. He settled in England and carried on business at 33 Cockspur Street, Charing Cross. He was made an honorary freeman of the Clockmakers' Company in 1781. He made fine watches, mostly with the cylinder escapement, but he is most famous for his pioneering of the lever escapement from 1782, being the first maker to use this escapement after Mudge. His lever escapements are described fully on pp. 120–21. He made about thirty lever watches from 1782 to about 1795. Before taking up the lever he made a few precision watches with a pivoted detent escapement and spiral compensation curb. Emery died in 1797 and was succeeded in business by Louis Recordon (*q.v.*).

FACIO DE DUILLIER, NICHOLAS, was born at Bàsle in 1664, settled in England in 1687 and died in Worcester in 1753. He was the first to succeed in piercing jewels for watch pivots, a method which he patented jointly with Debaufre (patent No. 371 for May 1704), although the patent was later defeated by the Clockmakers' Company. For the next eighty years jewelling was an entirely British monopoly, to the great advantage of home trade. Facio was a Fellow of the Royal Society.

FASOLDT, CHARLES, was born in Germany in 1818 and emigrated to New York in 1849 where he made high-quality watches with a curious form of lever escapement, having two concentric escape wheels. The larger wheel was concerned only with locking the two ordinary pallets of the lever, and gave no impulse. However, the lever was continued towards the escape wheel pivot and ended in an impulse jewel, which was impulsed by the second, smaller escape wheel. Impulse therefore was given only on alternate swings of the balance and generally operated very much like the Robin escapement, except for impulse being delivered via the lever, and not by the escape wheel direct. The escapement therefore combined most of the disadvantages of the lever and chronometer escapements. Fasoldt called his watches chronometers, and made about fifty of them before moving in 1861 to Albany, in the State of New York, where he set up a factory for making watches and clocks, employing some fifty men. He died in 1898.

FATTON, FREDERICK LOUIS, was one of Breguet's most eminent pupils. He settled in New Bond Street, London, and made many fine watches and clocks in the Breguet style, frequently signed 'Fatton élève de Breguet'. He is best known for his ink-recording chronograph watches, patented in 1822.

FRODSHAM. The numerous members of the Frodsham family were in business from some time before 1780. William Frodsham, the second generation of watchmakers, married Alice, a grand-daughter of John Harrison. Charles, the last member of the family to be active in watchmaking, died in 1871 but the firm still continues. From some time before the middle of the last century they began making lever escapement and chronometer escapement watches of the very highest quality, and continued to do so until the outbreak of war in 1939. For the 1851 Exhibition they introduced a three-quarter-plate calibre which they marked *AD.FMSZ*, which continued subsequently to be put on all their highest grade work. The significance of FMSZ is said to be found by putting the name Frodsham against numbers thus

F R O D S H A M Z
1 2 3 4 5 6 7 8 0

Thus FMSZ gives the date 1850.

During the first part of this century the firm was closely connected with Nicole Nielsen, who made most of the movements, including some of the most perfect tourbillons ever made, which can still perform with almost unrivalled accuracy. When other watches became increasingly ugly during the nineteenth and twentieth centuries, Frodsham watches were almost always elegant and well proportioned, and even as late as 1914, some of them, with engine-turned silver dials, were worthy of Breguet.

GRAHAM, GEORGE, was born in 1673 or 1674. In 1688 he was apprenticed to Henry Aske and was admitted to the Clockmakers' Company in 1695. He then worked for Thomas Tompion whose niece Elizabeth he married in 1696. In about 1711 Tompion took Graham into partnership and from then until Tompion's death in 1713 some or all of the work was signed 'Tompion & Graham'. Graham was elected a Fellow of the Royal Society in 1721 and Master of the Clockmakers' Company in 1722. In about 1726 he developed and rapidly almost perfected the cylinder escapement, and thereafter very seldom used any other for watches. These nearly all have an elegant enamel dial with blued steel beetle and poker hands. The plain watches frequently have a polished steel centre-seconds hand, but this is not found on the repeaters, which are also somewhat smaller. Graham continued Tompion's numbering, No. 4369 being a 'T. Tompion & G. Graham'. Baillie gives the following dates and numbers.

Plain Watches			*Repeaters*	
1715	No. 4660		1720	No. 480
1725	5260		1730	620
1735	5610		1740	790
1745	6180		1750	960
1750	6480			

The number is stamped on the pillar plate and the underside of the balance cock.

Apart from Graham's earliest work, he always used a solid, engraved cock-foot and the balance staff is jewelled with a big diamond end stone. There are many faked Grahams, but the genuine watches are of such standardised design that, having seen one, one can easily detect any fake.

Graham died in 1751 and was succeeded by his executors, Samuel Barclay and Thomas Colley; but this partnership did not survive long. The maker who effectively continued Graham's tradition and style of watchmaking was his most famous pupil Thomas Mudge (*q.v.*).

GRANT, JOHN & SON. John Grant senior was apprenticed to his famous uncle, Alexander Cumming, and became an honorary freeman of the Clockmakers' Company in 1781. He was probably the last of the band of English experimenters with the lever escapement and his work is of the very highest quality. He used a helical balance spring and a variety of complicated compensation balances. Working without draw, he evidently set great store on deep locking of the lever pallets and it was perhaps this that induced him, in about 1800 or a little later, to arrange the escape wheel with its arbor at right angles to the lever and escape wheel arbors. There are examples of this lay-out, commonly known as Grant's 'chaff-cutter', in the Ilbert and Guildhall Collections. The chaff-cutter escapement is well illustrated on page 75 of Chamberlain's *It's About Time*.

In the Guildhall is a movement with his earlier, more normal lever escapement, with two balances geared together.

Grant died in 1810 when he was succeeded by his son John who was also a fine maker, and was five times Master of the Clockmakers' Company between 1838 and 1867. Both father and son traded at 75 Fleet Street.

GRIGNION. Successive members of this family did fine work from about 1690 to 1825. Thomas (1713–84) introduced improvements to the cylinder escapement and prolonged the life of the cylinder by arranging the teeth of the escape wheel at different levels.

GUILLAUME, CHARLES-EDOUARD (1861–1938), was not primarily an horologist, but his metallurgical researches brought him in contact with the problem of middle temperature error, which he finally eliminated by the nickel alloy 'Invar' for pendulums and balance wheels, and 'Elinvar' for balance springs, which are virtually unaffected by changes of temperature.

HALEY, CHARLES, was admitted to the Clockmakers' Company in 1781 and died in 1825. He was a pioneer chronometer maker and made fine watches with duplex or detent escapements. In 1796 he patented a remontoir, or constant force escapement which, despite its complication, he contrived to compress into a watch movement of normal size, of which there is an example in the Guildhall Museum. The escapement is illustrated on page 167 of Chamberlain's *It's About Time*, copied from a drawing by Berthoud. Impulse is imparted to the balance by a helical spring which is rewound at every beat by the train. The escapement, which was the first of several of this kind devised by contemporary and later makers, was never reliable.

HARRISON, JOHN (1693–1776), despite his achievements in the field of marine timekeepers, is memorable in connection with watches only for the prototype of the successful 'Number four' made for him in 1753 by John Jeffreys, which has the first example of maintaining power and temperature compensation ever applied to a watch.

HAUTEFEUILLE, JEAN (1647–1724), was a physician and mechanic, and not primarily an horologist. Nevertheless, he made early experiments with the balance spring, and devised a straight line spring similar to Hooke's. With it he was able to defeat Huygen's application for a French patent for his spiral spring. In 1722 Hautefeuille also invented the forerunner of the rack lever escapement, but it did not become popular until taken up by the English Peter Litherland and the Preston makers over sixty years later.

HELE or HENLEIN, PETER, is the first recorded maker of pocket watches, in about 1520. He worked in Nürnberg and died in 1542.

HOOKE, ROBERT (1635–1703), worked for Robert Boyle and from 1662 was Curator of the Royal Society. A man of many and varied talents, he was also Professor of Geometry at Gresham College. From about 1658 he experimented with balance springs, probably in the form of a straight spring. In 1675 he was prominent in contesting with Huygens for the priority of invention of the balance spring. He also claimed the invention of the anchor escapement for pendulum clocks, but in neither case does there seem to be much solid evidence to support his claims. He did much to assist Tompion, with whom he co-operated in making the first English balance spring watches.

HOURIET, JACQUES FRÉDÉRIC (1743–1830), was one of the finest Swiss makers and he also worked for the foremost French makers, including Breguet. He made fine chronometers and tourbillons and in 1814 devised the spherical balance spring.

HOWARD, EDWARD (1813–1904), was a pioneer of high-quality watchmaking in America. He was apprenticed to the famous American clockmaker Aaron Willard, but in 1849 he went into partnership with A. L. Dennison to found the company which eventually became the Waltham Watch Company (*see also* A. L. Dennison).

HUYGENS, CHRISTIAN (1629–95). Although primarily a mathematician and optician, Huygens made more important contributions to horology than probably any other man, before or since. Huygens was Dutch and sometimes referred to (particularly by Hooke) as 'Zulichem', this being his home town. In 1657 he successfully applied a pendulum to the verge escapement and the first examples were made by Samuel Coster. In 1675 he brought the spiral balance spring to a practicable stage of development, the first examples being made in Paris by Jacques Thuret. Thus, although he actually invented neither, he was nevertheless the first man to bring to a practical form the pendulum and the balance spring, upon which two all precision timekeeping for the next 250 years was to be based.

INGOLD, PIERRE-FRÉDÉRIC (1787–1878), a pupil of Breguet, was a pioneer of interchangeability of parts in watch manufacture and invented the Ingold milling tool.

JAPY, FRÉDÉRIC (1749–1813), effectively founded the modern methods of mass-produced watch manufacture. He manufactured ébauches by machine tools as early as 1776.

JEANRICHARD, DANIEL (1665–1741), was the leading pioneer in setting up the Neuchâtel watchmaking industry.

JUMP, JOSEPH, was articled to Benjamin Lewis Vulliamy in 1827 and on Vulliamy's death in 1854 continued on his own account in Bond Street, and later Pall Mall, until his death in 1899. He made very fine watches with typical English three-quarter-plate movements, but externally in a style fully worthy of the Breguet tradition.

JÜRGENSEN, URBAN and JULES. Urban Jürgensen was born in Copenhagen in 1776. After preliminary training in his father's workshop, in 1796 he was sent to Houriet at Le Locle. In 1800 he moved to Paris and continued his training with Berthoud and Breguet. In 1800 or 1801 he moved to London where he studied with Arnold. Later in 1801 he married Houriet's daughter, Sophie Henriette. Urban then returned to Copenhagen and began to train his brother Frédéric. He also began to make the marine and pocket chronometers for which he is famous. It seems certain that he carried out a very much higher proportion of the work himself than did his contemporaries and with such a catholic training, coupled with his own genius, it is natural that his work should be of a very high order.

He made seventy-three chronometers and about 800 watches. The chronometers have two numbers, the first being those of the chronometer series and the second of the watch series. Nearly all have silver cases.

Urban made very fine balances and often made the weights of oval shape to reduce air resistance. In view of his practical experience of both types it is interesting to find him say: 'from my own experience I know very well that a soldered balance is just as good if not better than a cast balance and has all the desired sensitiveness and regularity of action'. He sometimes employed Arnold's type of gold helical balance spring and escape wheel and in general, the style of his work resembles the English more closely than the French. He died in 1830 and was succeeded by his sons Louis Urban and Jules Frederick. The latter was the better business man. He moved to Switzerland in 1834 and opened a factory at le Locle, although his watches continued to be signed 'Jules Jürgensen, Copenhagen'. On his death in 1877 he was succeeded by his sons Jules Urban and Jacques Alfred. Jules Urban died in 1894, but Jacques Alfred lived, and continued in business, until 1912. After 1912 the business was carried on by David Golay and after passing subsequently through several hands, still survives in America. The genealogy of the clockmaking members of the family is as follows:

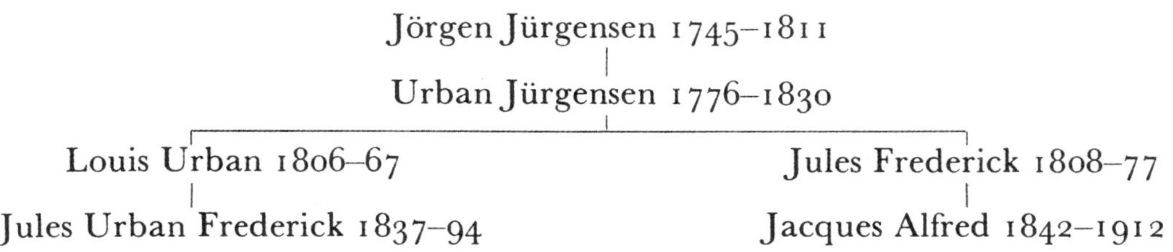

Jörgen Jürgensen 1745–1811

Urban Jürgensen 1776–1830

Louis Urban 1806–67   Jules Frederick 1808–77

Jules Urban Frederick 1837–94   Jacques Alfred 1842–1912

KENDALL, LARCUM (1721–95) is best known for his duplicate of Harrison's marine timekeeper No. 4 and two subsequent variants on the same theme. Watches by him are rare, but there is a very fine pivoted detent watch by him, hallmarked 1786, in the Guildhall Museum.

KULLBERG, VICTOR (1824–90) was a Swede. After working for the trade he moved to England in 1851 and set up on his own in 1856. He was overwhelmingly successful as a maker of marine chronometers, at first with a form of flat-rimmed balance and subsequently with a more normal pattern with auxiliary compensation. He also specialised in keyless-wound fusee, chronometer watches. To minimise the position errors to which spring detents are specially subject he arranged his pocket watches so that the detent spring is vertical in the 12-up position. His work was all of a very high order.

LEPAUTE, JEAN-ANDRÉ (1720–89), is best known as a clockmaker of great merit, and as watchmaker his fame rests upon his invention of the virgule escapement in 1753. He retired from business in 1774, being succeeded by two nephews, and the firm continued far into the nineteenth century under a variety of names, including 'Lepaute et fils' and 'Henry Neveu Lepaute'.

LEPINE, JEAN-ANTOINE (1720–1814). Although Lepaute invented the virgule escapement its development and wider use was attributable to Lepine. But his greatest importance as a watchmaker rests on his introduction of the cocked (as opposed to the plated) movement, often known as the 'Lepine calibre' in about 1770. The firm continued long after his death.

LEROUX, JOHN (1744–1808). Nothing is known about Leroux's personal history but he was an important and very early pioneer of the lever escapement in England, and differed from his contemporaries in putting all the lift on the teeth of the escape wheel. His lever spans only $1\frac{1}{2}$ teeth of the escape wheel. He used a compensation balance and a conical-shaped balance spring. Only two examples of his lever escapement are known: a movement in the Ilbert collection and a watch, hallmarked 1785, in the Guildhall Museum.

LE ROY, JULIEN and PIERRE. Julien Le Roy was born in 1686 and died in 1759. He was a very fine maker and raised French watch and clockmaking to a level at which it could successfully challenge the supremacy which British makers had established in the last quarter of the seventeenth century. In this he was much assisted by the English Henry Sully, and perfected the oil sinks invented by Sully. In about 1725 he introduced the adjustable potence for the escape wheel in verge watches and in 1727 obtained from George Graham an example of his cylinder escapement, which he subsequently often used himself. It is thought that Julien was the first to use wire gongs, instead of bells, in repeater watches, thus paving the way for Breguet's thin repeaters. He also used a small anchor escapement (also used later by Breguet) to regulate the speed of striking in a repeater.

Julien's son Pierre (1717–85) was almost entirely taken up with the invention of a marine timekeeper. As early as 1748 he invented a detached detent escapement of a sort, and by 1766 he had brought it to a considerable stage of perfection, in the instrument now preserved in the Musée des Arts et Métiers in Paris. Pierre did not take a great interest in watchmaking, although he made some watches with Sully's escapement and gridiron temperature compensation. One of these has a primitive form of detached lever escapement (see pp. 45 and 53).

LITHERLAND, PETER (1756–1804), was born at Warrington, and in 1791 took out his first patent for the rack lever escapement, which had been dormant since its invention by Hautefeuille in 1722. Litherland's form of it is so different from Hautefeuille's that he may well not have known of the latter. The rack lever as made by Litherland came to be made in great numbers in Lancashire. Despite its complete non-detachment it performed remarkably well, and probably paved the way for the introduction of Massey's detached lever escapement in 1815 and subsequently the development, mostly in Lancashire, of the single-roller lever, from the early 1820s.

Dr. Vaudrey Mercer has done much research into the work of the Litherlands, which is set out in *Antiquarian Horology* (the proceedings of the Antiquarian Horological Society), Volume 3, No. 11, for June 1962. He gives the following list of the successive styles of the firm, with dates:

1796	Peter Litherland,	21 Mount Pleasant
1800–3	Peter Litherland,	12 Commutation Road
1800–7	Litherland Whiteside & Co.,	Ranelagh Street
1810–13	Litherland Whiteside & Co.,	Church Street
1816–35	Litherland, Davies & Co.,	Church Street
1837–76	Litherland, Davies & Co.,	Bold Street.

McCABE, JAMES, came of a Belfast clockmaking family, but moved to London and was admitted to the Clockmakers' Company in 1781. He made many fine watches, with a variety of escapements, and in particular brought the duplex escapement to a high level of perfection. His best watches were signed 'James McCabe', and the second graded 'McCabe'. He died in 1811 but the firm continued until 1883, and Britten gives the following dates, styles and addresses:

James McCabe was first at 11 Bell's Buildings, Fleet Street, in 1778; 34 King Street, Cheapside, in 1783; 8 King Street, Cheapside, in 1788; 97 Cornhill, Royal Exchange, in 1804.

After James's death the business was carried on by successive members of the family as McCabe & Son, at 99 Cornhill, until 1820. McCabe & Strahan, 1825–6; J. McCabe, 97 Cornhill, until 1838, and then 32 Cornhill until 1883, when Robert Jeremy McCabe retired and closed the business.

However, the signature of James McCabe seems also to have persisted long after his death, as there survives a movement with the Savage two-pin escapement, which must be later than 1820, and by the style more like 1830, which is signed 'Jas McCabe, Royal Exchange London'.

MAIRET (alternatively JEANMAIRET), SYLVAIN (1805–90), was one of the finest Swiss watchmakers of the nineteenth century, alike on account of the quality of work, the technical interest of his escapements (particularly the lever) and the external elegance of his watches. Between 1830 and 1840 he lived for five years in London, working largely for B. L. Vulliamy, and also Hunt and Roskell. Most of his lever watches have all the lift on the escape wheel teeth.

MARGETTS, GEORGE, was born in 1748, admitted to the Clockmakers' Company in 1779 and died insane in 1808. He was an early maker of eight-day marine chronometers and one pocket chronometer by him is known. In the watch field he made a number of very complicated and decorative astronomical watches, and was also among the pioneers of the lever escapement in England, employing a fairly straightforward form of cranked roller with separate safety roller. His trains are finely cut and high-numbered, but the superficial finish is remarkably rough and the movements are seldom even signed. It is thought that no specimen of his lever escapement has survived. One was said to be in the Dennison collection, but this was so badly catalogued that the attribution is most unreliable, and it certainly did not appear in the sale of the collection at Christie's in 1961.

MARTINOT. This prolific family were watchmakers to the French court, on and off, for two centuries from about 1570 to 1770. As watchmakers they introduced no special innovations, but are mentioned because of their long history and the large number of watches by them, especially fine oignons, which have survived.

MASSEY, EDWARD (1772–1852) was a Coventry maker. He patented a form of pump-wind and, in 1815, a form of detached lever escapement. This is of great importance, being the first introduction of a simple, cheaply-made lever escapement into England, as opposed to the very complicated work of the earlier experimenters from Mudge to Grant. It is doubtful if any lever watches were made in England during the decade before Massey's patent in 1815 (and very few even then before 1820). Massey is often quoted as having invented the crank-roller lever escapement, which he did not. He did make a few watches with cranked rollers, but most watches by him (and very few actually bear his name) or others, described as having a 'Massey crank roller lever', have a nib which is an integral part of the roller. The peculiarity of his escapement lies in the safety action. This consists of slots cut into the roller, on each side of the nib impulse pallet, into which the horns of the lever must enter before allowing the escapement to operate. The escapement may therefore be more accurately described as 'Massey's Escapement'.

MOINET, LOUIS (1758–1853), was Breguet's business secretary and later set up on his own account, making watches in the Breguet style. Breguet never published anything except what amounted to catalogues in his lifetime, but it was known that he had compiled a lot of notes, and when these could not be discovered after his death, Moinet was widely thought to have purloined them. This became a certainty in 1848 when Moinet published his *Nouveau Traité Général Astronomique et Civil d'Horlogerie Théoretique et Pratique*. The Breguet family successfully applied to the courts for the return of the original notes, which still exist. Moinet's *Traité* is a most valuable work, finely illustrated.

MOTEL, JEAN FRANÇOIS HENRI (1786–1859), was one of Breguet's most eminent pupils and made watches of the highest quality, especially with a form of pivoted detent escapement. He used Houriet's spherical-shaped balance spring, and a cylindrical spring with conical-shaped ends, of his own.

MUDGE, THOMAS, was born in 1715 and apprenticed to George Graham. Although Thomas Colley was legally Graham's successor at his death in 1751, Mudge was so effectively. For the King of Spain he made a watch, set in the top of a walking stick, which showed true and apparent time, struck the hours, and was a minute repeater. In the Guildhall Museum is the movement of another Mudge watch which reputedly belonged to the King of Spain, but the evidence for this is obscure. It has a very sophisticated form of verge escapement, a spiral bi-metallic compensation curb and a remontoir wound once every minute. This is almost certainly the first time a remontoir was fitted to a pocket watch and indeed, it has very rarely been done at any time. The movement is believed to date from 1755.

For his ordinary watches Mudge used the cylinder escapement and in almost every way, including the dial, hands and case, closely followed the design laid down by Graham. In about 1755 he entered into partnership with William Dutton, another apprentice of Graham, and the work was then signed 'Thos Mudge Wm Dutton'.

During the 1760s Mudge had been experimenting with the detached lever escapement and probably the bracket clock with lever escapement in the Ilbert Collection dates from this decade. But in 1770 he succeeded (with great difficulty, on his own admission) in compressing it into the famous pocket watch which was sold to George III and is still in the Royal ownership. This was the first lever escapement watch, from which all modern watches have been developed. In 1771 Mudge retired from active business and moved to Plymouth to concentrate on the development of his remontoir escapement, marine timekeeper, but the name of the firm continued unchanged to the time of his death in 1794. Mudge, or Mudge & Dutton, cylinder watches are of very fine workmanship and appearance and are now moderately rare.

NOUWEN (alternatively NOWE, NOWAN, NEUWERS), MICHAEL, was active from about 1580–1613 as one of the first English watchmakers of the highest quality. Several specimens of his work survive, including a superb crystal watch in the British Museum.

OUDIN, CHARLES, was one of Breguet's best pupils and was in business on his own account from 1807–25. He produced a close copy of Breguet's souscription watch and frequently signed himself 'Oudin, élève de Breguet'. He invented a form of keyless work.

PENDLETON, RICHARD was active from about 1780 and died in 1808. He worked first for Emery and was generally known to have made Emery's lever escapement watches. Very few lever watches survive that are signed by him and only one is known to be in its original case, on which the hallmark is for 1797. From this, and the numbering of the others, it seems that they were made in the closing years of the century, immediately after Emery's death. They closely follow

Emery's general pattern, and the bridge balance cocks and 'double-S' balances are indistinguishable. They are signed 'Richd Pendleton, Pentonville, London'. In about 1794 he became one of the team employed by Thomas Mudge Jnr. to make marine timekeepers on his father's plan.

PENNINGTON, ROBERT, was active in Camberwell from about 1780–1816 as a chronometer maker, and with Pendleton was one of the team got together by Thomas Mudge Jnr. to make marine timekeepers on his father's plan. He made pocket watches of very high quality, including chronometers, and employed a much lighter and more modern form of bi-metallic compensation balance than most of his British contemporaries. At his death he was succeeded by his son, also Robert, who continued in business at 11 Portland Row, Camberwell, until 1842.

PERIGAL, FRANCIS, was active from about 1770 to the time of his death in 1794. He was admitted as honorary freeman of the Clockmakers' Company in 1781. A fine maker, he is particularly important by virtue of the very early lever watch by him which survives, certainly made before 1790 (unfortunately it cannot be dated by the hallmark as the case is not original). Since he was appointed watchmaker to the King in 1784 he may well have had an opportunity of examining the Mudge lever watch, which to some extent his resembles. He was succeeded in business by his son, also Francis, and there were several other makers of the same family.

PERRELET, ABRAM-LOUIS (1729–1826), worked at le Locle and was described by Breguet as 'un homme si bon et tout plein de talent'. In *La Montre Automatique Ancienne* by Alfred Chapuis and Eugène Jaquet (Paris, 1952) it is proved at length, and pretty conclusively, that the self-winding watch was invented by Abram-Louis Perrelet. The experiments went on during the 1770s and had attained to a satisfactorily-reliable result by 1780. The first examples of these are said to have been bought by Breguet and Recordon. Unfortunately no example of Perrelet's own work can be identified since the Neuchâtel makers did not sign their watches; but there are good grounds for supposing that a very ancient self-winding watch in the Le Roy collection may be by Perrelet. If so, it has the peculiarity that, unlike Breguet's and Recordon's oscillating weights, this one is centre-pivoted and revolves through 360 degrees.

PETO, a watchmaker whose Christian name is not known, was active from about 1780 to 1800 and worked for the Brockbanks. He is remembered by his cross detent variant of the spring detent escapement which he may have arrived at to combine the theoretical advantages of Arnold's and Earnshaw's lay-outs, but possibly only to evade Earnshaw's patent. Several pocket watches with Peto's escapement were made by Brockbanks, and long after the British makers had given it up, Breguet thought so well of it as to introduce it into at least two of his finest tourbillons made after 1810.

PHILLIPS, EDOUARD (1821–89), was a mechanical engineer and mathematician. From 1860 he also carried out investigations into the mathematics and properties of balance springs. Although terminal end curves to helical springs had been used empirically by Arnold from 1776, and Breguet introduced his overcoil for flat spiral springs some time around 1800, it remained for Phillips to lay down their mathematically correct formation. His best-known treatise is *Le Spiral Réglant*, published in 1861.

POUZAIT (or POUZZAIT), JEAN MOÏSE (1743–93), was a capable maker at Geneva where he was put in charge of the first school of horology in 1788. He experimented with the lever escapement and produced a model in 1786 which has no resemblance to the English types, so it may well have been arrived at quite independently. The escape wheel teeth project vertically from the face of the wheel and have chamfered impulse faces, as also have the anchor pallets. Pouzait was thus certainly the first to introduce divided lift, which was not taken up by Breguet, nor probably any other maker, until well after 1800. His escapement was, however, defective in having no true safety action, and he seems generally to have used very large, seconds-beating balance wheels the full diameter of the watch.

PREST, THOMAS, was John Roger Arnold's foreman and probably made most of his best watches. In 1820 he patented a form of keyless wind which was occasionally used by Arnold (No. 4501). It is applicable only to a going barrel and the hands have to be set by a key on the usual square on the cannon pinion. Prest died in 1855.

QUARE, DANIEL (& HORSEMAN) (*c.* 1648–1724). Quare was admitted to the Clockmakers' Company in 1671 and was Master in 1708. He was second to none in his day as a watchmaker, and his work is always handsomely proportioned, especially his early work and six-hour dials. He was conservative in case design, keeping a loose-ring pendent and square-ended hinges long after others had gone over to ringed pendents and curve-ended hinges. In 1680 he made a quarter repeating watch with which he successfully opposed the Rev. Barlow's application to James II in 1686 for a patent. Barlow's had two push pieces, one for the hours and another for the quarters, but Quare operated both trains with a single push piece. In 1718 he went into partnership with Stephen Horseman, after which their work is signed 'Quare & Horseman' and Horseman continued it unaltered after Quare's death in 1724, until 1733, when he became bankrupt. Known watches signed by Quare have numbers 233 to 4989, and known watches by Quare & Horseman have numbers 4677 to 5503 (indicating some overlap). Quare repeaters are known from 109 to 257 and Quare & Horseman repeaters are known from 843 to 1129. Quare was much faked in his lifetime, but some watches signed 'Quaré' are of high quality, quite consistent with his own work, and may be genuine, the accent having been added by him on export pieces.

RAMSAY, DAVID (*c.* 1590 to *c.* 1654), was born in Scotland but worked both in France and London. He was watchmaker to James I and, apparently, Charles I, at any rate until the royal appointment of Edward East. He was the first Master of the Clockmakers' Company, in 1632. He made watches of high quality, a few of which survive, including a particularly fine example which is part of the insignia of the successive Masters of the Clockmakers' Company.

RECORDON, LOUIS (active 1778–1824), worked in Greek Street, Soho, and in 1797 succeeded Emery at No. 33 Cockspur Street. He also acted as Breguet's London agent and in 1780 took out for him an English patent (No. 1249) for self-winding watches. He made fine watches with a variety of escapements, in a style standing somewhat between English and Breguet's. At some time about 1796 he entered into partnership with Paul Dupont when their work was signed 'Recordon & Dupont'. Recordon retired in 1796, but the name continued unaltered until about 1816 when the business was taken over by Peter Des Granges who continued at 33 Cockspur Street until 1842.

ROBIN, ROBERT (1742–1809), a very fine maker, and clockmaker to Louis XV and XVI, and subsequently to the Republic. He devised a pinwheel lever escapement for watches and also, in 1791, the escapement still known by his name. In this, a lever is used for locking only and impulse is given direct by the escape wheel to the balance. Impulse can therefore only be given on alternate swings of the balance. The escapement was used a good deal by Breguet, including in some of his perpetuelles, while his 'échappement naturel' is really only a double Robin. Robert Robin is not to be confused with a later Robin who operated in the Rue de Richelieu from about 1790–1825 and was clockmaker to Louis XVIII.

ROSKELL, ROBERT (active 1798–1830), worked both in Liverpool and London. He made many good quality rack-lever watches and, later, watches with Massey's lever escapement. Many of his watches have dials decorated in multi-coloured gold and he frequently used a small escape wheel to which was fixed a seconds hand revolving once in 15 seconds. Other members of the family also operated in Liverpool.

SAVAGE, GEORGE (active about 1808–55). He was born in Huddersfield where in 1808 he patented a remontoir. He later moved to London and in about 1820 invented a lever escapement which was used by a few makers until late in the nineteenth century, and is known as the 'Savage two pin'. In it, a very wide pallet on the balance arbor engages with two pins on the lever, but all or most of the impulse is given by the safety pin which engages with a very narrow slot on the roller. The escapement works well with a very brisk action, but it is difficult to make and never came into wide use. It is nevertheless important for the part it played in the revival of the lever escapement after 1820. Savage later emigrated to Canada where he founded a successful retail business in Montreal and died in 1855.

SULLY, HENRY (1680–1728), had a career of continuous failure, both in France and England. Soon after completing his apprenticeship with Charles Gretton he travelled abroad, where Julien Le Roy recognised his abilities, and he was persuaded by Law, a Scottish speculator living in Paris, to obtain the services of a number of skilled English watch and clockmakers and set up a factory at Versailles. This venture failed, but it did introduce to France a standard of craftsmanship which had a profound effect upon the whole French industry. Sully's most important single contribution to watchmaking was the invention of oil sinks, later perfected by Julien Le Roy.

After his unsuccessful French venture he returned to England, bringing some of his staff with him, but here he was equally unsuccessful and eventually returned to France where he became slightly more prosperous and in 1721 turned his attention to the problem of marine timekeeping. In this he was not successful either, but one of his attempts survives in the Guildhall Museum, with an escapement similar to Debaufre's (but having one escape wheel and two pallets, as opposed to Debaufre's two escape wheels and one pallet) which has, though very rarely, been applied to watches.

In 1714 he published in Vienna an important horological work, *Règle Artificielle du Temps*.

TAVAN, ANTOINE (1749–1836), worked in Geneva where he came to be considered the most skilled and inventive watchmaker; he was also much admired for his character and amiable personality. He invented a number of escapements of considerable ingenuity and complexity, at a time when the lever and chronometer escapements had not yet established themselves in their position of unchallenged supremacy. Several of these are illustrated in Chamberlain's *It's About Time*. In 1806 he was commissioned by Melly Frères to make models of ten escapements. These survive in the museum of the horological school at Geneva. They all measure 10 cm (4 in) in diameter and are of great beauty. The escapements, including some of Tavan's own, are: verge, virgule, lever, 'Arnold' (pivoted and spring-detent), constant-force (Tavan); pin-escapement for watches, lobster-claw (Tavan, a form of Robin escapement) and 'brise et à surprise' (Tavan, a variant of the lobster-claw, with less drop). Only one watch signed by Tavan is known to survive, and formed part of the Chamberlain collection. It has the 'brise et à surprise' escapement and may well be the watch which won for Tavan the gold medal at the Geneva Competition in 1819.

THURET, JACQUES. Thuret's dates are not known, though he succeeded his father as Horloger du Roi in 1694. His horological importance is that in 1675 Christian Huygens, then living in Paris, selected Thuret to execute a watch with his newly invented spiral balance spring. The first examples had a gearing between the verge and the balance wheel with its spring, whereby the vibrations of the balance were amplified. But this complication was soon given up. Several fine oignon watches with normal verge and balance-spring escapements, by Thuret, survive.

TOMPION, THOMAS (1639–1713), is thought to have been brought up by his father as a blacksmith, in Northill, Bedfordshire, where he was born. It is not known what influenced him to take up watch and clockmaking, nor to whom he was apprenticed. But he was admitted to the Clockmakers' Company in 1671 and became Master in 1704. While Tompion made some outstandingly fine clocks, his watches are no better than the best of his contemporaries. He was, however, much assisted by Robert Hooke in co-operation with whom—however unwillingly at times—he developed a practicable form of balance spring, although not in time to be able to claim the priority over Huygens and Thuret. Nevertheless, the superiority of his wheelwork, in conjunction with the new balance spring, attained an accuracy in portable timekeeping far exceeding anything that had been known or contemplated previously. And this in turn benefited the whole

British industry, giving it a clear pre-eminence over other Continental artists which it enjoyed for at least eighty years. Only one watch by Tompion without a balance spring, and therefore made before 1675, has survived, and that as a dial-less movement only. It is in the Guildhall Museum of the Clockmakers' Company.

In 1701 Tompion took Edward Banger (who had married his sister's daughter) into partnership and their work was then signed 'Thos Tompion Edwd Banger'; but the partnership seems to have been an uneasy one and it broke up in 1707 or 1708. He then took into partnership George Graham, who had worked for him since 1695, married his brother's daughter, and ran his business after he died in 1713. Pieces signed 'T. Tompion & G. Graham' are, for some reason, very rare, and the joint signature does not seem to have been used before about 1711. 1711.

Tompion started to number his watches in about 1680. The number was stamped under the balance cock and this is a useful point in establishing authenticity, since Tompion was much faked, both in his lifetime, and long afterwards; although the fakes are usually not difficult to recognise. He seems to have had three series of watch numbers. Baillie gives the following dates and numbers:

Signature	Year	Plain Watches	Repeaters
Thos Tompion	1701	No. 3292	No. 203
Thos Tompion Edwd Banger	1701–8	Nos 3252–4119	Nos 196–290
Thos Tompion	1709–13	Nos 4265–4312	Nos 359–392
T. Tompion & G. Graham	1711–13	Nos 4369–4543	

George Graham continued the series after Tompion's death and the earliest recorded number with his name alone is No. 4669 (plain) and 393 (repeater). There is a very early repeater by Tompion, in the Ilbert Collection, dating from about 1690, with the number 63. In addition to the above two series Tompion seems to have had a third series for special watches of one kind or another, in which the number was prefixed by an O.

Tompion had a nephew, also Thomas, his brother's son, who was apprenticed to Charles Kemp and admitted to the Clockmakers' Company in 1702, but he does not seem ever to have worked with his uncle and was evidently a bad character, as in 1720 he was sent to prison for pickpocketing.

TYRER, THOMAS, of Red Lion Street, Clerkenwell, did not invent the duplex escapement, but he brought it more or less to its established form and in 1782 took out a patent (No. 1311) in respect of it, described as a 'horizontal escapement for a watch to act with two wheels'. Apart from this nothing, not even his dates, is known.

VACHERON & CONSTANTIN. The Vacheron family were watchmakers from 1785 and the Constantins were in business by about the same date. They amalgamated in 1819 to form the firm which still flourishes under that name. Early work was of no more than average quality, but in 1839 they took into partnership George Leschot who introduced a large measure of machine tooling and interchangeability. Since this date the products of the firm have been uniformly of high quality.

VULLIAMY. Three successive members of this family were active from 1730 and 1854 and although they are most famous for their unsurpassed regulator clocks they also made some excellent watches.

Justin Vulliamy was in partnership with Benjamin Gray from 1730 to 1775. His son Benjamin, of Pall Mall, was active from 1775 to 1820 and his son Benjamin Lewis, of 68 Pall Mall, continued the business up to his death in 1854.

The most characteristic Vulliamy watches have the duplex escapement and the whole top plate is covered with a decorative pierced fret. These watches were made with little or no variation over a surprisingly long period, from before 1800 until well after 1820. Another, rarer class of Vulliamy watch could almost be taken for a Breguet lever repeater, except that the centre of the gold or silver dial is matted and not engine-turned. It is highly probable that these were made for Vulliamy by Sylvan Mairet.

WRIGHT, THOMAS, was admitted to the Clockmakers' Company in 1770 and died in 1792. He is only of importance in having taken out for Thomas Earnshaw a patent (in 1783, No. 1354) for his spring detent escapement and compensation balance, as Earnshaw could not afford to do so for himself. Earnshaw's earliest watches with this escapement were stamped 'Wright's Patent'.

Glossary of technical terms

This glossary is not intended to be comprehensive. Its aim is to define the technical terms which are not self-explanatory and are commonly met with in discussion about watches. A fuller description of many of the terms may be found in other parts of this book, by reference to the index.

An exhaustive list of horological terms is contained in *Watchmakers' and Clockmakers' Encylopaedic Dictionary* by Donald de Carle (London 1959).

Alarum A mechanism that can be set to sound a bell at a predetermined time.

All-or-nothing Piece The catch, in the repeating mechanism of a repeater watch, that prevents the mechanism from operating unless the push pendent or slide is moved to the end of its travel.

Anti-friction Roller The roller bearing sometimes used during the eighteenth century, instead of jewelled bearings, to support the balance axis of watches.

Arbor The axle of a watch wheel. The arbor of the balance wheel is usually called the balance staff.

Arc The angle through which a balance wheel swings; the arc of vibration. *See also* Escaping Arc *and* Supplementary Arc.

Auxiliary Compensation See Middle-temperature Error.

Back Plate In a plated, or part-plated movement; the plate seen from the back of the movement (as opposed to the 'front', or 'dial' plate).

Balance The means of controlling the speed at which the mainspring of a watch unwinds. In some very early watches it took the form of a dumbell, or foliot, but almost invariably it is an oscillating wheel, kept in motion by the escapement of the watch. The application of the balance spring to watches (*c.* 1675) secured the rate of oscillation as more or less isochronous, regardless of the extent of the arc. The effect of a balance spring alters with changes of temperature; these alterations are counteracted by various forms of 'compensation balances.' Since the development of metals for balance springs which are not affected by changes of temperature, compensation balances have become redundant. *See* Balance Spring.

Balance Cock The bracket in which the bearing for the upper pivot of the balance staff is fitted.

Balance Spring A thin wire spring used to control the period of oscillation of the balance. It is usually in the shape of a spiral of which the outer end is fixed to some convenient part of the balance cock or plate, and the inner end is attached by a collet to the centre of the balance. It was first experimented with by Robert Hooke, from *c.* 1658 onwards, but its successful application, in spiral shape, is almost certainly attributable to Christian Huygens in 1675. In chronometers and in watches where thinness is not a requisite, the spring may be of cylindrical, or helical shape, and there have been other shapes, not widely used, such as a pyramid, sphere and 'duo-in-uno'. The isochronism of a balance spring is largely determined by its terminal curves, which were developed by John Arnold. Broadly speaking, the terminals of a helical spring are curved inwards to as nearly as possible the centre of oscillation. A. -L. Breguet discovered that much the same effect can be obtained with a spiral spring, by bending the outer end of the spring upwards and over the coils of the spiral to a point of attachment as nearly as possible to the centre of oscillation. This is known as the 'Breguet overcoil'. Balance springs may be made of iron (originally), steel or gold (eighteenth and nineteenth centuries) or one of the proprietary metals (such as Elinvar, palladium, Nivarox and others) whose elasticity is not altered by changes of temperature.

Balance Staff The arbor of the balance wheel.

Band The middle ring of a watch case into which the movement fits. The back and front bezels are hinged, or snapped onto it.

Banking Pin A pin used to limit the motion of a moving component. In a watch it limits the balance arc.

Barley Corn The type of engine-turning most frequently used to decorate watch cases, so called because it resembles concentric rows of barley seeds laid end to end.

Barrel The circular box containing the mainspring. There are three kinds of mainspring barrel. The first is called a standing barrel. The barrel is fixed to the plate and the outer end of the spring is attached to it. The inner end is attached to the rotating arbor at the centre of the barrel. The first wheel of the train is fitted loosely on the arbor and is connected to it by a ratchet. The ratchet enables the spring to be wound onto the arbor. It then transfers the drive to the first wheel and thus drives the watch.

The second type is used in conjunction with a fusee. The central arbor is fixed and cannot turn. The barrel fits loosely on the arbor and is free to turn. A gut cord, or chain, is wound round the outside of the barrel and one end is attached to it. When the mainspring is run down, the whole chain is wound round the barrel. The other end of the chain is attached to the fusee (French: *fusée =* spindle). The fusee is roughly conical in shape with a spiral track cut round it and, thus, of reducing diameter from one end of the fusee to the other. The fusee is rigidly attached to its arbor. The wide end of the fusee is connected by a ratchet to the first wheel of the train. The winding key is attached to the fusee arbor. As it is turned, it winds the chain off the barrel onto the fusee. The ratchet enables this to happen without moving the first wheel. As the chain is transferred to the fusee it winds up the spring. When

pressure is taken off the winding key the spring operates to wind the chain back on to the barrel. It is impeded from doing so by the fusee ratchet, connected to the train of wheels and the escapement, thus allowing it to unwind only at the speed dictated by the escapement. A stop prevents further winding when the chain reaches the final turn of the fusee. In order to ensure that the watch will run out, some initial tension has to be imparted to the spring. This is achieved by mounting the fixed barrel arbor on a ratchet or wormwheel. This is wound up about a quarter of a turn when the watch is assembled and the tension so provided is called the 'set-up'. The purpose of the fusee is to equalise the torque, or driving-force of the mainspring between wound-up and run-down. This was vital with non-detached and therefore non-isochronous escapements. *See also* Maintaining Power.

The third kind is the going barrel which is now universal in watches. As in a standing barrel, the central arbor can be turned and is the means of winding up the spring. For this purpose it has a ratchet. The first wheel of the train is fixed to the barrel so that the force of the spring is constantly applied to it. The torque thus varies between wound-up and run-down and for this reason a going barrel is only applicable to isochronous escapements. To minimise the change in torque it is usual to fit as long and as weak a mainspring as possible into the barrel, and ensure by means of stopwork that only part of the potential duration of the spring is utilised to drive the watch.

Barrel Arbor The steel arbor at the centre of the spring barrel.

Barrow Regulator See Regulator. *See* figs. 102, 103.

Basse-taille A form of watch case decoration consisting of engine-turned gold covered with transparent, coloured enamel.

Beetle Hand The form of hour hand found in most eighteenth-century watches, so called from its fancied resemblance to a stag beetle. Usually accompanied by a 'poker' minute hand.

Bezel The rim into which the glass of a watch is snapped.

Bi-metallic See Compensation Balance.

Blueing The process of heating polished steel until the surface turns blue.

Bow The pivoted loop of a watch pendent used for hanging the watch or attaching it to a chain. Also, the piece of whalebone and horse-hair used by watchmakers for rotating the workpiece in the turns.

Breguet Hand The style of watch hands almost invariably used by Breguet. The hand terminates in a pointer carried on a circle pierced with an eccentric hole.

Breguet Key The form of watch key invariably supplied by Breguet with his watches. A ratchet within the shaft frustrates attempts to wind the spring in the wrong direction. Also known as 'tipsy-key'.

Breguet Overcoil See Balance Spring.

Bridge or *Bridge-cock* A bridge-shaped support, attached to the movement at both ends, containing the bearing for a pivot. *See also* Cock.

Bristle Regulator *See* Regulator.

Buffer Spring The spring in self-winding watches with oscillating weights, which limits the travel of the winding weight and absorbs the shock caused by violent movements on the part of the bearer.

Calibre The arrangement of the components of a watch movement.

Cannon Pinion The pinion to which the minute hand is fitted. It drives the motion work which in turn drives the hour hand.

Carriage The rotating frame carrying the escapement of a tourbillon or karrusel watch.

Centre Pinion The first driven pinion of a watch. It is usually at the centre of the movement but it is still so called even when not central.

Centre Seconds A seconds hand pivoted concentrically with the hour and minute hands. Also termed Sweep Seconds.

Centre Wheel *See* Centre Pinion.

Chapter Ring The circle of division on the dial indicating the hours.

Champlevé Enamel A form of enamelling much used for the dials of British seventeenth- and early eighteenth-century watch dials. The area to be enamelled is scooped out of the field and the hollow so formed is filled with enamel. The chapter ring is frequently treated in this way.

Chronograph Commonly called a stop-watch; but strictly it should be a stop-watch which makes a mark where the chronograph hand is stopped. Fatton (a pupil of Breguet) made ink-recording chronographs and Breguet made them under licence. But chronograph has come to be synonymous with stop-watch. It is a watch with a seconds hand (usually centre seconds) which, by pressing a button, can be started, stopped and returned to zero at will. A split-seconds chronograph has two such seconds hands which can be stopped and started independently, thus enabling two separate operations to be timed simultaneously. Chronographs also have a subsidiary dial recording the number of minutes elapsed from the time of starting the seconds hand. It returns to zero, together with the seconds hand.

Chronometer A term now applied to any precision portable timekeeper, but historically applicable only to watches with the detent escapement.

Clock-watch A watch which automatically strikes the hours, or hours and quarters.

Cloisonné Enamel A form of enamelling used for decorating watch cases. Used especially for floral patterns. Thin strips of gold are first soldered to the watch case and serve to divide the different colours of enamel. The dividing strips are known as cloisons.

Cock A bracket attached to the movement which contains the bearing for a pivot. *See also* Bridge.

Collet A small metal ring, usually split, used to attach the balance spring with a friction fit to the balance staff.

Compensation Balance A 'compensation' balance is one designed to counteract alterations in the elasticity of the balance spring under changes of temperature. The effect of such alterations is to make the watch go slow in high temperatures, and fast in low temperatures. Compensation of the balance is achieved by enlarging the effective radius in low temperatures and reducing the radius in high temperatures. This is achieved by making the rim of the balance of two laminae (usually) of brass on the outer edge and steel on the inner edge. The laminae are fused together. This rim may be in two, three, or four sections; but is usually in two. One end of each section is mounted on a spoke or arm of the balance. The other end is free. When the temperature changes, the different coefficients of expansion of the two laminated metals cause the free end of the rim to move towards or away from the centre of the balance. Its effective radius is thus altered, and a constant rate is achieved. *See also* Balance, Balance Spring *and* Middle-temperature Error.

Compensation Curb An automatic watch regulator, fitted during the late eighteenth and early nineteenth century to watches of high quality, which did not, however, possess a compensation balance. The curb may be of various shapes but always, effectively, of a laminated brass and steel strip. The movement of the free end of the strip operates so as to lengthen or shorten the balance spring by means of curb-pins at its outer end.

Complicated Work The addition to a watch of any mechanism other than repeating work.

Conical Pivot The pointed pivot used, principally, by Breguet for the balance staffs of his shock-proof watches. *See also* Parachute. *See* fig. 38.

Constant-force Escapement A watch escapement in which the impulse to the balance is given by a released spring. On completion of the impulse, the train re-tensions the impulse spring in readiness for the next impulse. *See* Remontoir Escapement.

Consular Case The form of watch case used by John Arnold and some of his contemporaries.

Contrate Wheel A wheel whose teeth are at right angles to the plane of rotation. Used to transfer a drive through ninety degrees, as in verge watches, to drive the crown or escape wheel.

Coqueret A polished steel plate, screwed to a cock, acting as an end plate to a pivot. Similar to an end stone. Used mainly in France before the introduction of jewelling.

Count Wheel Found in clock-watches before the introduction of rack-striking. The count wheel has twelve slots cut into its periphery. The slots are progressively wider from the first to the twelfth. A claw drops into each slot ensuring that the correct number of hours are struck. The watch, therefore, cannot avoid striking the hours progressively, regardless of the position of the hands.

Crown Wheel An alternative name (not much used) for the escape wheel in verge watches. Similarly Crown Wheel Escapement.

Crystal A flat watch glass with a bevelled edge. Also, any watch glass made of rock-crystal. Much used by A.-L. Breguet.

Curb-pins The pins fitted to a regulator index, or a compensation curb, which embrace the balance spring.

Cuvette The inner cover of a watch case. In key-wound watches it has a hole or holes for inserting the key. The English equivalent is 'dome'.

Cylinder Escapement The dead-beat, frictional rest escapement developed by George Graham in 1726. Also known as the 'horizontal escapement'. A cylinder, with a segment cut away, forms part of the balance staff. The escape wheel teeth have impulse planes cut into their acting faces. They import impulse to the cylinder as they pass the entry and exit lips. Between impulses, the point of an escape tooth rests alternately on the outside and inside faces of the cylinder; hence the name 'frictional rest'. The cylinder was usually made of steel and the escape wheel of brass, but from about 1760 some very high-grade watches had ruby cylinders and steel escape wheels which were less subject to wear. A.-L. Breguet invented a special form of ruby cylinder which is almost indestructible. The cylinder is mounted on a cranked extension of the balance staff, beyond the end-bearing of the staff. The escape wheel is simpler and more robust than the conventional Graham type.

Dart In the lever escapement, the dart-shaped projection attached to the end of the lever that provides the safety action. *See* Lever Escapement.

Dead-beat Escapement Any escapement in which there is no recoil. *See* Recoil.

Debaufre Escapement A frictional rest escapement invented in about 1704 by Peter Debaufre. The Sully and Ormskirk escapements were developed from it. A disk mounted on the balance staff has a segment cut out of it. An impulse plane is on the radial edge of the segment. Two escape wheels with ratchet-shaped teeth are mounted on a common arbor. This arbor is at right angles to the balance staff. The teeth of the two wheels are staggered in relationship to each other. One tooth escapes through the cut-out segment of the disk and gives impulse. The next tooth of the other wheel is then held on the surface of the disk until the segment reaches that tooth, which in turn escapes. This is the form of the escapement used by the Ormskirk makers and by Paul Garnier. Another form has one escape wheel, and two parallel, concentric disks, mounted on the balance staff. The cut-out segments are staggered so that as a tooth of the escape wheel passes through and impulses one segment, it is held on the second disk until it can pass through the segment of that disk. The next tooth is then held on the first disk, and so on. This is Sully's form of the escapement. *See* figs. 44, 45.

Deck Watch A large precision watch, usually too large for pocket wear, and contained in a box. Mainly used on board ship to note the time while taking observations to determine the ship's longitude. It is checked by the ship's chronometer.

Detached Escapement An escapement in which the balance is detached from it during the supplementary arc: e.g. the lever and detent escapements.

Detent Any piece used to lock a wheel by catching its teeth, in particular the spring- or pivot-mounted detent in the chronometer, or detent escapement.

Detent Escapement A detached escapement used in watches and marine chronometers (whence it is often called the chronometer escapement). Impulse is given to the balance only on alternate oscillations. During the supplementary arc the escape wheel is locked by a detent. It is the watch escapement possessing the highest degree of detachment. *See* figs. 11–20.

Dial Plate In a plated, or part-plated movement, the plate nearest the dial. Also referred to as the 'front plate'. *See also* Back Plate.

Differential Dial A very rare form of dial used in the late seventeenth century. There is only one hand, which is an ordinary minute hand. The hour chapter ring is on a small disk concentric with the hand. This rotates at eleven twelfths the speed of the minute hand so that the relevant hour numeral is always under the hand. *See* figs. 123, 132.

Dome The inside cover in the back of most watches, other than pair cases. The French term *cuvette* is more commonly used.

Double-roller Escapement A lever escapement with separate rollers for the impulse and safety actions.

Double-virgule Escapement *See* Virgule Escapement.

Draw In the lever escapement, a small angular displacement of the locking faces of the lever pallets as they are drawn up the point of the escape wheel teeth. It ensures that the lever moves firmly to the banking. First used by Leroux in 1785, it did not come into regular use until after 1820.

Drop The free movement of the escape wheel between the instants of impulse and locking.

Dumb Repeater A repeater in which the hammer strikes a block of metal instead of a bell or gong.

Duplex Escapement A frictional rest escapement with impulse only at alternate oscillations. An impulse pallet and locking roller are mounted on the balance staff. The locking roller is a jewel, of very small diameter, with a segment cut out of it to permit the locking teeth of the escape wheel to pass through it. As a tooth escapes, an impulse tooth imparts impulse to the impulse pallet. British duplex watches have one escape wheel with two sets of teeth for locking and impulse. Continental watches have two co-axial wheels. The escapement in its widely-used form was invented by Thomas Tyrer in 1782. It was popular during the first half of the nineteenth century and in England during that period surpassed the lever escapement in high-grade pocket watches, where the extreme precision of a chronometer, with its tendency to trip, was not desired. *See* figs. 8, 9, 10.

Dust Cap From about 1715 for the next hundred years, many watches, mostly English, had a dust-excluding cap which fitted over the movement and was held by a bayonet fixing.

Ébauche The watch movement, complete with its wheels, excluding the escapement, complications, case, dial, hands, etc.

Enamel A form of decoration for watch cases. A type of glass coloured by the addition of metal oxides. In the most usual form of enamelled cases the colours are not true enamels, but metal oxides mixed with sufficient flux to make them vitrifiable. *See also* Basse-taille, Champlevé *and* Cloisonné.

End Plate A steel covering for a pivot bearing used to limit end float.

End Stone An end plate set with a jewel to reduce friction.

Engine-turning A form of decorative machined engraving used on cases and metal dials, especially during the first half of the nineteenth century. The work is secured to a machine that is rotated and oscillated as the cutter is applied to the surface, producing a variety of patterns, of which barley-corn was the most popular. It was extensively used by Breguet after about 1795 for cases, and after about 1805 for dials. The French equivalent is guilloché.

Entry Pallet In a lever escapement, the pallet stone of the anchor on the advancing side of the escape wheel. The stone on the

293

departing side is called the exit stone.

Equation of Time The difference at any date between solar time and mean time.

Escapement The part of a watch interposed between the balance wheel and the train of wheels. At each oscillation of the balance the escapement is moved, permitting the mainspring to give impulse to the balance via the train of wheels.

Escaping Arc The part of the total arc of the balance during which unlocking and impulse take place. The remainder of the arc is the supplementary arc.

Escape Wheel The last wheel in the train which is alternately locked by and gives impulse to the balance, either directly or through an intermediary such as the lever.

False Pendulum For a time after the invention of the balance spring, an attachment to the balance staff was sometimes arranged so as to be seen through a radial slit cut out of the dial. It was supposed to look like the bob of a pendulum and thus to encourage its owner to think it was keeping good time. *See* fig. 114.

Foliot An alternative in some very early watches to a balance wheel. It consists of a bar resembling a dumbell.

Fork In a lever escapement, the forked end of the lever which engages with the pallet on the balance staff roller and effects locking and impulse.

Form Watch A watch case in the shape of some non-functional object such as a cross, star, skull, flower-bud etc.

Fourth Wheel The fourth wheel in the train of wheels from the great or first wheel.

French Silvering A process of applying silver chloride paste to a brass surface. Used to give a silver-white finish to dials, mostly of marine chronometers or deck watches. Rarely used in pocket watches.

Frame The plates and pillars of a watch movement.

Frictional rest Escapement The family of non-detached, dead-beat escapements, in which locking is achieved by a tooth of the escape wheel being held against a roller etc. on the balance staff: e.g. cylinder, Debaufre, duplex, Ormskirk, Sully.

Front Plate See Dial Plate.

Full Plate The form of watch calibre in which all the wheels are pivoted between the front and back plates.

Fusee The conical-shaped wheel with a spiral grave cut round it, connected via a gut or chain to the spring barrel. The great wheel is attached to the fusee. The fusee acts as an infinitely variable gearing to even out the torque of the mainspring. *See also* Barrel.

Gathering Pallet A single-toothed pinion used in striking and repeating work. At each rotation the pinion gathers a single tooth of a rack to count the number of hours struck. When the last tooth is gathered the rack re-locks the striking train.

Grande Sonnerie A form of quarter-striking where the preceding hour is struck before each quarter, as in quarter-repeating.

Great Wheel The first wheel in the train of wheels of a watch.

Guard Pin See Dart.

Hair Spring See Balance Spring.

Half-chronometer A widely-used but meaningless expression, sometimes applied to the Robin, or union chronometer, escapements.

Half-plate A calibre in which all the train wheels except the fourth and escape wheels are pivoted between plates.

Helical Spring See Balance Spring.

Hoop Wheel In early clock-watches, prior to the invention of rack-striking, a wheel with a hoop fixed to it, which holds up the count wheel detent during striking.

Horizontal Escapement An alternative for cylinder escapement.

Hour Rack In a repeating watch or clock-watch, the toothed rack which determines the number of hours to be struck.

Hour Wheel The wheel to which the hour hand is fitted.

Hunting Case A watch case with a hinged protective metal cover over the glass covering the dial. If the cover has a small window in its centre, with a chapter ring round it, through which the hands may be seen, it is known as a 'half-hunter'. The French term is *Savonette*.

Impulse and Locking The two incidents in the action of dead-beat escapement. Impulse is the action by which the escape wheel (driven by the mainspring) restores energy to the balance. It is then locked (by different means, according to the escapement) until the balance is ready again to receive impulse.

Isochronism The ability of a balance fitted with an appropriate balance spring to traverse large or small arcs in the same amount of time.

Jump Hour An hour hand which only moves once an hour. Much used by Breguet in his repeating watches.

Karrusel A revolving carriage patented in 1894 by Bahn Bonniksen, in which (as opposed to the tourbillon) the carriage and escape wheel are driven separately. The usual period of rotation is 35 or 52.5 minutes. *See also* Tourbillon.

Leaf (of Pinion) The tooth of a pinion.

Lepine Calibre A lay-out in which, in place of a back-plate, each wheel is pivoted in a separate cock. Each cock is screwed to the front plate. Developed by Jean-Antoine Lepine (1720–1814) in about 1770 and perfected by Breguet.

Lever Escapement A detached escapement invented by Thomas Mudge in about 1754 and first used by him in a watch in 1769. The escapement is now used in all mechanical watches. A centrally-pivoted lever is interposed between the escape wheel and the balance. At the escape wheel end is a cross piece looking like an anchor. At each end of the anchor is a pallet-stone with a locking face (cut radial to the escape wheel) and an impulse plane. The radial faces alternately lock the escape wheel. At the opposite end of the lever is a fork. A pin (the impulse pin) is mounted on a roller on the balance staff. As the balance swings, the pin enters the fork of the lever and carries it along with the pin. This moves the lever so that the locking pallet is drawn away from the escape wheel tooth. As the tooth reaches the impulse plane it imparts impulse to the lever which transfers it via the impulse pin to the balance. The other pallet of the anchor then relocks the escape wheel until the process is repeated on the return swing of the balance. There is, in addition, a safety action to prevent a severe jolt causing the lever to jump out of engagement at the wrong time. This consists of the dart and safety roller. *See* figs. 22–39.

Lift The action of an escape-wheel tooth when traversing the incline of a pallet.

Litherland Escapement A form of rack-lever escapement patented by Peter Litherland in 1791. *See* Rack Lever.

Locking and Impulse See Impulse and Locking.

Lunette Glass The form of watch glass first used by Breguet. The edges are sharply radiused, leading to a lightly curved central section.

Mainspring The spring, coiled into a barrel, which supplies the motive power in a watch or any other portable timekeeper.

Maintaining Power In fusee watches, the train of a watch is reversed during winding so that the escape wheel cannot advance. To overcome this, a small supplementary spring continues to drive the train during winding.

Matting The finely-speckled finish used principally for the centre of watch dials from about 1650–1750, but also occasionally later, into the nineteenth century.

Mean Time The time as recorded by watches. The average length of the day throughout the year, as opposed to solar time. *See* Equation of Time.

Middle Temperature Error A bi-metallic compensation balance is not perfectly matched to the change in elasticity of the balance spring with changes of temperature. It will be accurate at two temperatures, but between those two it will gain. This is known as middle temperature error and is counteracted by a variety of auxiliary devices.

Minute Repeater See Repeater.

Mock Pendulum See False Pendulum.

Moon Hand An alternative name for the Breguet Hand.

Motion Work The gearing by which the minute hand drives the hour hand.

Movement The complete mechanism of a watch.

Nib The projection of the balance staff that engages the two-pin form of fork in Breguet's early lever escapements, and some others.

Oil-sink The small recess round a pivot bearing made to retain the oil. Invented by Julien Le Roy.

Ormskirk Watches Lancashire-made watches of the early nineteenth century with a frictional-rest escapement. *See* Debaufre Escapement.

Overcoil See Breguet overcoil *in* Balance Spring.

Pair Case Two cases for one watch. The inner case contains the movement and is removed from the outer case for winding. The outer case may be decorated. Almost universal in British watches from about 1650 to 1800; less common on the Continent.

Pallet A small jewel stone used for locking the escape wheel and receiving impulse.

295

Parachute A safety device invented by Breguet to prevent damage to balance pivots if the watch is dropped, etc. The pivot bearing is spring-mounted.

Passing Spring The thin flexible spring attached to the chronometer detent, lying in the path of the impulse roller on the balance. On the impulse swing, it carries the detent with it to unlock the escape wheel. On the return swing it flexes away from the detent which remains in its locking position. It is usually made of gold and sometimes called the gold spring. Some Continental makers used steel.

Pendent The part of a watch to which the bow is fitted. In key-wound repeaters it usually forms the plunger which operates the repeater. It is sometimes hollow and the plunger passes through it.

Perpetual Calendar A calendar watch which records February 29th at leap years.

Perpetuelle The name given by Breguet to his self-winding watches.

Pillar The distance piece between the front and back plates of a plated movement.

Pinion On each arbor of a watch train is a large brass wheel and a small steel wheel. The latter is a pinion. Its teeth are called leaves, thus: 'a six-leaf pinion'.

Pin-wheel Lever Escapement Tavan's form of lever escapement in which the escape wheel teeth are pins standing vertically from the plane of the wheel.

Pirouette A geared-up balance. Usually applied to the cylinder escapement. The balance and balance spring are not fitted to the staff containing the cylinder. The latter staff has fitted to it a segment of a large wheel which drives a pinion fitted on the balance staff itself. Huygen's first balance spring watches had a pirouette. The pirouette has rarely been fitted to other forms of frictional rest escapement.

Pivot The end of an arbor reduced in size to turn in a bearing. In the interests of robustness Breguet, and subsequently others, frequently made them conical in shape.

Pivoted Detent A chronometer escapement with a detent pivoting on an arbor, as opposed to a spring detent which is one mounted on a spring.

Poinçon The French equivalent of a British hallmark on gold and silver.

Poise The balances of watches are poised to prevent changes of rate with changes of position of the watch.

Poker Hand The usual form of minute hand in British watches during *c.* 1690–1760, so called for its resemblance to a poker. Usually in company with a beetle hour hand.

Pump-wind An early form of keyless-winding where the pendent is moved in and out (pumped) to wind the watch.

Puritan Watch A fairly small oval-shaped pair case watch, without decoration, typical of the mid-seventeenth century Commonwealth period.

Quarter-rack The rack of a repeating-, or clock-watch, that releases the train and counts the quarters to be struck.

Quarter Repeater See Repeater.

Quarter-snail The four-stepped cam fitted to the cannon pinion against which the quarter rack buts.

Rack See Quarter-rack.

Rack Lever A predecessor of the detached lever escapement. Invented by the Abbé Hautefeuille in 1722 and brought into common use by Peter Litherland after his patent of 1791. The lever has an anchor similar to a detached lever (without draw). There is a toothed segment at the opposite end of the lever, instead of a fork. This meshes with a pinion on the balance staff. The balance usually makes two full turns. *See* fig. 30.

Rack-striking The form of striking which, from 1675, largely supplanted the system of lock-plate striking. A toothed segmental rack drops on a snail which determines how many hours will be struck. Rack striking is the basis of repeating work.

Rate The performance of a watch. If a watch has a steady losing or gaining rate of x seconds a day, any departure from this average is described as a loss or gain on its rate.

Recoil The backward rotation of the escape wheel in a verge escapement which takes place towards the end of each swing of the balance. A recoil escapement is the opposite of a dead-beat escapement.

Regulator The movable lever on which are mounted the pins which embrace the balance spring of a watch. Moving the regulator lengthens or shortens the effective length of the spring and so alters its rate. In clock terms, however, 'regulator' means a precision clock which does not strike and which has a compensation pendulum. During about 1680–1780 almost all watches had the form of regulator developed by Tompion. The pins are mounted on a segmental rack which moves in a guide and is geared to a pinion. The pinion is fitted with a square and turned by a key. Some very early balance spring watches had a straight section at the outer end of the spring along which the regulator pins moved. They were mounted on a carriage moving along a long key-operated worm. This is known as a Barrow regulator after its inventor, Nathaniel Barrow. Before the balance spring, the only easy way of regulating a watch was by altering the set-up. *See* Set-up Regulator. Stackfreed watches were often regulated by a bristle mounted on a movable index, and standing in the path of the foliot balance. By limiting the swing of the foliot it had some effect on the rate of the watch. It is known as a 'bristle regulator'. From it the balance spring was eventually developed.

Remontoir A supplementary spring interposed between the mainspring and escape wheel. The train is normally locked, but it is released at fixed intervals (usually about 15 seconds) by the escape wheel. It then winds up the remontoir spring. The remontoir spring drives the escape wheel. It is this that isolates the escape wheel from the variable torque of the mainspring between fully wound and run down. It thus obviates the need for a fusee.

Remontoir Escapement When the remontoir spring is wound by the escape wheel, and itself directly impulses the balance, it is called a remontoir escapement. *See* Remontoir *and* Constant Force Escapement.

Repeater A watch which can be caused at will to strike the time on a bell, block of metal, or one or two gongs. The repeating mechanism is set in motion by depressing the pendent in most key-wound watches, or by moving a slide round the band in most keyless-wound watches. The repeating train is derived from rack-striking and was developed, notably by Daniel Quare, in about 1685. The repeater may signal the time as:

Hour repeater, striking only the last hour.

Quarter repeater, striking the last hour; and then the quarters, with a double stroke at each quarter.

Half-quarter repeater, in which, after the quarter is struck, a single stroke follows after 7·5 minutes of the quarter have elapsed. This arrangement was much used by Breguet.

Ten-minute repeater, in which a double blow is struck for each ten minutes. This is an inconvenient arrangement, possibly only used, very rarely, by Breguet.

Five-minute repeater. After the last quarter a single, detached blow is struck for each five minutes, until the next quarter.

Minute repeater. The final refinement of repeating work. After each quarter a single blow is struck for each minute up to the next quarter.

Reversed Fusee Reversing the path of the chain passing between the fusee and spring barrel. This has the effect of halving the pressure on the fusee pivots. Probably first used by Thomas Mudge.

Roller Usually a disk on the balance staff; either the impulse roller or safety roller.

Ruby Cylinder A cylinder escapement in which the cylinder is formed of a ruby instead of steel. *See* Cylinder Escapement.

Safety Roller In the lever escapement, a small roller on the balance staff out of which a crescent is cut, allowing the safety dart to cross it. If the lever is jolted out of engagement with the escape wheel, the dart will rub against the edge of the roller until it is able to pass through the crescent, after which normal operation will continue. The detent escapement has a similar safety action. *See also* Setting.

Savonette French term for hunting case (*q.v.*).

Screws, Balance The screws in the rim of a compensation. They are movable in the rim to vary the poise and the effects of the compensation.

Self-winding Watch A watch containing a weight (either oscillating or revolving) which is caused to move by the activity of the wearer. In so doing it winds up the mainspring through a ratchet and pawl. Probably developed in about 1780 by Abram-Louis Perrelet and perfected by Breguet who called it a perpetuelle. Self-winding wrist watches were first developed by John Harwood in 1928.

Setting A watch is said to have set when it is stopped by a violent movement or jolt, causing the safety pin to pass to the wrong side of the safety roller, where it jams.

Set-up The residual power maintained in a mainspring by the stop-work, after the watch has run down. *See also* Barrel.

Set-up Regulator Before the introduction of the balance spring the only easy way of regulating a watch was by altering the set-up. This was done either by a ratchet and wheel, or worm and wheel on the back plate.

Siderial Time The time elapsing between two successive transits of a fixed star across the meridian. A siderial day is 3 minutes 55·5 seconds shorter than the mean time day.

Six-hour Dial A rare form of dial used in the late seventeenth century. There is only one hand which turns once in six hours. The chapter ring is marked in Roman numerals from I to VI. Superimposed on these are Arabic numerals from 7 to 12. *See* figs. 104, 108.

Snail A cam that rises uniformly through 360 degrees and then reverts to the starting radius. *See also* Quarter-snail.

Spade Hand A watch hand terminating in a pointer shaped like a playing-card spade. Used regularly by Arnold and Emery.

Spring Barrel The barrel containing the main-spring.

Spring Detent A detent mounted on a spring. Generally applied to the spring-detent escapement.

Stackfreed An early, crude alternative to the fusee, used in Germany during the sixteenth and early seventeenth century. Geared to the great wheel is a wheel turning once in twenty-four hours. Mounted on the same arbor is a cam against which a follower is pressed by a strong spring. During the hours after winding, when the torque is strong, the follower works against the mainspring. When the mainspring approaches the end of its run and the torque is weak, the follower helps it forward. The torque is thus roughly equalised throughout the day. The going period of a stackfreed watch is usually twenty-six hours. Stackfreed watches were regulated by a bristle regulator. *See* Regulator. *See* fig. 60.

Star Wheel A star-shaped wheel to effect the intermittent rotation of a striking or calendar mechanism.

Steady Pins Pins in the underside of a cock which register with holes in the plate, thus locating it precisely.

Stop-watch See Chronograph.

Stop-work A device in going-barrel watches to control the number of turns of winding and unwinding. The mainspring is therefore not used in its fully-wound or unwound positions, thus ensuring a reasonably equal torque throughout the day. In fusee watches the stop work prevents further winding after the chain or gut is fully wound onto the fusee.

Straight-line Lever A lever escapement in which the escape wheel arbor, the lever pivot, and the escape wheel arbor are set out in a straight line.

Sully's Escapement A frictional escapement developed by Henry Sully in connection with his marine timekeeper. *See* Debaufre Escapement. *See* fig. 45.

Sun and Moon Dial. A form of dial found on some late seventeenth-century watches. The minute hand is of normal type but the hour is indicated through a semi-circular aperture round the rim of which the hours are marked from VI to VI (with XII at the top centre). A disk revolves behind the dial plate, on one half of which is a gilt sun on a polished ground. On the other half is a silver moon surrounded by stars on a blued ground. Only half of the disk being ever visible, the sun and moon point alternately to the hour on the chapter. The wearer is thus assisted in concluding whether it is day or night-time. *See* fig. 112.

Supplementary Arc The continuing vibration of a balance after impulse and locking are completed.

Surprise Piece The part of a repeating train which comes into operation just before the hour is reached and ensures that the correct hour just either side of it is struck.

Sweep Seconds A seconds hand pivoted concentrically with the hour and minute hands. Also termed centre seconds.

Tact (Montré à Tact) A watch with a revolvable hand fitted to the front (if savonette) or back of the case. It carries a pawl which registers with a single-toothed ratchet on the hour wheel. When the hand is turned it will be brought to a stop when the pawl reaches the ratchet. The time may then be determined by touch in relationship to studs set around the edge of the cover. The arm may be freely turned anti-clockwise. It was devised by Breguet as an alternative to a repeater. Sometimes called a blind man's watch. *See* fig. 228.

Temperature Compensation Any device for compensating the effects of changes of temperature on the timekeeping of a watch. *See* Compensation Balance *and* Compensation Curb.

Terminal Curve A change in the radius of curvature of the end turns of a balance spring. Usually applied to help make an escapement isochronous. *See also* Balance Spring.

Third Wheel The wheel between the centre and fourth wheels of a watch.

Three-quarter Plate A calibre in which all the wheels, up to and including the fourth wheel of the train, are pivoted in the plates. The escape wheel and balance wheel are separately cocked.

Timing Screws The screws in the balance rim of a compensation balance for adjusting the daily rate but not the compensation. They are set at the root of each compensation rim.

Tipsy-Key See Breguet Key.

Tompion Regulator See Regulator.

Top Plate The plate of a watch next to the dial.

Total Arc The sum of the escaping arc and supplementary arc of a balance.

Tourbillon A revolving carriage invented by Breguet to carry the escapement of a watch. By continuously turning the escapement through 360 degrees the effects of poising errors in the balance are nullified. The usual period of rotation is one minute, but Breguet made tourbillons with up to a six-minute period. *See also* Karrusel.

Train The wheels and pinions of a watch.

Trial Number The marks awarded to a marine chronometer submitted for trial at Greenwich Observatory. Various systems of marking were employed at different times (see Appendix I of *The Marine Chronometer* by R. T. Gould). The system of marking used for watches submitted for trial at the National Physics Laboratory at Kew is similar. These last six weeks and test the watch for temperatures and positions. The theoretically-possible maximum is 100 marks, but this has never been achieved.

Tripping A term used to describe the action of an escapement when it allows two or more teeth of the escape wheel to pass during one vibration of the balance.

Tulip Hand A form of hour hand somewhat resembling a tulip flower, found mostly in British watches during the late seventeenth century.

Under Sprung A balance spring fitted under the balance.

Unlocking Pallet The jewel stone in a detent escapement that flexes the detent to unlock the escape wheel.

Verge Escapement A recoil escapement, the first escapement applied to mechanical timekeepers, in use continuously from 1300 to 1900. The balance staff has two flags or pallets one at each end, set at about 100 degrees to each other. These register with the ratchet-shaped teeth of the escape or crown wheel, whose arbor is at right angles to the balance staff. The pallets escape alternately past teeth on opposite sides of the escape wheel. *See* fig. 1.

Virgule Escapement A rare frictional rest escapement invented by Jean André Lepaute in about 1780 and found in Continental watches during the late eighteenth and early nineteenth century. Its operation is something between a cylinder and a duplex. It is a development of an escapement by Tompion in the late seventeenth century. Impulse takes place only at alternate oscillations of the balance, but in the double virgule, two virgules are superimposed, with two sets of teeth on the escape wheel. Impulse then takes place on both oscillations. The double virgule seems to have been developed by P. A. Caron (later Beaumarchais). *See* figs. 6, 7.

Wandering-hour Dial A rare form of dial used mostly during the late seventeenth century. The dial plate is solid and an annular slit covering 180 degrees is cut in its upper half. There are marked numerals from 0 to 60 on the outside of this slit which are the minutes of an hour. A revolving plate behind the dial is seen through it. There are two round windows in it which pass alternately along the slit. Behind the revolving plate are two smaller revolving plates which are turned by the train and have the hours marked on them. On one plate are the even numbers and on the other the odd numbers. Thus, the hour appears in the window, through the slot, starts at 0 minutes, and traverses the slot up to 60 in the course of an hour. The next window, registering the next hour, then appears at 0. *See* fig. 134.

Wheel Engine A machine used to divide and cut teeth on wheels.

Bibliography

The collector of horological books has been well served in recent years by the publication in facsimile of many important books which had long been out of print. Since these and new books cover almost every aspect of the subject, it has not been thought necessary to cite books which are unduly rare or unobtainable.

For a general history of clocks and watches, F. J. Britten's *Old Clocks and Watches and their Makers* has long held an unchallenged position. It was completely re-written by Messrs Baillie, Clutton and Ilbert for the Seventh Edition and is currently in the Eighth Edition, published by Eyre Methuen (1973). Of the earlier editions, the Sixth, originally published in 1932, has been re-published in facsimile by the E.P. Publishing Company (Wakefield, 1971). Britten is largely famous for its list of early makers. This list is greatly exceeded in G. H. Baillie's *Watchmakers and Clockmakers of the World*, (N.A.G. Press, 1947), but Britten's remains useful for the amount of information it contains. Another important list of makers is the *Dictionnaire des Horlogers Français*, written and published by Tardy (Paris, 1972).

One of the most important books on horology is G. H. Baillie's *Clocks and Watches: an Historical Bibliography*, which gives a précis of every known horological book up to 1799. Originally published by N.A.G. Press it has been republished in facsimile by Holland Press (1978). Equally comprehensive and useful is the *Watch and Clock Encyclopaedia* by D. de Carle (N.A.G. Press, 1950).

A reference book on hallmarks is essential to any collector and the most compact is the *Guide to Marks of Origin on British and Irish Silver Plate from the mid Sixteenth Century*, last revised in 1978 and published by Frederick Bradbury. More detailed is *Chaffer's Handbook to Hall Marks on Gold and Silver Plate* by C. G. E. Bunt (William Reeves, Seventh Edition, 1945). French *poinçons* are quite chaotic and it is seldom possible to date a French watch to within ten years; but Tardy's *Les poinçons de Garantie Internationaux pour L'Argent* and a matching volume . . . *pour L'Or* contain the relevant mass of mutually conflicting information. A. G. Grimwade's immense *London Goldsmiths* (Faber, 1976) is useful on watch case makers.

Catalogues of public collections are useful as reference works. Eagerly awaited is a catalogue, now in course of preparation, of the British Museum's incomparable collection. The London Science Museum's useful collection is described in the Museum's *Time Measurement: Science Museum* by Dr F. A. B. Ward. Useful for its pictures is the *Collections du Museé International d'Horlogerie La Chaux-de-Fonds Suisse*, but the cataloguing itself is inaccurate. One of the most important collections is that belonging to the Worshipful Company of Clockmakers on display in the London Guildhall Museum. A detailed illustrated, critical catalogue by Cecil Clutton and George Daniels is published by Sotheby Parke Bernet Publications (1975) and there is a matching catalogue of the Company's library compiled by John Bromley (1977). The horological section of the Conservatoire des Arts et Métiers in Paris is chiefly valuable for its collection of Ferdinand Berthoud's marine timekeepers. It is published by the Museum with the subtitle 'Catalogue du Musée Section JB. Horlogerie', but is at present out of print.

There are two books on private collections which may be found useful: *Antique Watches* by T. P. and T. A. Cuss is written around the highly important Cuss Collection and is noteworthy for the very high quality of its illustrations which hardly duplicate those in *Watches*. It was published by the Antique Collectors' Club (Woodbridge, 1976). A much smaller, highly selective collection, is the subject of *Collector's Collection* by Cecil Clutton (The Antiquarian Horological Society, 1974). The same Society has also sponsored the following books, all of which are of great value to collectors and amateurs: *Paul Philip Barraud* by Cedric Jagger (1968); *John Arnold and Son* by Vaudrey Mercer

(1972); *Edward John Dent and his Successors* by Vaudrey Mercer (1977); *Pioneers of Precision Timekeeping: A Collection of Important Essays* (1965). The Proceedings of the Society, a quarterly entitled *Antiquarian Horology* is available to members.

A book no longer in print, but not yet in a state of serious rarity, is *The Swiss Watch* by Eugène Jaquet and Alfred Chapuis. It is a large and comprehensive work, complete to the date of its publication in 1953.

There seems as yet to be no definitive book on the important and wide-ranging subject of American watches, but pending the publication of such a book, *English and American Watches* by George Daniels is a useful short work (Abelard-Schuman, 1967).

Many books have been written about Breguet including that by Sir David Salomon, about his own collection, now long out of print. But while all have their own value the whole subject is now covered definitively in *The Art of Breguet* by George Daniels (Sotheby Parke Bernet, 1975).

Carriage clocks are so closely allied to watches that collectors of watches are likely to have an interest in them. There was no book of any kind on this wide and somewhat confusing subject until the publication of *Carriage Clocks, their History and Development* by Charles Allix and Peter Bonnert (Antique Collectors' Club, 1974).

The following list gives the most important classical works, so far as watches are concerned, which have been republished in facsimile. These are given in alphabetical order by name of author:

Ferdinand Berthoud. *Histoire de la Mesure du Temps par les Horloges* published in 1802. (Facsimile edition by Berger-Levrault, 2 vols. Paris, 1976.) *Essai sur L'Horlogerie*, the Second Edition, of 1786, with a long preface in English by Cecil Clutton, was also published by Berger-Levrault in 1978. Also in two volumes, this is the outstanding eighteenth-century work on the practical art of clock and watchmaking. In course of preparation, and of great value in the study of the history of marine timekeepers, is *Histoire De la Mesure du Temps ou supplément au Traité des Horloges Marines et à l'Essai sur l'Horlogerie*, originally published in 1787. It is to be hoped that there will eventually be a facsimile of Louis Berthoud's small and delightful *Entretiens*

sur *L'Horlogerie*, first published in 1812.

Paul Chamberlain's *It's About Time* is a collection of essays by this famous horological historian published in 1941, soon after his death. It contains descriptions, with Chamberlain's superb drawings, of a large number of highly eccentric escapements and some invaluable biographical essays. It is one of the most entertaining of all horological books and is published in facsimile by Holland Press (1978).

William Derham's *The Artificial Clockmaker; a Treatise of Watch and Clockwork* was first published in 1696. It has been reproduced by the Thames Facsimile Company (1962).

Diderot drew the plates for the gigantic French *Encyclopédie Raisonné* published in Paris between 1751 and 1780. The horological plates are reproduced in facsimile by Magikal (Milan) in about 1963 and, to an extent, they overlap the plates in Berthoud's *Essai*.

R. T. Gould's *The Marine Chronometer, its History and Development* was first published in 1923 and for historical interest, entertainment value and excellence of style, this book is unrivalled in horological literature. The facsimile edition is published by Holland Press (1976).

T. Hatton's *Introduction to the mechanical parts of Clock and Watchwork*, first published in 1773, was the first book in English to compare with Berthoud's great works. The facsimile edition is by Turner and Devereux (1977).

M. L. Moinet's *Nouveau Traité d'Horologerie* was published in 1848 and it was subsequently established to have been based extensively on notes by Breguet, intended for publication, and purloined by Moinet. It is an important essay on the art of horology at the middle of the nineteenth century, with a separate volume of superb plates.

Thomas Mudge's *A Description with Plates* is the somewhat uninviting title of a book published by Mudge's son in 1799; but it is a book of great historical value. It is in four parts, of which the first is a turgid narrative of Mudge's unhappy relationship with the Board of Longitude. At the end of the book is a set of beautiful drawings of Mudge's marine timekeeper, prepared by Robert Pennington. For today's amateur, however, the value and enjoyment of the book is to be found in the voluminous collection of letters which are published in it from Mudge to Count von Bruhl over the period 1772 to 1787. There is also a short and valuable section containing Mudge's *Thoughts on the Means of improving Watches and more particularly those at Sea*. The facsimile edition is by Turner and Devereux (1977).

Rees's *Cyclopaedia of Literature, Science and the Arts* was published serially in 47 volumes, terminating in 1820. It contains a number of excellent articles relating to horology, with fine plates, and these have been collected into one volume, entitled *Rees's Clocks Watches and Chronometers* published by David & Charles (Newton Abbot, 1970).

Claudius Saunier's *Treatise on Modern Horology* was published in 1869 and the English translation came out in 1882. It is considered by those technically well informed to be invaluable and was republished in facsimile by Foyles in 1952.

John Smith published his curious little *Horological Dialogues* and *Horological Disquisitions* in 1675 and 1694, respectively. Like Derham's *Artificial Clockmaker* these are interesting because of their early date. They were published *en suite* with the Derham book by the Thames Facsimile Company in 1962.

R. W. Symonds was a well-known authority on old clocks, and his book, *Thomas Tompion: His Life and Work,* published in 1951, has a useful section on his watches and includes an account of his life generally. It was published in facsimile by the Hamblin Publishing Group (1969).

Anthoine Thiout's *Traité de L'Hologerie Mecanique et Pratique* may be regarded as a shorter precursor of Berthoud's *Essai* published in 1741. It is a useful work with good plates, and was reproduced in facsimile by Editions du Palais Royal (Paris, 1972).

Acknowledgements

The authors have expressed their appreciation for permission to illustrate watches from private collections whose owners wish to remain anonymous. It remains to thank the following for their help in obtaining information and photographs for inclusion in this enlarged volume: Mr Geoffrey De Bellaigue, Surveyor of the Queen's Works of Art, Mr Charles Allix, The British Museum, The Worshipful Company of Clockmakers, The L. A. Mayer Memorial Foundation (Jerusalem), Messrs. Asprey & Co., Bobinet & Co., Camera Cuss & Co., Omega Watch Company, Patek Philippe & Co. (Geneva), Christie Manson and Woods, Sotheby Parke Bernet & Co. and especially Miss Tina Millar.

1979

Index

Figures in italic relate to monochrome illustration numbers. Figures in roman numerals relate to colour plates. The Glossary of technical terms, being itself in alphabetical order, is not included in this Index.